≡ LA FRONTERA

A book in the series
Radical Perspectives: A Radical History Review book series
Series editors: Daniel J. Walkowitz, New York University
Barbara Weinstein, New York University

History, as radical historians have long observed, cannot be severed from authorial subjectivity, indeed from politics. Political concerns animate the questions we ask, the subjects on which we write. For over thirty years the *Radical History Review* has led in nurturing and advancing politically engaged historical research. *Radical Perspectives* seeks to further the journal's mission: any author wishing to be in the series makes a self-conscious decision to associate her or his work with a radical perspective. To be sure, many of us are currently struggling with the issue of what it means to be a radical historian in the early twenty-first century, and this series is intended to provide some signposts for what we would judge to be radical history. It will offer innovative ways of telling stories from multiple perspectives; comparative, transnational, and global histories that transcend conventional boundaries of region and nation; works that elaborate on the implications of the postcolonial move to "provincialize Europe"; studies of the public in and of the past, including those that consider the commodification of the past; histories that explore the intersection of identities such as gender, race, class, and sexuality with an eye to their political implications and complications. Above all, this book series seeks to create an important intellectual space and discursive community to explore the very issue of what constitutes radical history. Within this context, some of the books published in the series may privilege alternative and oppositional political cultures, but all will be concerned with the way power is constituted, contested, used, and abused.

☰ LA FRONTERA

*Forests and Ecological Conflict
in Chile's Frontier Territory*

THOMAS MILLER KLUBOCK

Duke University Press | Durham and London | 2014

Library of Congress Cataloging-in-Publication Data
Klubock, Thomas Miller.
La frontera : forests and ecological conflict in Chile's frontier territory /

Thomas Miller Klubock.
p. cm. — (Radical perspectives:)
Includes bibliographical references and index.
ISBN 978–0–8223–5598–4 (cloth)
ISBN 978–0–8223–5603–5 (pbk.)
1. Forests and forestry—Environmental aspects—Boundaries—Chile.
2. Chile—Boundaries—Environmental conditions.
3. Chile—Boundaries—Social conditions.
I. Title. II. Series: Radical perspectives.
SD583.K58 2014
333.750983–dc23   2013045337

Duke University Press gratefully acknowledges the support of the University of Virginia Dean of the College of Arts and Sciences, and the Vice President for Research and Graduate Studies, which provided funds toward the publication of this book.

Contents

Acknowledgments

I have incurred many debts while writing this book. I am happy to be able to finally recognize all those who have contributed to it in so many different ways over the years.

First and foremost, without the research assistance of Alberto Harambour Ross, Marcos Fernández Labbé, and Karen Donoso Fritz, all fine historians in their own right, this book could never have been written. I am deeply indebted to all three for their indispensable research at different times during the years I worked on this project.

My colleagues and graduate students at Stony Brook University, where I worked when most of this book was researched and written, provided invaluable intellectual comradeship. I thank especially Brooke Larson, Paul Gootenberg, Christopher Sellers, Alberto Harambour Ross, Consuelo Figueroa, Brenda Elsey, Hernán Prudén, Ana Julia Ramírez, and Enrique Garguín. The New York Area Latin American history seminar provided a robust space for intellectual conversation and I benefited from presenting my work there; I would like to thank Sinclair Thomas, Pablo Piccato, and Nara Milanich, as well as the Stony Brook cohort named above, for their suggestions and comments on pieces of this research.

I began thinking about writing this book when I was at Georgetown University. I am grateful to Georgetown for an important research leave that allowed me to embark on what would turn out to be years of research. I am also thankful for the lively intellectual climate provided by the Georgetown history department. Conversations with John Tutino, Erick Langer, and John McNeill were especially helpful.

I benefitted enormously from the feedback I received during presentations of this work in Chile. I would like to thank especially Jaime Flores Chávez and Jorge Pinto at the Universidad de la Frontera in Temuco for their helpful comments and suggestions. I would also like to thank the Universidad Diego Portales history department and Facultad de Ciencias Sociales e Historia, who invited me to give talks on my research at the Norbert Lechner Seminar and the Jornadas de Historia de Chile. Discussions with colleagues who attended these talks helped clarify many issues and have made the book better than it would have been otherwise.

For many years of conversation and friendship I thank my good friends historians Julio Pinto Vallejos and Verónica Valdivia Ortiz de Zárate. For invigorating discussions that have made me think hard about the questions raised by this book and for their support over the years I also thank my friends and colleagues John French, Elizabeth Hutchison, Gil Joseph, Temma Kaplan, Karin Rosenblatt, Heidi Tinsman, Barbara Weinstein, and Peter Winn.

I was fortunate to receive extraordinarily detailed and critically engaged comments from the two readers for Duke University Press, Brian Loveman and Florencia Mallon. This book is immeasurably improved by their generous and meticulous critical readings, although of course they bear no responsibility for whatever weaknesses remain in the text.

The staffs at the Archivo Nacional de la Administración, Archivo Nacional, the Sala de Fundadores of the Biblioteca Nacional, the Biblioteca del Congreso Nacional, the Museo Histórico Nacional, and the Servicio Agrícola y Ganadero in Santiago, and the Archivo Regional de la Araucanía and Archivo General de Asuntos Indígenas in Temuco provided indispensable assistance in locating often difficult to find, at times uncatalogued, materials. Without their help I would never have been able to track down many of the archival sources used in this book. I would like to thank especially María Eugenia Queupumil Burgos, who went out of her way to assist me with gaining access to documents held by the Archivo General de Asuntos Indígenas, now located in the Archivo Regional de la Araucanía. In Concepción, the Vicaría Pastoral Obrera facilitated contacts with forestry workers and provided me access to their collection of documents on the labor movement in the forestry industry during the 1980s. In Valdivia, the office of the Corporación de Promoción y Defensa de los Derechos del Pueblo (CODEPU) provided me copies of their important studies of military repression in Neltume, the core of the former Complejo Maderero Panguipulli, and helped me to make contact with former residents of Neltume.

I am very grateful to Felipe Orrego, a wonderful photographer, for allowing me to include his powerful photographs of forests and the effects of deforestation in southern Chile, a number of which were included originally in the environmentalist group Defensores del Bosque Chileno's indispensable volume on Chile's forests, *La tragedia del bosque chileno*. I would also like to thank Bill Nelson for producing the maps included in this volume.

I hold an unpayable debt to many individuals in Chile who agreed to be interviewed for this book and who received me not only with warmth and hospitality, but also with an acute engagement with the questions and themes which animate this book: in Concepción, the leadership of the Confederacion de Trabajadores Forestales, especially Caupolicán Pávez, and the Federación Liberación; in Panguipulli, Luis Rosales, Pedro Cardyn, Moises Durán, and Mauricio Durán, as well as members of the Parlamento de Coz Coz, especially Angela Loncoñanco and Humberto Manquel; in Valdivia, Ida Sepúlveda and Alex Rudloff; in Lonquimay, members of the Quinquén and Bernardo Ñanco communities, especially Segundo Carilao, and forester Frida Schweitzer; in Santiago, a number of CONAF foresters and union leaders who I interviewed at the Corporación Nacional Forestal (CONAF) workers' union, the Federación Nacional de Sindicatos de CONAF (FENASIC).

I received essential assistance in the form of research grants and leaves provided by Stony Brook University between 2001 and 2007 and by the University of Virginia between 2008 and 2013. I would like to thank the University of Virginia Office of the Vice President for Research and the College of Arts and Sciences for grants which helped me to complete this book. An American Council of Learned Societies (ACLS) Burkhardt fellowship at the National Humanities Center provided a wonderful year of writing and intellectual community in North Carolina. I am especially appreciative of the staff at the National Humanities Center who generously helped me in gaining access to materials from far-flung libraries across the country.

I owe thanks to my editor at Duke University Press, Valerie Millholland, for her support for this book from early on. I am also grateful to Sue Deeks for her meticulous and intelligent copy editing of the manuscript. I thank Barbara Weinstein and Danny Walkowitz for the honor of including the book in their Radical Perspectives series where it enjoys fine company.

It is impossible to put into words my gratitude to the members of my family. Without the support of my father and mother Daniel Klubock and Dorothy Miller, and my sister Katharine Klubock, I certainly would never have been able to finish this book. They also provided passionate intellectual and political discussions over the years, which have certainly made their mark on this project. It is a great pleasure to be able to thank Sandhya Shukla for her love, companionship, and brilliance. Like it or not she has been a true partner in this project. I have learned and continue to learn a great deal during our life together. Our sons Kiran and Ishan have provided many moments of joy and invaluable perspective on this work.

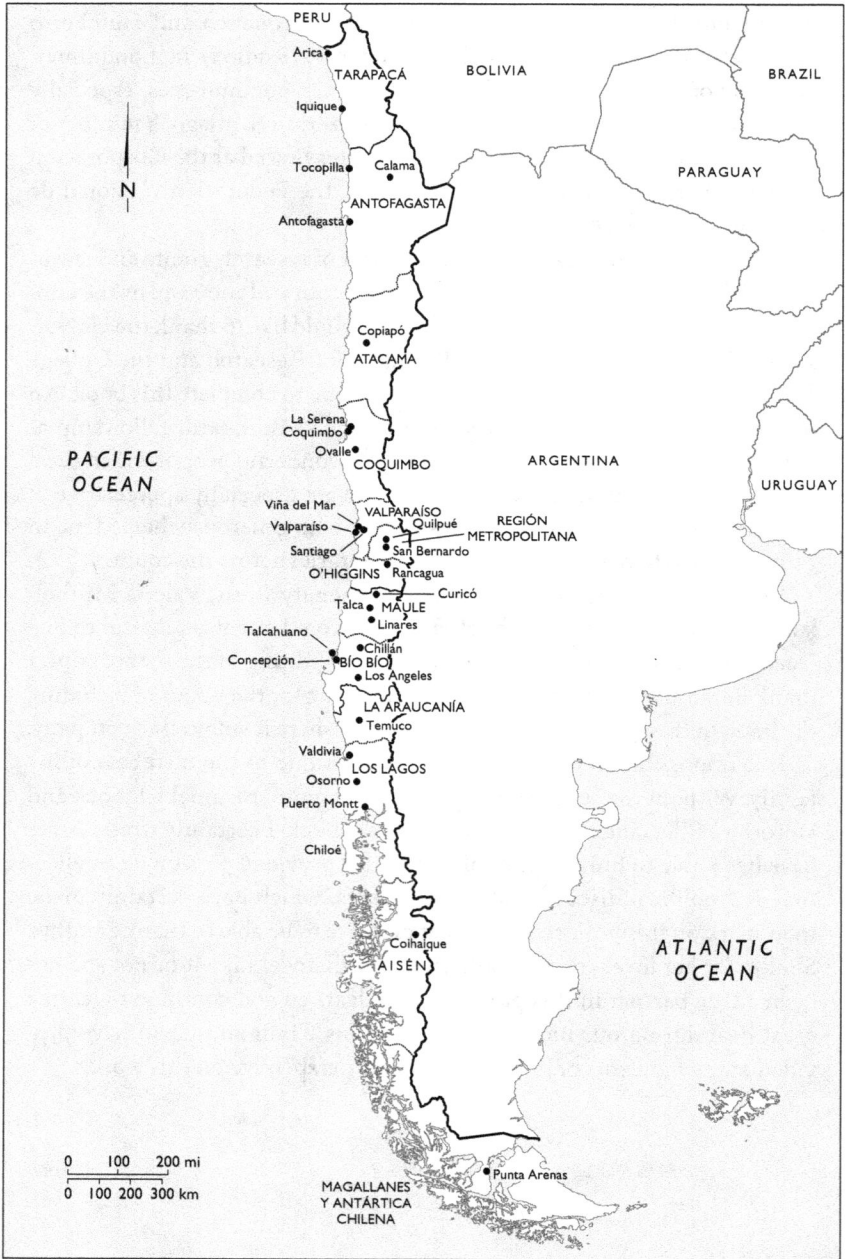

PERU

Arica

TARAPACÁ

BOLIVIA

BRAZIL

Iquique

Tocopilla

Calama

PARAGUAY

ANTOFAGASTA

Antofagasta

Copiapó

ATACAMA

PACIFIC
OCEAN

La Serena
Coquimbo

Ovalle

COQUIMBO

ARGENTINA

URUGUAY

Viña del Mar

VALPARAÍSO

Quilpué

REGIÓN
METROPOLITANA

Valparaíso

Santiago

San Bernardo

O'HIGGINS

Rancagua

Curicó

Talca

MAULE

Linares

Talcahuano

Chillán

Concepción

BÍO BÍO

Los Angeles

LA ARAUCANÍA

Temuco

Valdivia

LOS LAGOS

Osorno

Puerto Montt

Chiloé

Coihaique

AISÉN

ATLANTIC
OCEAN

| 0 | 100 | 200 mi |
| 0 | 100 200 | 300 km |

Punta Arenas

MAGALLANES
Y ANTÁRTICA
CHILENA

N

Map 1. Chile

Map 2. Southern Chile

Map 3. Southern Chile (Araucania)

BÍO BÍO

Concepción

Lota

Bío Bío

Los Angeles

Curanilahue

Santa Barbara

Lebu

Nacimiento

□ Reserva Nacional Ralco

Mulchén

Reserva Nacional
Malalcahuello

Cañete

Angol

Collipulli

Reserva Nacional
Nahuelbuta

Purén

Ercilla

□ Reserva Nacional Malleco

Reserva de la
Biósfera Araucarias
(Quinquén)

Lumaco

Victoria

Lonquimay

Imperial

Traiguén

Lautaro

Curacautín □

□ Reserva Nacional Alto Bio Bio

LA ARAUCANÍA

□ Reserva Nacional Galletué

Carahue

Temuco

Melipeuco

Saavedra

Nueva Imperial

Cunco

□ Reserva Nacional Villarrica

Pitrufquén

Toltén

Villarrica

Pucón

Curarrehue

Toltén

Loncoche

Coñaripe

Parque Nacional
Villarrica

Calafquén

Valdivia

Panguipulli

Liquiñe

Valdivia

Riñihue

Neltume

Puerto Pirihueico

Reserva Huilo Huilo

La Unión

Lago Ranco

Rio Bueno

PACIFIC
OCEAN

ARGENTINA

Puerto Montt

Ancud

LOS
LAGOS

Castro

N

Introduction

When you arrive in the southern Chilean city of Concepción by air, magnificent views of pine stands stretch out beneath you in evenly spaced rows, covering the undulating foothills of Chile's coastal cordillera and running down to the very edge of the Pacific Ocean. Although it is difficult to discern from the window of an airplane, Concepción's pine forests are really not forests at all. They are, rather, plantations of the North American conifer known as Monterey pine (radiata pine or, in Chile, *pino insigne*), with none of the life that characterizes a forest. They contain no underbrush, vines, or trees other than pine, none of the intermingling of tree and plant species that characterizes forest ecosystems. The pine plantations bear no traces of the native forests that held a multitude of species endemic to Chile, such as the ancient araucaria pine, and covered much of the coastal cordillera and its foothills just a century ago. They sustain as little diversity of fauna as they do flora. This is true in large part because of the exertions of the large forestry companies that own them. Systematic aerial spraying has purged the pine plantations of all competing insect, fungal, or vegetable species. Fences and forest guards keep out straying mammals that might feed on young saplings. They are there, too, to prevent any denizen of the countryside from entering the tree plantations to collect firewood or forest products or, perhaps, to fell a tree or two. These are forests without people, completely uninhabited.

Pine plantations now cover extensive stretches of eroded soil left by deforestation and intensive agriculture, replacing wheat and livestock on large estates, as well as the cereals and garden crops cultivated on small peasant plots, from the Bío Bío River south to the Valdivia River and Los Lagos (the Lake Region). They also occupy land where Chile's frontier forests, large expanses of undisturbed native forests characterized by biodiversity, once stood. A century ago, one might have found stands of araucaria pine and different types of Chilean beech—both deciduous varieties such as raulí and roble and perennial species such as coigüe—in mixed stands with a wide variety of trees native to Chile's temperate forests, intermingling with vines, underbrush, and wild bamboo in the Andes and the coastal cordilleras and their valleys. Farther south lay stretches of the broadleaf evergreens of the Valdivian temperate rain forest; stands of the gigantic conifer alerce, a member of the cypress family, which rivals California's redwoods in age and size, along the coast; and raulí and coigüe forests in the Andes cordillera. Now uninterrupted waves of Monterey pine (about 85 percent of all tree plantations in Chile) and eucalyptus cover vast areas from Llanquihue and Valdivia provinces to Cautín, Malleco, Arauco, and Bío Bío provinces farther north. Chile's frontier forests, which still compose one-third of the world's intact stands of temperate forest, appear to have been swept away by a vast wave of monocultural pine plantations.[1] Today, Chile has the largest expanse of tree plantations in Latin America, and forestry exports are a significant source of foreign revenue for the Chilean economy, in third place behind mining and industry. For many boosters of Chile's dynamic forestry economy, plantations of Monterey pine are the signature success story of the country's recent free-market "miracle."[2]

During the 1990s, indigenous Mapuche communities throughout southern Chile initiated a series of land invasions of large estates covered with Monterey pine and owned by a handful of Chile's largest financial conglomerates. Mapuches couched their claims to land on forestry estates in terms of their historic occupation of frontier territory. Under the rows upon rows of pine trees, they argued, lay long histories of usurped land they had occupied "since time immemorial" or land stolen from communities with legal titles granted by the Chilean state at the beginning of the twentieth century. In addition, Mapuche communities wielded environmentalist arguments to challenge the prevailing triumphalist narrative of pine's miraculous transformation of the southern landscape. They underlined that pine plantations had a destructive effect on the ecology of southern Chile's forests and soil. They contended that rather than a "green" motor of development, literally and figuratively, pine produced soil acidification and dried up rivers and streams. The chemical sprays employed by the forestry companies poisoned wild game and Mapuche and non-Mapuche peasants' livestock, destroyed their crops, and made their children sick.

They also pointed out that forestry companies often substituted the native forests peasants relied on for firewood, game, and forest products with more profitable pine plantations. On large estates, pine replaced native forests, as well as resident estate laborers and seasonal workers, throwing increasing numbers of land-starved peasants, many of them Mapuches, into the swelling ranks of the rural unemployed. For the Mapuche communities that broke down the fences encircling pine plantations and disrupted forestry companies' logging operations, the spread of Monterey pine had led to a new moment of ecological degradation in southern Chile, uprooting increasing numbers of poor peasants from the countryside.[3]

This book brings together the social and environmental histories of the southern frontier territory to examine the origins of Chile's recent forestry boom and uncover the roots of today's bitter conflict between forestry companies and Mapuche communities. The history of Monterey pine's movement from the United States Northwest down the Pacific coast to southern Chile is inextricably tied to the history of colonization and settlement in the region known as *la frontera* (the frontier), or the Araucanía, roughly the territory that lies between the Toltén and Bío Bío rivers. I examine the ecological crises produced by colonization in southern Chile's native forests from the late nineteenth century, when the military conquest of la frontera was completed and settlers set fire to forests to clear land for crops and livestock, until the late twentieth century, when pine plantations established their dominion over southern soil during the military dictatorship of Augusto Pinochet (1973–90) and the transition democratic governments of the center-left Concertación de Partidos por la Democracia (Coalition of Parties for Democracy, 1990–2010). My goal is to write this history of ecological change in southern Chile's temperate forests "from the bottom up," as it was experienced by the rural poor, members of Mapuche communities, Mapuche and non-Mapuche squatters (*ocupantes*), settlers (*colonos*), seasonal laborers (*peones* or *gañanes*), and full-time resident estate laborers (*inquilinos*)—roughly the broad population of laboring rural poor often referred to as peasants or campesinos. I combine social history's interest in rural land and labor relations with environmental history's focus on the ecological changes wrought by economic development, settlement, and colonization.[4]

I ask three basic questions about the environmental and social history of the frontier. First, what were the origins of today's forestry miracle? While the ecological triumph of Monterey pine has been interpreted by both boosters and critics as the product of the radical free-market reforms designed by students of Milton Friedman at the University of Chicago (*los Chicago boys* in Chile) and implemented at gunpoint by the Pinochet dictatorship, I demonstrate that the spread of the North American conifer throughout southern Chile was largely the result of state-directed development programs and forestry policy before 1973.[5] I show that in response to ecological crises provoked by deforestation, including drought, climatic changes, and soil erosion, during the early twentieth century, land and colonization officials imposed the authority of the state on the frontier's natural and social landscapes. They reined in the practices of both large landowners and campesinos, laying the foundation for a strong state role in ordering rural property relations and regulating the exploitation of natural resources. State officials looked to scientific forestry management to build an industrial forestry economy rooted in plantations of Monterey pine while establishing new conservationist restrictions on the destruction of

Chile's remaining native forests. My argument is that conservation, forestry, and forestry science served as tools for extending state governance into a frontier territory often referred to as Chile's "Wild West."

Second, I ask how forestry development remade southern Chile's social landscape. Most literature on forestry in Chile and the contemporary "Mapuche conflict" attributes struggles over forests and forestation to the free-market restructuring introduced by the Pinochet dictatorship after 1973. However, my research makes clear that neoliberal economic "shock therapy" exacerbated, rather than initiated, the social dislocations produced by several generations of government forestry policy in Chile. As early as the 1940s, state-fomented forestation with pine had led to poor peasants' exclusion from public forest reserves and to their eviction from privately held estates. Government officials charged with land and colonization in southern Chile came to believe that reorganizing the frontier's natural landscape around industrial forestry was essential to ordering both ecological and social relations. For the state, forestry replaced the often environmentally destructive practices of the large estate, or hacienda, transforming an unproductive property into a scientifically managed and ecologically sustainable modern business that would be the motor of frontier development. In addition, government officials believed industrial forestry would mold an often rebellious and itinerant population of landless rural laborers into a stable and settled workforce employed in forestry industries and logging. Forest and land officials defined the frontier's campesino population as both socially disruptive and ecologically destructive—a cause of deforestation, drought, and soil erosion—and sought to transform campesinos' relationship to nature by incorporating them into the forestry economy. Pine plantations offered a technocratic solution to southern Chile's chronic social and ecological crises. They provided both an alternative to reforming the frontier's unequal system of property ownership and a means of redressing the impact of indiscriminate logging in native forests. For land and colonization officials, scientifically directed commercial forestry would civilize the frontier's social and natural worlds, introducing the rational management of people and forests.

Third, I ask how southern campesinos confronted changing environmental and social conditions on the frontier, tracing their shifting relationship to the tree plantation economy and modern systems of forest management. Mapuche communities' recent struggles with forestry companies make up a chapter in a century-long history of protests by both Mapuche and non-Mapuche peasants of logging, modern forest management, and tree plantations. Campesinos in the frontier territory viewed logging and deforestation as belonging to a broader pattern of injustice rooted in profound land inequalities and their exclusion from the resources offered by southern Chile's temperate forests during the first decades of colonization. They articulated a moral economy—or, in the words of historian Karl Jacoby, "moral ecology"—that defined the south's frontier forests as a commons to denounce the accumulation of land in the hands of large estates and the loss of forests on which they depended due to logging and fire.[6] As state officials imposed new forest regulations and promoted forestation with pine in response to the ecological catastrophes that beset the southern frontier only a generation after colonization had begun, campesinos confronted new restrictions on their customary uses of the forests. Forestry science and conservationist policies defined campesinos as a threat to the forest and restricted their access to the basic forest resources necessary to subsistence. For campesinos, both forestation with pine and conservationist regulations on exploiting native forests constituted forces leading to proletarianization, or their transformation into a landless labor force.

However, beginning in the 1960s, as the Chilean state initiated one of Latin America's most important agrarian reforms, campesinos appropriated conservationist ideology and made it their own to characterize large estates as irrationally exploitative of nature and to demand their expropriation. They questioned estate owners' property rights by underlining the ecological damage caused by logging and burning forests. Conservation and forestry science gave campesinos a language to make claims to frontier forests they believed to be public or theirs by rights conferred by generations of occupation. During the agrarian reform, state officials backed campesinos' claims and employed environmental laws to appropriate forestry estates they defined as failing to follow modern practices of forest management. By the 1980s, the Pinochet dictatorship had dismantled agrarian reform and introduced radical free-market economic restructuring, handing over extensive tracts of native forests, as well as tree plantations and paper and pulp plants developed with significant state investment, to a handful of financial conglomerates. Mapuche and non-Mapuche campesinos drew on environmentalist arguments to critique the industrial forestry economy and, by implication, the free-market economic model maintained by the governments of the Concertación. They drew on older understandings of the moral ecology attached to their use of the southern frontier forests, but phrased their claims to forestry estates in the language of modern environmentalism. Their goal, they contended, was to produce a more just social order rooted in the biodiversity of Chile's remaining temperate rain forests.[7]

Whether pine is an ecologically and socially sustainable crop that constitutes a green strategy of agricultural modernization animates heated debate today. On the one hand, Monterey pine is a pioneer species that evolved to invade open areas in conditions that are inhospitable to other vegetation. It spreads efficiently on cleared land degraded by agriculture, logging, and ranching. Monterey pine therefore is well suited to reclaiming the eroded lands of southern Chile's central valley and coastal piedmont. The region's wet and temperate climate provides an ideal environment for Monterey pine, similar to its home along the coast of northern California. When cultivated in plantations along southern Chile's coast, Monterey pine trees can be harvested after only two decades, a fraction of the time it takes for commercially valuable native species such as the Chilean raulí beech to mature. Monterey pine has another advantage: it supplies the prime material for producing long-fiber cellulose, used in manufacturing paper. This means it is both a valuable agricultural commodity and linked to the profitable pulp and paper industries. In addition, pine plantations can take pressure off native forests, as many proponents of pine have long argued, by supplying to the timber industry an inexpensive, quick growing, easily managed substitute for native woods.[8]

On the other hand, critics point to a number of weaknesses in pine-fueled forestry development. Pine plantations, unlike forests, have short life spans, and because the trees are harvested after only twenty years, they do not regenerate naturally. While they do very well on already eroded soil, it remains unclear how many rotations of monocultural pine plantations forestry companies can cultivate and what the impact on soil and the wider environment will be. Pine plantations return few nutrients to the soil because they are harvested as they mature, and the plantations do not allow for the decomposition of vegetation, be it trees or underbrush. In addition, as Mapuche communities contend today, studies have demonstrated that pine plantations contribute to increased soil acidity, undermining the conditions for the regeneration of any native vegetation or agricultural crops in the regions they dominate. Pine's detrimental impact on soil is exacerbated by its absorption of water. Whereas southern Chile's dense native forests regulate and preserve rainfall, pine trees retain water in their needles, facilitating evaporation before water hits the soil. Pine plantations lack the low plants and bushes that grow in Chile's native forests and that help to conserve rainwater and humidity in the soil. In addition, the very concentrated nature of pine plantations, their biomass, means that they absorb several times the amount of water consumed by native forests. Lack of water, combined with diminished sunlight in densely planted plantations, also leads to a decline in the decomposition of organic material and prevents the formation of a layer of nutrient-rich humus. As opposed to native forests, which maintain a thick layer of humus filled with nutrients from decayed vegetation, the soil of pine forests stays dry even during rainy winter months.[9]

Critics of pine plantations also point out that, as with any monocultural crop, genetic uniformity makes pine vulnerable to plagues and infestations. To combat possible competing species, such as underbrush, weeds, or animals that might find forage in pine saplings, as well as potential plagues, forestry companies employ an arsenal of chemical herbicides and pesticides. Chemicals sprayed aerially and indiscriminately poison estuaries, streams, and watersheds. Defoliants destroy agricultural crops on land that neighbors plantations and prevent the regeneration of native vegetation. Finally, environmentalist critics have noted that while pine plantations can, and sometimes do, take pressure off native forests, their promotion by the Chilean state over the years has often led to a process of substitution or conversion. Landowners and forestry companies have found it profitable to clear native forest, both old-growth frontier forests and second-growth forests, and plant pine to take advantage of market demand and state subsidies.[10]

Pine plantations have had social as well as ecological costs. The Mapuche land invasions of the 1990s made it clear that, while pine generates jobs in forestry and the paper and pulp industries, its most significant impact has been to expel campesinos from the countryside, swelling the ranks of southern Chile's underemployed and unemployed. Indeed, while Chilean governments from across the political spectrum have seen in pine an engine of economic development for the frontier, one of the major consequences of pine's expansion across the southern Chilean countryside has been campesinos' dispossession. During the second half of the twentieth century, campesinos—from the members of Mapuche communities to smallholders and resident estate laborers who historically have exchanged their labor for small plots within the borders of large landed properties—surrendered their land to pine plantations from the Valdivia River to the Bío Bío River farther north as large estates, backed by state subsidies and incentives, turned from cultivating cereals and pasturing livestock to planting trees. Campesinos' uprooting by pine was exacerbated by the ecological impact of the plantations. Surrounded by plantations, many were forced to sell their land and labor to forestry companies. Pine, like many commercial crops before it, has served as the wedge that separates rural people from their land and from the natural resources essential to their subsistence, turning them into a population of deracinated, landless laborers—an inexpensive labor reserve that often is employed by the forestry companies.[11]

While the appropriation of the small plots of campesino agriculturalists during the expansion of Chile's dynamic forestry economy was shaped by the particularities of Chile's environmental and social history, it is possible to view this process as part of a broader global history that began with the process of enclosure that emerged with the first forest laws and the development of modern forestry practices in Europe and its colonies and then radiated out to other parts of the world with the expansion of the capital-ist market.[12] Chile's history of forestry development belongs to a transna-tional history shaped by the circulation of forestry science and foresters trained in Europe and North America and the influence of international organizations such as the US Forest Service, the United Nations Food and Agriculture Organization, the International Monetary Fund (IMF), the World Bank, and the International Bank for Reconstruction and Devel-opment (IBRD), all of which participated in projects to promote forestry industrialization in Chile in response to growing global demand for paper pulp after the Second World War. In Chile, the imperative of modern for-est management and forestry development based on the forestry science originating in Europe and the United States drove both the expansion of state authority over frontier territory and campesinos' exclusion from once public frontier land and forests.

Chilean campesinos' experience of the development of modern forestry during the twentieth century reproduced many of the contradictions that bedeviled rural people's earlier opposition to forestry and forest regula-tion. Proponents of pine-fueled forestry cast Mapuche campesinos who burn logging trucks or invade forestry estates as the opponents of envi-ronmentally sustainable strategies of development. Even as Mapuche or-ganizations have increasingly adopted the mantle of environmentalism themselves, forestry companies and government officials have defined their opposition to pine plantations as antimodern or premodern and eco-logically destructive. Today, Mapuches in southern Chile risk incurring the charge that they are irresponsible stewards of nature, an accusation leveled consistently by estate owners and state officials over the course of the twentieth century to justify the eviction of both Mapuche and non-Mapuche campesinos from the frontier's forests. In addition, the Chilean state has responded to peasants' protests of modern forestry practices with draconian measures. The democratically elected center-left coalition governments of the Concertación largely backed the forestry companies during the 1990s, applying national security and antiterrorism laws, some on the books since the 1930s and others handed down by the military dic-tatorship, to quell an increasingly militant Mapuche movement for land recuperation. As late as 2010, a number of Mapuche activists remained in jail, held under antiterrorism laws decried by international human rights organizations and the United Nations.[13]

The violence produced by forestry development in recent decades in Chile belongs to a long history of conflict between rural people and sci-entific forest management around the globe. For example, more than "acts of terror," a phrase frequently employed by both the Chilean state and for-estry companies, Mapuches' attacks on logging trucks and pine plantations echo the crimes of the eighteenth-century British peasants known as "the Blacks," who intruded on forest preserves and parks to poach deer; collect firewood; cut down trees; and set fire to haystacks, barns, and houses to protest the enclosure of forest commons, or the actions of the nineteenth-century French male peasants known as the *demoiselles* of the Ariège, who invaded fenced-in forests and attacked forest guards to protest new regulations on the extraction of forest products.[14] Peasants' confrontations with modern forest codes and forestry traveled from Europe with imperial expansion during the late nineteenth century. In the early twentieth century, for example, Himalayan peasants set fire to forests that were regulated, enclosed, and managed under the British Raj to promote both commercial forestry and forest conservation. Much like their Chilean counterparts today, they had seen their customary access to forests curtailed by scientific forestry and logging.[15] The Chilean state's disproportionate response to minor acts of violence against property since the 1990s recalls the cruel legislation employed to punish violators of forests laws in eighteenth-century and nineteenth-century Europe.[16] During the late 1990s and early 2000s, much of forested southern Chile looked like occupied territory, with the massive presence of *carabineros* (uniformed police who compose one branch of the Chilean armed forces) protecting tree plantations and encircling Mapuche communities who found themselves under siege by both Monterey pine and the military.[17]

The Frontier in Chilean History

For many centuries, the southern region of Chile dominated by temperate forests was known as *la frontera* (the frontier). The term designated a territory sandwiched between the Bío Bío and Toltén rivers on the north and south and the Pacific Ocean and Andes cordillera to the west and east, today the provinces of Arauco, Bío Bío, Malleco, and Cautín. Until the 1880s, the frontier region composed not an edge or a border but a space that split Chilean national territory. As frontier, the territory stood outside the reach of first the colonial and then the modern nation-state and was, instead, governed by independent Mapuche groups.[18] While the natural borders of two rivers, a mountain cordillera, and an ocean made a lot of sense, history and geography disrupted these natural, as well as political, demarcations. Both human and ecological movement made the frontier's boundaries ever-shifting. During the late eighteenth century and early nineteenth century, the frontier region expanded both ecologically and socially. Mapuche groups extended their territorial reach south of the Toltén and across the Andes cordillera well into Argentina's pampas, building extensive military and commercial networks that linked the frontier to Argentina and central Chile; to the capital city, Santiago; to the port of Valparaíso; and even to Lima, Peru, where the products of Mapuches' livestock economy made their way on Chilean ships.[19] By the mid-nineteenth century, even territory south of the Toltén River, in the region around Valdivia and into Llanquihue Province—today known as the Lakes Region—still theoretically ruled by the Chilean state, was largely ungoverned territory where Mapuche groups maintained significant autonomy and large stretches of Valdivian rain forest proved an impenetrable barrier to settlement. While the region around the city of Valdivia had been settled and incorporated into the Chilean colony before the seventeenth century, by the mid-nineteenth century, as Benjamin Vicuña Mackenna, Valdivia's deputy in Congress, observed in 1868, the province's "civilized" population had diminished; abandoned settlements throughout the province had disappeared beneath "large stretches of new forests and second-growth forests."[20] The frontier and its forests, rather than receding, appeared to be extending their reach southward. The frontier discussed in this book was not a fixed space; through the twentieth century it was shaped by both ecological and human movement contained by neither natural nor political borders.

In the mid-nineteenth century, the southern frontier contained a wide diversity of ecological zones. Like central Chile, the territory south of the Bío Bío River is defined by the fertile and narrow level floor of the central valley walled in by the Andean cordillera to the east; it is lower than farther north, with altitudes that reach just over 3,000 meters, and the smaller coastal cordillera to the west, with altitudes that reach 2,000 meters at the thirty-eighth parallel and only 800 meters at the forty-second parallel. Because of the mountain barriers created by the Andes, the northern Atacama Desert, and the shift in climate north of the Bío Bío River, Chile's southern forests remained relatively isolated, a "biogeographic island" that contains numerous endemic plant and animal species, including some of the world's oldest conifers.[21] As in the North American northwest, the influence of the ocean produces both a temperate climate and frequent rainfall moving in from the west. In contrast to the relatively homogeneous temperate forests of the northern hemisphere, however, Chile's southern forests contain high levels of biodiversity due to the heterogeneity of the environments in which they grow. As Rodrigo Catalán Labarías and Ruperto Ramos Antiqueo note, over the millennia variations in temperature and climate at different latitudes and altitudes, as well as ecological disturbances caused by earthquakes, volcanic eruptions, and the movements of glaciers, made Chile's temperate forests a repository for an unusual quantity of indigenous flora and fauna.[22] Chile's southern frontier appeared to nineteenth-century travelers to possess a true anarchy of intermixed species of trees, underbrush, and ubiquitous creepers, vines, and wild bamboo.[23] As the German botanist Rudolph Philippi wrote in the 1850s, "The vegetation of the jungles of Brazil cannot be more varied than those of the Valdivian forests."[24]

Although both travelers to the south and government officials frequently described these forests as uniformly dense and virgin, contributing to the myth of a pristine nature, Chile's temperate forests were shaped by a long history of human intervention and ecological change. In the south, the Spanish conquest did not provoke the destruction of the southern forests. Rather, the arrival of the Spanish created conditions for the forests' regeneration and expansion throughout certain southern zones. Before the conquest, indigenous groups had engaged in a variety of practices, ranging from hunting and gathering to sedentary agricultural cultivation, that made their mark in the forests.[25] The biodiversity of southern Chile's temperate forests provided abundant possibilities for collecting forest products, food, medicinal herbs and plants, and hunting birds and mammals.[26] In coastal areas, fishing and collecting shellfish were a staple of Mapuches' subsistence and regional trading networks. In addition, before the conquest, indigenous groups engaged in significant levels of agricultural production. Mapuches used a mobile slash-and-burn agriculture, or swidden, burning patches of forest to cultivate crops such as maize, squash, potatoes, and beans for two or three years, then moving on to another patch and letting the forest grow back over the cleared land so the soil could recover its fertility. Fire was a key agricultural method of capturing nutrients from vegetation and returning them to the soil. In some areas, Mapuches pastured domesticated animals, primarily alpacas and llamas, although the scale of their pastoral economy was limited. Before the Spanish conquest, plains, prairies, and clearings cultivated with crops in the central valley interrupted the expanse of dense forest that covered the coastal and Andes cordilleras.[27]

The protracted wars with the Spanish produced a catastrophic demographic decline among Mapuches in Chile. Chronic warfare, slavery, and forced labor—and, most important, the spread of European epidemics—caused the indigenous population of central and southern Chile to collapse in a century to about 10 percent of its preconquest size. While Mapuche groups in the south maintained their independence from the Spanish, recurrent military conflict and low population levels undermined their sedentary agricultural activities and reinforced a tendency to raise livestock and hunt and gather. Pastoralism was also strengthened by the introduction of European cattle and sheep, which quickly became the cornerstone of Mapuche groups' economic activities. The radical drop in the indigenous population and Mapuches' extensive pastoral economy allowed forests to reclaim cleared land.[28]

In southern Chile, the Andes cordillera, which separates Chile and Argentina, is lower than in central Chile, gradually descending as it moves south from the Bío Bío River. Mountain passes allow the movement of people from one side of the cordillera to the other. By the nineteenth century, the Mapuche had expanded their military and trading networks across the Andes cordillera to the Argentine pampas and to Buenos Aires Province, leading to an "Araucanization" of nomadic indigenous groups such as the Pehuenches in Argentina and the Chilean cordillera.[29] In addition, during the second half of the nineteenth century, many Mapuches, fleeing warfare or dispossessed of the fertile land of the central valley and coastal piedmont by large estates and colonization concessions, settled in the mountain valleys of the coastal and Andes cordilleras, where they found land they believed to be public, meadows to pasture their livestock, and forest resources such as the piñon (the seed of the araucaria pinecone) to provide for their subsistence. By the beginning of the twentieth century, they were often joined there by Chilean campesinos seeking public land to settle. The Andes cordillera constituted not so much a simple line or national border between the Chilean and Argentine states as a space inhabited by Mapuche and non-Mapuche campesinos, who moved back and forth from one country to the other in search of work, pasture, and forest products and as part of a booming trans-Andean livestock trade.

While many Mapuche groups maintained their sovereignty over frontier territory until the late nineteenth century, they efficiently absorbed European crops and livestock. This did not amount to simple "ecological imperialism," in Alfred Crosby's famous words.[30] Rather, Mapuche groups selectively adopted some crops and species while limiting their use of others. Independent Mapuche territory possessed hybrids of indigenous and European species and agricultural systems. Indeed, Mapuches maintained their pre-European patterns of limited cultivation, combined with hunting and gathering in the forests, but folded new commodities into these economies. Mapuches' adoption of horses, cattle, and sheep facilitated their mobility throughout the frontier region south of the Bío Bío River, aiding their military capacity to confront the Spanish. However, pasturing livestock did not produce radical changes in indigenous groups' economies. Raising livestock permitted Mapuches to take advantage of the southern forests' biodiversity, pasturing their herds in different ecological zones in the mountains, with winter pasture in valleys at lower altitudes (invernadas) and summer pasture high up the cordilleras in high-altitude meadows (veranadas) and forests.[31] Mapuches supplemented livestock rearing with hunting and gathering. Nineteenth-century travelers' accounts contained exhaustive lists of the extensive buffet provided by the southern forests. One early twentieth-century botanical study listed more than sixty products Mapuches gathered in the forests, including wild strawberries, the fruit of the *copihue* (Chilean bellflower), berries of the *maqui*, an astonishing variety of mushrooms, several species of potato, other roots and tubers, and food from other farinaceous plants and legumes, including grains and edible oils. Mapuche groups made seasonal migrations into the Andes and coastal cordilleras to harvest piñones, which they made into nutritious flour.[32]

While Mapuche-Pehuenche groups in the Andes cordillera, which depended almost entirely on collecting piñones and pasturing livestock, engaged in almost no agricultural activity, more sedentary groups in the frontier's river valleys and around the shores of lakes cultivated indigenous crops such as squash, quinoa, potatoes, and beans, combined with European crops such as wheat, oats, and barley.[33] In the plains and valleys, Mapuches also cultivated orchards of apple trees, whose fruit could be made into *chicha*, a fruit liquor.[34] However, even in these more sedentary zones, seasonal migrations between microenvironments dictated the rhythms of Mapuches' economic life. While many Chileans viewed the Mapuches as a vagrant population of barbarian nomads with no fixed relationship to territory or land, their mobility was part of an extensive agricultural economy that reflected a dynamic engagement with southern Chile's ecological landscape. Movement from habitat to habitat and ecological zone to ecological zone allowed Mapuches to enjoy a certain degree of abundance, while minimizing both their labor and their impact on the land.[35]

Nonetheless, Mapuche agriculture and livestock rearing did make an imprint on the southern landscape. In addition to clearing land in the valleys to cultivate crops and maintaining patterns of slash-and-burn agriculture, Mapuches continued to use fire to manage southern Chile's environment. In low-altitude mountain valleys and on plains covered with native grasses, especially *coirón*, a perennial grass that grows in the frontier's mountain valleys, Mapuches burned over the meadows and pasture seasonally.[36] They employed fire in the mountains, where livestock were brought to pasture in the araucaria forests, as well. As the ethnobotanist David Aagesen argues, fire was used to burn the araucaria and araucaria beech forests' understory to facilitate the collection of piñones. However, these fires were largely low-level and controlled. In addition, because araucaria pines have a thick, fire-resistant bark, the fires would have inflicted more damage on the beech trees, which are more vulnerable to fire. One possibility is that by eliminating beeches, fires opened spaces for the hardier araucaria pines to spread. In this case, fire would have enabled both the regeneration and expansion of the araucaria forests.[37] Thus, despite the dense forests that covered the coastal and Andes cordilleras, at the end of the nineteenth century, the southern landscape had been marked by centuries of human management.

It is commonplace to view frontier spaces as on the edge of or outside both nation-states and global economies. But as Patricia Limerick has argued about the United States, frontier societies and economies historically have been integrated into, and dependent on, linkages to national and international webs of trade and investment, as well as political ties to nation-states.[38] In the case of Chile's southern frontier during the late eighteenth century and nineteenth century, Mapuche commerce in livestock, salt, and ponchos was incorporated into trade routes that carried these commodities to Santiago and Valparaíso and farther north to Lima. Mapuche groups established extensive trading networks both across the cordillera into the Argentine plains and across the Bío Bío River with Spaniards and Chileans during the late colonial and early republican periods. During the late eighteenth century and early nineteenth century, Mapuche groups met with Chilean traders at markets to exchange their cattle, ponchos, and salt for silver, clothing, tools, wine, and aguardiente.[39] Emblematically, even piñones, harvested by Mapuche groups in high-altitude araucaria forests, circulated in markets in Valparaíso and Lima during the late eighteenth century and early nineteenth century.[40] Trade among the coast, plains, and mountains, as well as with Spanish and then Chilean merchants, combined with seasonal migrations as a strategy for gaining access to the products of different ecological zones, producing significant prosperity for independent Mapuche groups. Nineteenth-century travelers to the frontier frequently wrote admiringly about the abundance produced by the Mapuches' extensive commercial, agricultural, and pastoral activities.[41]

Primitive Accumulation and the Conquest of the Frontier

After the 1850s, the Chilean state initiated the incremental military conquest of the frontier, a process that slowly compressed independent Mapuche territory as lines of military forts pushed southward from the Bío Bío River, eventually integrating Chilean national territory during the intermittent military conflicts known as the "Pacification of the Araucanía." On the Argentine side of the border, a parallel military campaign during the 1870s, the "War of the Desert," pushed the frontier's eastern border back to the Andes cordillera. As the historian Consuelo Figueroa has argued, the southern frontier and its "pacification" have played a minimal role in Chile's national imaginary compared with the incorporation of the northern frontier provinces of Antofagasta and Tarapacá during the War of the Pacific (1879–83) with Peru and Bolivia. The heroic exploits of the Chilean soldiers and then of the nitrate miners in the north—embodied in the figure of the *roto chileno* (literally, "broken one," a derogatory term for lower-class Chileans)—came to symbolize Chile's distinctive historical trajectory and national cohesion. The *roto* moved from a despised distillation of the supposed vices of Chile's vagrant rural laboring classes to an emblem of national strength and masculine virility memorialized in poems, songs, and monuments. As Figueroa notes, the violence defined as military glory that accompanied the conquest of the northern frontier and its arid but nitrate-rich deserts has been celebrated for more than a century in nationalist iconography, historical narratives, school textbooks, and a parade of monuments to the heroes of the war with Bolivia and Peru. The defeat of independent Mapuche groups and the colonization of the southern frontier, however, have never found similar representational power. Judging from the absence of monuments, museum exhibitions, and military histories, the Pacification of the Araucanía included few acts of heroism or national greatness.[42]

In this book, I argue that one of the reasons for the absence of commemorations of the Pacification of the Araucanía derives from the incomplete history of colonization in the southern frontier territory. Unlike in the resounding military victory over Bolivia and Peru, the Chilean state's hegemony over its southern frontier remained limited even after the conquest of Mapuche groups was completed in 1883. The state's weak presence in the south meant that it resorted to violence to exercise authority in more naked ways than in central Chile. Indeed, many of Chile's worst episodes of state-directed repression before the Pinochet dictatorship were located in the south, where countless acts of low-level violence against campesinos, including the infamous massacre of hundreds of rebellious rural laborers in 1934 in Ránquil, high up in the Andes near the headwaters of the Bío Bío River, marked the state's presence in frontier territory. Because of the state's restricted reach in the south, social conflict also took on a more violent cast. Whereas in central Chile, landowners, backed by the state, held the power of coercion and built paternalist social relations with inquilino resident estate laborers, on the frontier campesinos frequently challenged both landowners' and the state's authority in many local-level rebellions, land invasions, and protests. It is no accident of history that waves of land occupations by southern campesinos impelled the radicalization of Chile's agrarian reform between 1967 and 1973, and that some of the worst massacres after the military coup of 1973 occurred in the southern countryside. A common question asked of recent Chilean history is how a country famous for its exceptional political stability and multiparty democracy during most of the twentieth century could descend into the terrible state terror unleashed by Pinochet's military regime, one of the longest dictatorships in twentieth-century Latin America at almost seventeen years. Viewed from the southern frontier, the violence of the post-1973 period appears less surprising.

In this book, I build on the work of a number of recent histories of the frontier and Mapuche groups to argue that a view of Chilean history from the southern frontier provides a privileged lens onto the violence at the heart of state formation, as well as the violence involved in primitive accumulation, the dispossession of peasants from their land, the formation of modern property regimes, and the disciplining of workers to the necessities of wage labor.[43] While historians have focused most recently on the conflicts surrounding Mapuche communities' historic land claims, I demonstrate that the history of exclusion and violence they describe so well also included poor non-Mapuche settler (colono) and squatter (ocupante) smallholders, as well as landless laborers (peones or gañanes) and resident estate workers (inquilinos). Ethnicity, rooted in the fixed place of the indigenous community and articulated in distinct linguistic, cultural, and religious traditions, established separate social identities and potential tensions between Mapuche and non-Mapuche campesinos. However, common experiences of dispossession often laid the foundation for collective movements that included both Mapuche and non-Mapuche rural laborers who often toiled together on large estates, squatted as neighbors on both estate and public frontier land, followed the same routes of migration into the mountains to pasture livestock or across the cordillera to Argentina, and exchanged labor through sharecropping relationships.

In Chile's southern frontier territory, primitive accumulation took place through military campaigns to defeat independent Mapuche groups, Mapuches' settlement on constricted allotments of land termed *reducciones* (from the word "reduction," a term handed down by the Spanish empire), land auctions and colonization concessions that led to the lease and sale of Mapuche land now deemed "empty" and "public" (*terrenos baldíos* and *terrenos fiscales*), and the expulsion of Mapuche and non-Mapuche campesinos from the now public land acquired through military conquest. During the first decades of colonization, legal measures of appropriation of frontier land, accompanied by systematic fraud and violence that was often backed by local and national authorities, led to the accelerated formation of large estates and the dispossession of southern campesinos. This book places the story of the violent appropriation of campesinos' land on the frontier at the center of Chile's national history and the history of Chile's southern frontier forests.[44]

The frontier provides an especially privileged view of the role of the extraction of nature's wealth, or what some theorists have termed "ground" or "forest" rent in capital accumulation.[45] As David Harvey argues, frontiers have been places where the history of "accumulation by dispossession"— the forcible appropriation of land and natural resources as a condition for the constitution of private property regimes and the expansion of capitalist markets—is replayed.⁴⁶ Central to this process has been the enclosure and appropriation of the natural resources found in frontier commons, often in the developing world, during the nineteenth and twentieth centuries. In southern Chile, settlers burned extensive stretches of native forest to clear land for pasture and crops. They also set massive forest fires, guided by the teachings of botanists and agronomists, to change southern Chile's climate and reduce the heavy rainfall that made cultivating cereal crops such as wheat a difficult proposition. Beginning in the mid-nineteenth century, landowners also ignited forest fires to fertilize their soil, an expansion and distortion of Mapuches' slash-and-burn agriculture now applied to vast extensions of forest, and denied any process of rotation that might allow the forests to regrow.

Southern Chile's temperate forests constituted an ecological safety valve; as central Chile's cereal-producing estates confronted chronic crises due to deforestation, soil erosion, and drought, settlers on the frontier extracted "forest rents" from their properties, extending their cultivations of cereals and pasture onto land fertilized with the ashes of frontier forests. In addition, incinerating forests raised land values quickly on cheaply, often fraudulently, acquired properties that could then be resold by land speculators. Just as the extraction of nitrates on Chile's northern frontier propelled nineteenth-century growth, the appropriation of nature's wealth through burning forests produced an economic boom as the southern frontier region became the country's leading producer of cereal crops and livestock at the end of the nineteenth century. Deforestation had a social dimension, as well; setting forest fires to clear soil, establish land claims, raise property values, and introduce climatic changes beneficial to cereal cultivation also shaped the formation of large estates on the frontier and the expulsion of campesinos from their land. Large estates required access to extensive stretches of forest to fertilize soil for cereals and pasture. In addition, by the first decade of the twentieth century, and continuing for decades in a downward spiral, soil erosion, flooding, and drought due to deforestation forced many campesinos to sell their land to large estates and to join a swelling population of chronically underemployed, landless rural laborers.

Forestry, Campesinos, and the Modern State on the Frontier

Despite the long history of violence directed by estate owners at southern campesinos with the complicity of government officials, a history of exclusion is insufficient to explain the process of nation-state formation on the frontier. In this book, I argue that ecological disaster in the south drove Chilean governments' efforts to establish governance in frontier territory. During the first decades of the twentieth century, government colonization, land, and forest policies established many of the institutions and attributes of the modern nation-state, in large part because of the frontier's status as public property. The initial bonanza harvests produced by burning forests during the first years of colonization gave way to drought and widespread soil erosion within a generation. For many officials in charge of the colonization of the frontier, loss of soil and forests represented a threat to both the regional economy and the interests of the state. By the first decade of the twentieth century, colonization officials had begun to assert their authority over both natural resources and property relations, elaborating forest regulations to stimulate the development of scientific forest management and commercial forestry. These restrictions on logging and burning native forests represented some of Chile's first state interventions to regulate the extraction of natural resources and limit the rights of property owners.

Officials in charge of colonizing the frontier hoped to employ forestry science to remake the southern Chilean landscape by developing plantations of exotic species such as Monterey pine and eucalyptus. For land officials, a scientifically managed timber industry would substitute for the stagnant monocultural wheat economy, combat the ecological degradation produced by deforestation during the first decades of colonization, and provide an important source of revenue for the state. Federico Albert, a German forester, was appointed head of the new Forest Department in 1910 and quickly became instrumental in designing forest regulations. Much like Gifford Pinchot in the United States, a similarly European-trained forester who was the first head of the US Forest Service between 1905 and 1910, Albert promoted the extension of central state authority over natural resources, particularly woodlands in frontier areas, designing forest codes and establishing forest reserves and national parks. Albert's conservationism was not shaped by a preservationist sensibility. Rather, he drew on the same European *dirigiste* tradition as Pinchot to argue that rational forest management based on both cultivated native forests and plantations of exotic species was essential to national strength and the interests of the state.[47] After 1930, a growing corps of foresters and agronomists following Albert worked to impose state regulation of logging and clearing forest for agriculture with fire (*roces,* as this action is called in Chile) in response to ecological crises caused by deforestation.

Endemic social unrest on the southern frontier also propelled the state to assert a strong role in regulating both forest exploitation and property relations. By the early twentieth century, land fraud had deprived the state of hundreds of thousands of hectares of valuable frontier forests, which were quickly organized in large estates, undermining programs to populate the frontier territory with foreign settlers. Colonization officials abandoned their fantasies of settling southern Chile with European immigrants and turned to projects of "Chileanizing" the frontier by discouraging the continual migrations of Chilean laborers through mountain passes to Argentina and settling them on the Chilean side of the Andes cordillera. They argued that the frontier's transient campesino population threatened the sovereignty of the state over national territory and damaged the regional economy. In addition, this itinerant labor force tended to be disobedient and rebellious. Estate owners faced the ever-present danger of land invasions by squatters, settlers, and members of Mapuche communities who claimed frontier land. For land and colonization officials, the ecological catastrophes produced by rapid deforestation belonged to a more general crisis on the frontier, where state officials held little sway and social relations approached, in the language of many government reports, the level of "anarchy" and violent "class conflict."

From the beginning of the twentieth century, southern campesinos had organized militant movements to claim land. For campesinos, land on the frontier was public, and large estates' claims to private property were illegitimate. Because the forests, lakes, streams, and pasture of southern Chile had only recently been incorporated into national territory and parceled into private property, they composed, for many campesinos, a collection of natural resources that belonged to the public domain. Mapuche and non-Mapuche campesinos' views of this frontier commons did not always converge. Mapuches nourished an abiding sense of rights to land based on long histories of possession since "time immemorial." In addition, Mapuches drew on the legal status of indigenous communities and indigenous ethnic identity to lay claim to frontier land. Non-Mapuche campesinos employed a different language of rights. They invoked the frontier's status as public, drawing on land rights embedded in colonization laws. Despite these divergent conceptions of the frontier commons, campesinos' shared sense that private property organized in large estates was illegitimate drove chronic conflicts with landowners. They couched their claims to frontier land in terms of a moral economy grounded in their belief that the forests represented a public resource, a commons usurped illegitimately by estate owners to which they enjoyed use rights because of years of occupancy and labor. Campesinos pointed to estate owners' destruction of native forests to underline the illegitimacy of their property claims and to justify land invasions that spanned the spectrum from squatting to violent rebellion.

For the Chilean state, establishing authority over frontier territory required reducing the complex ecosystems of the native forests and the disorderly social order embedded in these apparently chaotic natural landscapes, to the scientifically managed landscape of tree plantations.[48] State colonization and land officials, following Albert, promoted forestation with Monterey pine as the solution to southern Chile's environmental and social woes. For government officials, Monterey pine constituted a valuable crop that could replace the frontier's dwindling harvests of wheat on soil degraded by overgrazing and monocultural cereal cultivation. Monterey pine also provided an excellent alternative to native forests, whose extraordinary heterogeneity rendered them difficult to log profitably. Monterey pine could be grown cheaply and quickly to supply expanding national and international markets for lumber and paper pulp. In addition, while little was known about the ecology of Chile's temperate forests, a great deal of forestry science was devoted to cultivating and managing pine plantations.[49] Pine held both the promise of development and the imprimatur of forestry science; in the context of unregulated deforestation in the frontier territory, it provided a new method of rational land use and forest exploitation.

Pine also offered a technocratic resolution to the tension between state efforts to exercise governance of frontier territory and large landowners' interest in extracting quick profits from the forests. The state's assertion of authority over forests and public land established a new logic for placing public interest over private property rights. However, rather than remake southern property relations or introduce sustained-yield timbering in native forests, colonization officials in Chile after 1930 focused on industrial forestry rooted in pine plantations on large estates and state-managed forest reserves to promote economic development while they looked to preserve frontier forests, primarily in inaccessible zones of the Andes cordilleras, in protected national parks.[50] Within government-administered reserves, foresters supervised leases to timber companies logging native forests and then assisted with their reforestation with Monterey pine. On privately held large estates, the Forest Department also oversaw logging in native forests and provided subsidies for reforestation with pine. Plantations of Monterey pine, rather than the managed exploitation of native forests, came to define sustainable logging both on estates and within public forest reserves. This forestry compromise benefited both landowners and timber companies for whom pine plantations held a series of benefits, including state incentives, as well as a means to recover soil degraded by agriculture, livestock, and logging. Monterey pine also operated to free landowners from the social and economic costs of unruly laborers, since pine required far fewer workers than agricultural crops such as wheat.

Modern forestry production regulated by the state did not, however, serve the interests of the south's large population of landless campesinos. State officials viewed campesinos as a threat to forestry development and sought to transform their relationship to the forests by turning them into trained and settled forestry workers. Governments dedicated to social reform, from the Popular Front coalitions of the 1940s to the Christian Democratic government of Eduardo Frei (1964–70), looked to pine plantations and industrial forestry to settle southern Chile's itinerant rural laborers by transforming them into full-time forestry workers, in effect severing them from their access to a makeshift peasant existence and resolving their often violent movements to wrest land from estates and colonize public land.

Foresters' focus on reforesting with Monterey pine led them to favor large logging interests over campesinos, since, they believed, campesinos possessed neither the capital nor the cultural resources to cultivate and manage tree plantations. The state leased logging rights to native forests in public reserves to large timber companies on the condition that they reforest with Monterey pine, rejecting campesinos' frequent petitions for public land to colonize. In national parks, by contrast, land officials viewed campesinos as a threat to the preservation of remaining stands of old-growth temperate forests and expelled them from plots to which they hoped to win legal titles as colonos or, in the case of many Mapuche communities, land they had occupied for generations. In response to these conservationist restrictions, throughout the twentieth century campesinos often invaded and squatted on land within national parks and forest reserves. This nurtured the stereotypes wielded by estate owners and state officials that campesinos constituted a threat to both native forests and their programs to produce forestry development based on cultivating pine plantations. Chile's history of conservation was thus rooted in the often violent expulsion of campesinos from land covered with native forests and land designated for reforestation with pine.

Nonetheless, the state's efforts to establish its governance of frontier land and forests during the first half of the twentieth century set the stage for a sweeping agrarian reform throughout Chile between 1967 and 1973. Decades before the agrarian reform, colonization officials sought to take back public frontier land and settle it with colonos, either purchasing land from estates or, in some cases, using powers of expropriation encoded in colonization and forest laws. The agency established to oversee the colonization of the frontier, the Caja de Colonización Agrícola (Agricultural Colonization Fund), for example, would become the agrarian reform agency of the 1960s, the Corporación de la Reforma Agraria (Agrarian Reform Corporation). In addition, laws that established state control of forests in reserves and parks also established a basis for asserting public over private property rights. In southern Chile, the agrarian reform drew on forest laws that had been on the books since the 1930s in some cases to justify legally the expropriation of large estates and their incorporation into national parks or forest reserves. In other instances, the state established centralized centers of logging and forestry production directed by trained foresters on expropriated forestry estates with the goal of building a scientifically directed, sustainable exploitation of the native forests based on cultivating tree plantations. Rather than rupture, the agrarian reform, even at its most revolutionary during the socialist government of Salvador Allende, represented basic continuities with state forest and land policies that dated back to Albert and the organization of the Forest Department.

The imperative of agrarian reform during the 1960s in southern Chile was driven by increasingly acute ecological crises. In the south, ecological degradation was a key component of the more general crisis of agricultural production that hampered Chilean economic development during the 1960s. Declining cereal yields were the result of estates' inefficient strategies of land use, but they were also caused by chronic soil erosion, flooding, and drought produced by the destruction of southern forests on campesinos' small plots. Over the course of the twentieth century, campesinos' subsistence came to depend increasingly on forest resources and often led to the degradation of the forests they inhabited. For many ocupantes and aspiring colonos, deprived of legal rights to land and pushed into mountain forests by the expansion of large estates, cutting wood from the forest was a vital source of subsistence. Likewise, reduced plots on marginal agricultural lands, uncertain land tenure, and the threat of appropriation at the hands of large estates pushed many Mapuche campesinos to exploit small patches of native forest to produce lumber, railroad ties, and firewood. This reinforced the opinion of foresters who worked with the agrarian reform agencies that campesinos could not be entrusted with the care of the forests.

A set of competing and contradictory imperatives shaped the agrarian reform in the south. The Frei and Allende governments sought to remedy ecological crises in southern Chile by investing in both pine plantations and cellulose plants. They also looked to address southern Chile's acute social inequalities by providing jobs for the large population of land-starved campesinos in new forestry industries. However, while both governments promoted forestation and forestry industrialization, they also worked to meet campesinos' demands for land. Land redistribution to meet a swelling campesino movement to take back frontier land from large estates frequently conflicted with state-directed programs to forest with pine and rationalize forestry production. Southern campesinos' resistance to proletarianization and their continuing struggles to maintain a peasant economy that relied, in part, on access to forest resources placed them in tension with agrarian reform organized around forestry industrialization. This was especially the case for Mapuche communities, which mobilized during the agrarian reform to recuperate land that had been usurped by large estates since the beginning of the century.

Nonetheless, many campesinos, Mapuche and non-Mapuche, embraced forestry development during these years. With state support they forested their small, eroded plots with pine and worked in government forestation projects, laboring to turn expropriated estates into forestry cooperatives guided by foresters working in the agencies of the agrarian reform. Supplied with subsidies, secure land tenure, and basic infrastructure such as seeds, saplings, technical assistance, and markets, campesinos looked to forestation as a means to recuperate eroded soil. They viewed forestry as a viable component of mixed agriculture in which tree plantations shared land with customary garden crops, livestock, wheat, and oats. This was true especially in regions where native forests were fragmented or cleared and where soil erosion had advanced, such as the foothills of the coastal cordillera. The agrarian reform allowed a large measure of local control and expanded access to land, pasture, and forests, making it possible for campesinos to combine forestry and forestation with their more traditional agricultural and pastoral activities.

In addition, while conservationist ideology, wedded to forestry development policy, served as a tool for excluding campesinos from forests during the first half of the twentieth century, by the 1960s campesinos had begun to employ environmentalist arguments in their attacks on the southern frontier's landed estates. As campesinos went on strike, invaded estates, and demanded agrarian reform, landowners' unchecked destruction of the southern forests stood as evidence of their backward and "feudal" status as *latifundistas* and reflected the illegitimacy of their property claims. During the Popular Unity government, campesinos, allied with agrarian reform officials, turned to both environmentalism and the reigning forestry ideology to justify the expropriation of large estates. Landowners who failed to manage their forests sustainably were deemed the opponents of modernization, feudal landlords responsible for the underdevelopment and ecological degradation of Chile's rural sector. Campesinos drew on an older moral ecology, rearticulated in modern environmentalist language, to demand forestry estates' expropriation.

After 1973, the Pinochet dictatorship dismantled the agrarian reform, auctioning off forestry estates covered with pine plantations and pulp and paper plants, often developed with significant state inputs, to a small group of large financial conglomerates, following the free-market policies dictated by the Chicago boys. Southern campesinos experienced another moment of dislocation, often expelled at gunpoint from land they had recuperated during the agrarian reform. They also faced the realities of a countryside that was increasingly dominated by large estates owned by major forestry conglomerates and planted with pine, as forestry companies replaced land once devoted to crops and livestock or covered with native forest with monocultural tree plantations. In the context of the vi-

olent repression unleashed by Pinochet and the radical socioeconomic inequalities produced by his dictatorship's free-market policies, pine came to
represent an alien commodity that was responsible for campesinos' loss of
land and livelihood. The brief moment when Monterey pine and forestry
held the promise of equitable development vanished, along with peasants'
small plots, engulfed by a swelling sea of tree plantations held by the most
powerful financial groups in Chile.

Over the course of the twentieth century, then, environmentalism's social meanings shifted. During the early decades of the twentieth century,
conservation belonged to the realm of forestry science and agronomy and
to the state officials who presided over the colonization of the frontier.
It often served both the state and the landowners in their efforts to produce a modernized forestry economy and rationalized social and natural
landscape. In the 1960s, however, modern environmentalism and forestry
science offered campesinos a new language of rights and an important
weapon in their arsenal to attack the dominance of large estates and logging companies. During the 1980s and 1990s, rural laborers and Mapuche
communities began to cast their struggles to recuperate land as a defense
of biodiversity and the remnants of southern Chile's frontier forests. In
addition, they called attention to the ecological and social disequilibria
in Chile's free-market industrial forestry economy, proposing instead a
model of environmental management that allowed for both ecological and
social sustainability rooted in local uses of native forests. They detached
environmentalism from the industrial forestry ideology to which it had
been wed for much of the twentieth century. After the end of the Pinochet dictatorship, Mapuche communities drew on generations of struggle
to organize a powerful movement to recuperate land and forests usurped
during the colonization of the frontier from the large forestry estates that
dominated the southern countryside. In doing so, they proposed an alternative to free-market forestry development in which the viability of local
campesino communities was linked to the sustainable use of southern
Chile's temperate forests.

La Frontera charts the shifting social and ecological relationships
that shaped over a century of frontier history. It draws on a wide variety
of sources, many of them never before examined by historians. To trace
government forest policy, I use a diverse group of government studies and
reports that document changing and often contradictory colonization
and forest policy from the Oficina Mensura de Tierras (Office of Land
Measurement), Inspección General de Tierras y Colonización (General
Inspectorate of Land and Colonization), and the offices of the Protectorado Indígena (Indigenous Protectorate), located in the Biblioteca del
Congreso Nacional (BCN) and the Biblioteca Nacional (BN) in Santiago.

In addition, I supplement my review of these government records with extensive research in the largely uncatalogued and hitherto unexamined archives of the Ministerio de Tierras y Colonización (Ministry of Land and Colonization), housed in the Archivo Nacional de la Administración in Santiago (ARNAD), and the archives of the frontier's regional governments (intendancies) located in the Archivo Nacional in Santiago (AN) and the Archivo Regional de la Araucanía (ARA) in Temuco. The government records for indigenous communities and the uncatalogued records of the indigenous courts, Juzgados de Indios, located in the Archivo General de Asuntos Indígenas in the Archivo Regional de la Araucanía, are an important source of information about the history of the formation of indigenous reducciones and often provide documentation of the communities' struggles over land and forests.

The files of the Agrarian Reform Corporation (Corporación de la Reforma Agraria, CORA), located in the Servicio Agrícola y Ganadero (Agricultural and Livestock Service, SAG) in Santiago, and the Corporación de Fomento de la Producción de Chile (Chilean National Development Corporation-CORFO), located in the Archivo Nacional de la Administración, contain reports from surveyors, agronomists, and foresters working to implement agrarian reform during the 1960s and 1970s and, following the military coup, to reverse the agrarian reform, as well as petitions from landowners and campesinos. These records provide a lens onto the highly conflictive process of agrarian reform and counter-agrarian reform, as well as the process of forestry development from the 1960s through the 1980s, offering a picture of agrarian reform as a process shaped by ecological change and state environmental policy.

While the documents in government archives—from the records of regional governments to reports issued by ministries in Santiago—often reflect the point of view of state authorities, they also provide a wealth of information reflecting the complex texture of local histories. In these files, petitions from Mapuche communities, colonos, and ocupantes, as well as from estate owners, and reports from local officials in the land and colonization offices, paint a rich picture of local conflicts over land and labor throughout the frontier. In addition, the reports by local offices of central government agencies offer details about social and ecological processes on the ground. The reports written by surveyors, foresters, and indigenous protectors working for regional branches of state agencies or ministries provide an invaluable perspective on the intersection of state formation, ecological change, and social experience. While they were often the agents of central state authority, these government officials more often than not reflected back to the state local interests that were not always in line with the national-level policy makers in Santiago. Surveyors, foresters, agronomists, and indigenous protectors, for example, labored to establish the authority of the state over both nature and society on the frontier and at times worked to further the interests of logging companies and large estates. But campesinos also frequently found allies in officials working to measure property boundaries, adjudicate land disputes, and regulate logging. In general terms, the archival data generated by these engineers of state formation reflect the social and environmental process of state building on the frontier. In addition, rather than depicting solely the top-down imposition of state policies, the work of these state officials who labored to untangle and resolve land disputes and adjudicate competing claims to natural resources help us to see state formation as a negotiated process. These archives provide both a view of the ways in which state forests and forestry policy were designed to remedy the frontier's ecological catastrophes and resolve the region's notorious social "anarchy" and provide insight into the local conflicts on the ground that helped shape the course of the frontier's social and environmental histories.

While there is a perhaps inevitable tension between the story these archives tell of state-directed forestry development, on the one hand, and the particular stories of local communities in southern Chile, on the other, we can read these state-generated archives "against the grain," in the famous words of the historian Ranajit Guha, to arrive at a sense of campesinos' actions and intentions as they disputed public land, land held by private estates, or forests they deemed theirs because of use rights grounded in custom.[51] Petitions and letters from workers on estates and in timbering enterprises and from ocupantes, colonos, and Mapuche communities fill the files of these government archives and allow us to trace the ways in which the frontier's laboring poor, often landless or land-starved campesinos, experienced industrial forestry development during the twentieth century and built their own ideas about the just uses of the forests. These records give us a clear sense of state formation as management of nature, but also of state formation as an incomplete process, negotiated on the ground by the state's agents with local campesino communities wielding their own understanding of nature and rights to natural resources.

The frontier's diversity of ecological zones and shifting social formations makes writing the history of the region a difficult business. Campesinos' social histories in the frontier are as complex and varied as the region's heterogeneous and ever-changing ecology, complicating the historian's work of discerning patterns and establishing a general narrative of a singular frontier history. Even within ecological zones, social histories often follow diverse trajectories, making simple reduction, or even correlation, of social to environmental history impossible. In one ecological zone, for example, a number of Mapuche communities might have widely varying histories of relationships to local estates, the state, one another, and the ecology of the native forests, despite shared geographical location. A historical narrative of state formation on the frontier as centralization and simplification confronts, on the ground, an array of local ecological and social histories that defy reduction to a story of progress or declension (from whichever side you choose to approach it) toward the scientifically directed landscape of tree plantations and commercial forestry, a stable regime of private property, and ordered social relations produced by proletarianization. Despite the best efforts of state officials, from Federico Albert to the Chicago boys, to redraw southern Chile's social and natural landscape, local realities defied simple characterization. The tension between state projects for engineering society and nature and the diversity of local social and environmental histories define the history of frontier as multiple.

La Frontera is not an ethnography or case study of one community or even of a handful of communities.[52] I use oral history interviews with forestry workers, labor militants, foresters, and members of Mapuche communities. But the texture and depth of social and cultural experience that only a more ethnographically inclined community study can provide are largely sacrificed to a larger-scale examination of the social experience of the environmental changes produced by colonization and forestry development among a wide variety of campesino communities throughout the frontier's different ecological zones over the course of the century following the final defeat of the Mapuche in 1883. Between 2000 and 2012, I interviewed a number of forestry workers and union leaders, members of Mapuche communities and organizations, campesinos, and foresters. In Concepción, I interviewed directors of the union federations the Confederación de Trabajadores Forestales (CTF) and the Federación Liberación, as well as agronomists working with the Catholic Church's Departamento Campesino. In Valdivia, I interviewed former residents of the Neltume logging camp and foresters employed by the Corporación Nacional Forestal (National Forestry Corporation, CONAF) who also worked in the Complejo Maderero Panguipulli. In Panguipulli, I interviewed former and current workers from Neltume and the Complejo Maderero. Also in Panguipulli, I interviewed members of the Mapuche organization the Parlamento de Coz Coz. In Lonquimay, I interviewed members of the Mapuche-Pehuenche Bernardo Ñanco and Quinquén communities, as well as foresters involved in forestry education programs in the region. In Santiago, I interviewed a number of foresters employed by CONAF. While many of the people who agreed to be interviewed are cited by name, I have left others anonymous to protect their privacy and safety, identifying only their locations and more general affiliations.

The oral history interviews illustrate the broader patterns and local texture of social and environmental change outlined in the book, especially in Concepción, Panguipulli, Valdivia, and Lonquimay, all major centers of forestry and locations of enduring land and labor conflicts. Yet inevitably, in tracing these different regional frontier histories, some of the complexities and particularities of individual local community histories are lost to the broader social and environmental processes I hope to elucidate. In the end, my goal is to paint a broad portrait of the frontier's environmental and social history, accounting for local variations by focusing on the histories of different regions and ecological zones to reveal broader comparative processes and patterns within the frontier's ever-shifting geography and history.

1

Landed Property and State
Sovereignty on the Frontier

In 1912, the Comisión Parlamentaria de Colonización (Congressional Colonization Commission) published a lengthy report of its investigation into the settlement of the southern frontier. The report provided a detailed indictment of the failures of colonization, underlining that the state had established only a feeble administrative grasp of the territory acquired through military conquest from independent Mapuche groups during the second half of the nineteenth century. It noted that early on, the state of war between Mapuches and Chileans had led to unsuccessful implementation of colonization laws, especially regarding the protection of indigenous and public land, since much of the frontier territory remained "unexplored." Then, as the railroad followed the military, advancing to the provinces farthest south, it brought "pacification, industrial progress, and rising land values." But it also brought increasing interest in acquiring property and the "abusive deforcement of public property, the dispossession of indigenous property, the indeterminancy of boundaries, the fraudulent altering of these boundaries, . . . the preparation of artificial property titles, . . . fictitious contracts, simulated partitions, fraudulent land registrations etc."[1] For the state, this proved to be disastrous in terms of establishing its dominion over the frontier and its forests, since "without a mapping of the territory, and often without inspection of locations, it is often very difficult for [the courts] to render correct decisions about the ownership of immense extensions of land." On the frontier, large estates had been formed, often with "vague titles, which signal as a boundary a river whose name often varies, a string of hills that is exactly the same as the strings of hills that extend in different directions in an uninterrupted line." According to the Congressional Colonization Commission, the state had lost large "extensions of land that undeniably belonged to it."[2]

The commission's journey through the southern provinces was provoked by a series of violent social confrontations between landless squatter ocupantes, aspiring colonos, and Mapuche communities, on the one hand, and large landowners who, more often than not, had assembled large estates through fraud and speculation, on the other. The state's inability to exercise its authority over public property and establish social peace in the frontier territories pushed members of Congress to seek solutions to the problems produced by colonization. Deputies and senators from the Conservative and social-reformist Democratic and Radical parties introduced reforms of colonization laws to favor "national colonization," the settlement of the frontier with Chilean smallholder citizens, as opposed to foreign immigrants. They also pushed for expanded state authority over public land and natural resources lost to enormous estates.[3]

Following the military defeat of the last independent Mapuche groups, the Chilean state implemented a series of often contradictory colonization policies to incorporate the new territory and promote its economic development. During the 1850s, the Chilean government had successfully settled a number of German colonists in the provinces of Valdivia and Llanquihue, burning vast extensions of southern forest to clear land and improve the rainy climate. As the line of forts demarcating independent Mapuche territory gradually moved southward from the 1860s to the 1880s, Chilean governments sought to reproduce this first experience of foreign settlement in the frontier provinces to the north, loosely the territory known as "la Araucanía." Yet the foreign colonization project quickly ran up against a number of obstacles. Frontier space was not composed of empty, fertile land awaiting the hand of European immigrants to transform it into an agricultural paradise. First, a large population of Mapuche groups controlled most of the territory. Second, by the late eighteenth century, non-Mapuche Chilean campesinos had begun to move south across the Bío Bío River, living at times amicably and at times in conflict with their Mapuche neighbors. Frontier land the state hoped to settle with European immigrants more often than not was occupied by Mapuche and non-Mapuche campesinos who were unwilling to cede their land to foreign settlers. Third, as the southern military line pushed southward from the Bío Bío, Chilean elites, often former military officers in the conquering army or high-level political officials, began to acquire large tracts of land, forming haciendas, or landed estates, in the Araucanía. With property on the frontier under the dominion of large landowners, finding territory to settle foreign immigrants became increasingly difficult during the late nineteenth century and early twentieth century.

These impediments to settling foreign colonists were further exacerbated by the weakness of the state's presence on the frontier. Indeed, the colonization of the frontier was fraught with tensions not only between Mapuche and Chilean understandings of property, but also between landowners' claims of private property rights and state claims of territorial sovereignty over "public" land won from the Mapuche during four decades of military campaigns.[4] As the Chilean state attempted to incorporate the new frontier territories, it found its capacity to exercise hegemony in the frontier undermined by Mapuche groups and Chilean squatters who often occupied public land without title and opposed the division and titling of new properties. More important, the formation of extensive landholdings by wealthy elites who rooted their property rights in often dubious land sales and land auctions rife with corruption and outright fraud, impeded the state's ability to establish political rule and a stable social order on the frontier. By the first decade of the twentieth century, the Chilean state's governance of frontier territory was defined by internal division and feeble administrative power. The state's inability to exercise governance in the frontier was particularly acute when it came to the region's temperate forests. As colonization policy turned from promoting clearing forests with fire to managing forests to build a commercially viable forestry industry, Chilean governments encountered a social landscape defined by violent conflict and rampant fraud that undermined any attempt to regulate either property relations or the destruction of the forests.[5]

Reducción and the Organization of Mapuche Communities

In 1866, the state passed its most important legislation on settling the indigenous subjects of military conquest on the southern frontier.[6] The law sought to impose state control of Mapuches by settling them on reducciones, literally reducing them to narrowly circumscribed sedentary communities. Rights to the land could be acquired only by proof of five years of permanent settlement for a group of households and one year for a single household. "Unoccupied land" belonged to the state. Mapuches' extensive networks of exchange, trade, and migration between different ecological niches meant that, for the most part, they did not establish permanent settlements. Their extensive understanding of territory collided with new laws that required an entirely different relationship to land based on settlement and intensive agriculture. The law of 1866 effectively reduced Mapuches' once extensive expanses of territory to small plots cultivated with crops or pasture. Implicit in this definition of property was an understanding that forests, representing unmanaged nature, were indications of lack of occupation and improvement and thus available for sale, auction, and colonization. The law essentially limited Mapuches' access to the forest resources they needed for subsistence while establishing incentives to clear forests to demonstrate property rights.

Colonization dictated a complete transformation of Mapuche groups' engagement with the southern frontier's natural landscape, a fundamental

change in how territory, property, and nature were understood and managed. Colonization officials reasoned that by reducing them to permanent
settlements with clear borders, Mapuches would be easier to govern, since
movement had always been Mapuches' most effective weapon in resisting
conquest. Settlement within reducciones also allowed the Chilean state to
restrict Mapuches' far-flung and dispersed social and economic networks,
reducing the territory they held to small reducciones and opening up huge
swaths of now public land, often covered with forests, for settlement. On
the new reducciones, Mapuches would become tied to a fixed place and
turn to peasant agriculture, primarily cultivating wheat and garden crops.
Settled and sedentary agriculture would civilize Mapuche "nomads," as
they were often called. Mapuches were concentrated together according to
putative family kinship networks, the land on the reducción granted to a
lonko (male head of the kinship line) and the members of his extended family. These new indigenous communities received only limited quantities of
land, while the state acquired huge new tracts of "empty" public land.

The law of 1866 had an essential component that would shape Mapuche communities' relationship to the state for more than a century: land
on the reducciones was held commonly and could not be alienated. The
law created a Protector de Indígenas (Indigenous Protector), whose job
was to represent the legal interests of the new Mapuche communities. The
deeply paternalist rationale for the law was that Mapuches required state
guidance and legal protection to shield themselves from the shysters and
land thieves who quickly invaded the frontier, following the military line
south. As a report from the Oficina Mensura de Tierras (Office of Land
Measurement) put it, "What was the reason for these proscriptions? [Was
it only] the protection of the interests of the *indígenas*, whom the state has
wanted to protect from the fraud and trickery of which they were victims?
No. The state has also attempted to preserve and protect land that falls under its dominion, [land] that, as the conquest proceeded, acquired value
and was coveted by the audacious." The report pointed out that differing
understandings of property had allowed speculators to engage in fraud,
usurping land that rightfully belonged to the state: "since indigenous land
had no drawn boundaries, and since the *indios* believed themselves to be
the owners of not only the land they occupied but also as much land as
stretched to the horizon, [the audacious] tried to acquire titles from them
to make legitimate their claims to unlimited extensions of land that was
public, which should have remained empty following the delimitation of
indigenous property."[7] Proscribing the alienation of land held by Mapuches was essential to the state's efforts to retain its control of the newly
acquired frontier territory, especially land described as *tierra baldía* (unoccupied, empty, and uncultivated).

The state's requirement that Mapuches demonstrate material occupation led to an extraordinary loss of land as Mapuche groups were reduced to small, circumscribed settlements. While perhaps a third of Mapuches did not even receive land through the process of reducción, those who did had to content themselves with very limited allotments.[8] A surveyor employed by the Indigenous Protectorate reported on the hundreds of lots of indigenous lands measured during 1905. Most reducciones had 150–300 hectares and were composed of numerous families, corresponding, on average, to 4–5 hectares per person. He noted that "they are given very little land since there are many families of four or five that only have three or four hectares of land. . . . [W]ith so little land, they can barely cultivate crops and have no place to pasture their animals, so they are obligated to take them to Argentina or to the canyons of the cordillera." According to the report, Mapuche communities possessed fewer animals than they had before they were settled because their small plots could sustain only limited herds.[9] Another report noted that, on average, the new land grant communities received no more than 5 hectares per person, which was "insufficient for their basic needs," and "the distribution of land to the indigenous, the original owners of this soil, is unequal compared to the distribution of lands to colonos who, at minimum, obtain 40 hectares and 20 [hectares] more for each male child older than ten."[10] A surveyor observed that the small allotments had been sufficient as long as the newly formed communities had "free access" to mountain valleys where they could take their livestock to pasture. However, as these valleys were auctioned off or granted as concessions and organized in large estates, many communities had to turn from raising cattle to cultivating crops intensively on their small plots.[11]

In 1907, a group of more than forty Mapuche lonkos from Cautín petitioned the government to expand the land allotments granted to the new indigenous communities. They argued that the number of hectares assigned to each community was insufficient, "taking into account that the indigenous dedicate themselves more to raising livestock than any other branch of agriculture." For the lonkos, no fewer than 10 hectares per person were required to "save our families and descendants from misery." The indigenous protector for Cautín backed their claims, noting, "This office has denounced repeatedly the limited extensions of land that are granted to the indigenous as being insufficient to satisfy their basic needs." Many Mapuche communities lost access to fertile lowland valleys and were settled on hillsides and in the foothills of the coastal and Andes cordilleras, on rocky soil vulnerable to erosion and that offered only limited amounts of pasture.[12] Reduction deprived Mapuches of access to the large stretches of land and multiple ecological niches that had defined their largely transhumant economy before 1883.

The small extensions of land granted to Mapuche *reducciones* were reduced even further by usurpations that often occurred simultaneously with the process of settlement. As land on the frontier acquired value with the extension of the railroad southward, the newly formed Mapuche communities faced constant threats to their property. The situation was exacerbated by the fact that private individuals knew perfectly well that "the Indigenous Protectorate counts on no elements for their defense; it does not have enough staff; it does not have the funds to pay the costs of judicial processes and for this reason the legal cases suffer long delays; it does not have an engineer [surveyor] who can work in defense [of indigenous communities]."[13] The many reports by indigenous protectors were unanimous in the sentiment that lack of funding and infrastructure made it impossible for them to pursue the task of defending indigenous *reducciones* from usurpation. In 1918, for example, the annual report of the Inspección Jeneral de Colonización e Inmigración (General Inspectorate of Colonization and Immigration) noted thousands of complaints lodged by Mapuche communities in the indigenous protectorates in Cautín, Malleco, and Valdivia and the obstacle to resolving the problem of the usurpation of indigenous lands posed by the judicial system. Mapuche demands for restitution were frequently held up in lengthy legal proceedings caused by the "appeals . . . entered by the private parties with the sole object of continuing the case for a long time during which they continued to control the lands usurped from the indios."[14]

These weaknesses in state institutions designed to protect Mapuche land from usurpation also undermined the process of settlement. Land surveyors often faced obstacles in mapping borders of Mapuche *reducciones*, resulting in loss of land to large estates. In 1901, for example, a government surveyor traveled to the Lonquimay and Bío Bío valleys to settle a number of Mapuche communities on *reducciones*. The surveyor was unable to measure the communities' borders, however, because wealthy landowners, who had purchased pasture at auction, "opposed [the settlements] with force." The landowners had obtained orders to suspend the settlement process in the region from the Inspección Jeneral de Tierras i Colonización (General Inspectorate of Land and Colonization). They made the case for their rights to pasture and forests historically occupied by Mapuche-Pehuenche groups in the region by arguing that the communities were not "Chilean" but, rather, "nomads" who had crossed the cordillera from Argentina. In this case, divisions between state institutions served the interests of landowners at the Mapuches' expense. The land surveyor and the land settlement commission had their authority to measure and grant land to Mapuche communities overturned by the land inspectorate, the agency charged with directing the colonization of the frontier.[15]

In Valdivia Province, laws prohibiting the alienation of indigenous land arrived late. An Indigenous Protectorate office for Valdivia and Llanquihue was established only in 1906. As Valdivia's indigenous protector noted in 1912, by the time the laws were passed, private parties already possessed thousands of land titles. Because the settlement commission began the work of placing Mapuches on reducciones and granting *títulos de merced* (land grant titles) in 1906, the land grab had proceeded unimpeded for years. When Mapuches in Valdivia began to petition for titles, usurpers already either occupied portions of the land in question or held titles to it. In Valdivia and Llanquihue, the protector reported that "private parties have sacked the southern territory, usurping land protected with the most shameful impunity. They have invented estates whose size fluctuates from 15,000 [to] 40,000 hectares. . . . Thus, the state has lost hundreds of thousands of hectares in the southern provinces, and the Araucanian race has been swindled of its lands." The protector argued that this generalized land fraud had led to the organization of large and unproductive estates, since many estate owners acquired the land for speculation.[16] Valdivia's indigenous protector summed up the history of land fraud in the province this way: "in the name of civilization and the progress of the Republic a few, yes a few chosen ones, have burned, left orphan homes, and cultivated tears on indigenous land. For the good of the country, the same fortunate few, have taken from the dominion of the state hundreds of thousands of hectares. Why if barely to work them? They simply maintain them in an unproductive state and wait for better times when they can sell them for healthy profits and then evict barbarously those who were born and raised there."[17] The land inspectorate noted in 1906 that only 4.5 percent of land titles in the province had fixed and clear boundaries. The other 95.5 percent had boundaries that were "completely illusory."[18] This was true partly because in Valdivia, land surveyors ran up against the often violent opposition of landowners when they attempted to chart property boundaries. A surveyor employed by Valdivia's Indigenous Protectorate reported in 1911 that on a number of occasions when he attempted to map indigenous property, landowners prohibited him from measuring the land they had usurped, threatening him "at gunpoint."[19]

In numerous cases, land assigned to Mapuche reducciones was lost when it was included in parcels auctioned by the state. Often the new communities found themselves surrounded by the land of a larger farm or estate purchased at auction and discovered that the landowners claimed reducción land, as well. As an indigenous protector argued in 1903, land auctions often preceded the settlement of Mapuches in land grant communities. This meant that the state auctioned off property and then settled communities within its borders, leaving newly formed Mapuche communities "encircled" by the new estates.[20] In addition, once they acquired land at auctions in Santiago, landowners frequently pushed the borders of their new property onto Mapuche reducciones, "mistaking" the boundaries and engaging in "fence running."[21] Auctioning parcels contiguous to indigenous or public lands became "a good business on the frontier," since after the auction, the new property owners could advance the boundaries of their lots onto lands that belonged to the state or to Mapuche communities, thus producing "a nice *finca* [hacienda] at little cost." Many communities had no access to roads to transport their crops, and when they attempted to cross land claimed by an estate, they were accused of trespassing or theft. In addition, it was difficult for small Mapuche communities hemmed in by surrounding farms to defend their land from constant encroachment by their neighbors. These transgressions of Mapuche reducciones were, for the most part, the crime of the "large wheat and timber producers."[22]

In 1903, a report by Cautín's indigenous protector described the various methods employed by private parties to steal indigenous lands.[23] Local tax collectors imposed excessive taxes on reducciones, at times fabricating tax bills and collecting up to ten times the legal tax. In addition, municipalities exaggerated the value of the land on reducciones to raise tax revenue. Drowning in debt, communities lost their crops, tools, and animals and even their land.[24] In addition, moneylenders provided loans to Mapuches, who required money to pay taxes or purchase goods and who used their land, livestock, and crops as collateral. Many lost their horses, oxen, and small harvests to municipal tax collectors and moneylenders.[25]

False *inquilinaje* contracts were another common method of acquiring Mapuche land. Landowners entered into contracts, often with members of neighboring Mapuche communities, in which the Mapuche workers agreed to serve their neighbors as *inquilinos* for a year or more. The contracts frequently designated the boundaries of the land to be worked by the inquilinos and in so doing expanded the neighboring estate to include the land that belonged to the Mapuches. The contracts were then used as evidence for the landowners' rights to the indigenous inquilinos' land. As a government report noted, "They pasture animals on the victim's own land, give him a small sum of money as an advance, and charge him with caring for the animals." Landowners acquired at once indigenous land and inquilinos to work it. Landowners often employed inquilinaje agreements to prove that Mapuches did not have rights to their plots since legally they were not members of reducciones but inquilinos.[26]

Mapuches continued to draw on earlier understandings of labor exchange and property rights to confront post-reducción realities. Reciprocal exchanges of labor and commodities had drawn the dispersed individual household units of Mapuche society together during the nineteenth century. Mapuches maintained these arrangements, often folding non-Mapuche peasant ocupantes or colonos into earlier patterns of land and labor exchange in rental or sharecropping arrangements. For Mapuche communities, renting land and sharecropping were important ways to supplement income, despite prohibitions in the new indigenous settlement laws.[27] Frequently, however, aspiring colonos took possession of Mapuche land by first entering to work the land "*a medias*" (on share) with the indigenous owners, renting the land or occupying it as sharecroppers and incorporating themselves into traditional Mapuche systems of reciprocal labor exchange. Once they could demonstrate that they were in material possession, they petitioned for titles as colonos. A surveyor in the office of an indigenous protectorate observed that "[the Mapuches] do not enjoy secure possession of their land since it is claimed by aspiring colonos nacionales [national colonists] who take possession of it, pretending to begin working on shares with its owners, renting it, or [working] simply as inquilinos. But, once they have cleared the land of forests, introducing improvements or occupying it for a time, they believe they have the right to petition for it as colonos." [28]

In one typical case, members of the Carileuf community near Temuco petitioned in 1907 to have a stretch of land occupied by Lorenzo Cayupil, a member of the reducción, restored. Cayupil had given permission to Otto Reusch to install a sawmill to exploit the community's forests. The agreement was to last one year, but following its expiration, Reusch claimed the land and petitioned the government to be granted a title as a colono. The community members complained that Reusch had threatened to use his contacts in the municipal government to acquire the land, prevented them from cultivating their own plots, and fenced a section of land within the reducción, leading to a number of "violent confrontations." The situation was exacerbated by the fact that since the community had been granted 250 hectares, its original population of sixty-nine "had grown considerably." In addition, the community had already lost a 20-hectare plot, which the government had separated from its original land grant and assigned to a colono.[29] Cayupil's action indicated a right to individual land ownership; the fact that he had entered into the contract as an individual reflected the persistence of claims individual families held to land within the new reducciones based on traditional forms of property rights rooted in individual households rather than an established community. The community's appeal to its status as an indigenous community whose lands were owned communally articulated a new property right that had been created during the process of reducción. Cayupil's right to rent land—a long-standing right and practice—was superseded by the community land rights enshrined in the título de merced. In this case, the logging contract with an individual member of the community served as a wedge Reusch could use to usurp forests for his sawmill. However, the community employed the land grant title to back its legal claims: Cayupil had had no right to enter into the contract, and the community's title had legal weight.[30]

In the end, the indigenous protector for Temuco backed the community's claims and reported that he had notified Reusch that he had to return the land. Nonetheless, the minister of land and colonization intervened to halt proceedings against Reusch, a frequent occurrence in which the judgments of indigenous protectors were overturned either by local officials or by the central authorities in Santiago. Finally, the General Inspectorate of Land and Colonization and the Land Measurement Office petitioned the Ministry of Land and Colonization to allow Reusch's expulsion. This lengthy bureaucratic internal conflict gave Reusch time to continue his illegitimate occupation, mill wood from the community's forest, build several houses on the land, and install a number of inquilinos on the plot he now claimed as a colono.

While the process of reducción has often been described as the transformation of Mapuches from pastoralists into campesino agriculturalists, Mapuche communities maintained many of their nineteenth-century economic activities, organized around patterns of mobility and migration. In large part, this was a response to narrow land allotments that simply did not permit a campesino subsistence economy. Migration to the mountains in search of pasture or forest products remained essential to many communities' survival. As the Cautín lonkos' petition of 1907 emphasized, raising livestock was the cornerstone of many Mapuche communities' subsistence, but limited land made it impossible to maintain this way of life. As a result, many communities stretched their economic activities beyond the borders of the reducciones, sending members across the cordillera to trade for livestock or find work on *estancias* (ranches) on the Argentine side of the border. A report from the Office of the Indigenous Protectorate noted that "it is not feasible to dedicate two, three, four hectares to crops and pasturing animals on bad-quality soil. . . . Many for this reason are obligated to take their animals to pasture in Argentina."[31] In addition, Mapuches found part-time seasonal work on the large properties of Chilean landowners. Although the reducción circumscribed Mapuche land occupation, it failed to contain networks of migration and trade that Mapuches had been building since the late eighteenth century. Indeed, gathering forest products and pasturing cattle in mountain valleys were more important than ever, since reducción land could not provide for their subsistence.

However, frequent travel to Argentina or into the mountains also abetted the usurpation of land on reducciones. In a number of cases, individuals rented plots from Mapuches who were traveling to Argentina to engage in the cross-border livestock trade or to work on Argentine estates as seasonal laborers.[32] The Office of the Indigenous Protectorate reported that many Mapuches violated the proscription against renting reducción land, leading eventually to the occupation and appropriation of the land by *arrendatarios* (renters): "sometimes to get around the laws that prohibit these [rental] contracts, they disguise the infraction, and the arrendatario figures as someone who is merely in charge of the land while the owner is absent. As you know, the indios travel continuously to Argentina, where they take their animals to graze."[33]

Mapuches often turned to the institutions of the Chilean state to protect their land. During the first decades of the twentieth century, numerous Mapuche lonkos made the long trip to Santiago to present petitions to the government for titles to land they occupied or to have usurped land restituted. Indigenous protectors and, to a lesser degree, the offices of the land inspectorate received incessant petitions from indigenous communities and families demanding titles or the restitution of usurped land. These petitions reflected the new role played by the land grant community in defining Mapuches' relationship to land and territory, despite continuing patterns of migration into and across the cordillera. When a congressional commission traveled throughout the south to investigate the state of colonization, thousands of Mapuches traveled to meet it and petition its members to intervene to grant land titles, expand reducciones, or restitute usurped land.[34]

In the end, however, Mapuche engagement with the state brought mixed results. Indigenous protectors intervened in hundreds of cases to defend newly formed communities and demand the restitution of usurped land. But the communities' efforts to recuperate land often came up against the obstacles presented by the courts and the limited authority of the indigenous protectors who represented them in administrative and judicial processes. As one indigenous protector reported in 1903, "The actions of this office have only weak results.... When individuals who, through illegal rentals, fraud, or force, take control of Indian land refuse to obey the administrative notification that I make through the provincial or departmental governments, there is no means of making them respect these property rights." Laws protecting indigenous land and establishing state authority to defend indigenous land rights were a "dead letter for the most part."[35] Similarly, the land inspectorate noted, "In the rural communes where the Protectorate has only a weak presence the mayor, the sub-delegate and the judge of the sub-delegation conspire together to exploit the Indians, stealing their lands."[36]

In many cases, land on the frontier was subject to incessant litigation and conflict because of superimposed borders. Which borders counted—

the borders described in titles registered in the offices of the Conservadores de Bienes y Raíces (Property Registries) after an auction or the borders inscribed in a Mapuche community's land grant title—had to be determined by the land office and in court. In the end, both the restricted state presence on the frontier and internal conflicts among different levels of government authority limited the efficacy of appealing to state authorities to protect reducción land. Land office officials and courts countermanded the orders of indigenous protectors and ignored decrees emanating from Santiago. In Santiago itself, conflicts among different spheres of authority, such as the Ministry of Land and Colonization, the Land Measurement Office, and the Indigenous Protectorates, also undermined Mapuche communities' efforts to use the indigenous laws and institutions to defend their land. In many cases, the Ministry of Land and Colonization simply refused to implement the decisions made by indigenous protectors or to respond to denunciations of land fraud made by surveyors working for the Land Measurement Office. In part, this was because land and colonization ministers viewed it as their job to promote the settlement of the frontier with European immigrants, not to protect Mapuche land. At the local level, courts and regional officials mostly sided with large landowners, undermining the authority of indigenous protectors.

European Immigration

Following the defeat of the Mapuche, the Chilean state looked to foreign colonization to incorporate the frontier into the national polity and economy. The government established colonization agencies in European capitals and subsidized steamship passage for European immigrants to Chile. A law enacted in 1874 eliminated colonization by Chileans and established that "in the colonies founded by the state in indigenous territory, only colonos who are immigrants from Europe or the United States of North America will be admitted."[37] As late as 1899, the General Inspectorate of Land and Colonization argued that "it is a fact beyond all doubt that the increase of the population in the states that possess extensive territories produces a proportionate development in public wealth. . . . This is why foreign colonization is important and why the state must give it the most decided support. . . . Today . . . the moment appears to have arrived to direct ourselves to the empty land located to the south of the Bío-Bío with the goal of populating it."[38]

Yet the general results of foreign immigration to Chile were unimpressive. Between 1857 and 1897, only 38,528 immigrants settled in Chile.[39] Foreign colonization encountered two fundamental obstacles. First, the frontier was far from "empty." Thousands of campesinos already occupied and cultivated small plots throughout the southern provinces. During the 1850s and 1860s, Chile experienced a major economic boom based on exporting wheat to Pacific markets in Australia and the United States. As landowners expanded wheat cultivation on central Chilean estates, replacing the herds of livestock that had once dominated Chile's agricultural economy, as well as any remaining stands of hillside forest, they often expelled their resident inquilino laborers, transforming them into a landless and itinerant labor force that wandered the countryside in search of work. Many moved south of the Bío Bío River in search of unoccupied frontier land. In addition, many of the soldiers who fought in the wars with Mapuche groups stayed in the south and sought land from the state and legal recognition as colonos. These campesinos often put up active resistance to their displacement from land auctioned off or granted to colonization societies. In 1905, the land inspectorate observed, for example, that land granted to foreign settlers was "generally occupied against the will of the government."[40]

Second, immigrants encountered hostility among local elites. Estate owners monopolized frontier land, often through fraud and speculation, leaving little place for settlers to go. According to a report by the Ministry of Land and Colonization in 1899, many southern landowners left large tracts uncultivated and waited for land values to rise, investing little in their property and undermining the state's project of settling immigrants and developing agriculture on the frontier. The report concluded with the futile exhortation: "let the agriculturalists understand that the most secure and important business for them at this moment is the subdivision of their property in order to settle them with Europeans."[41] As land speculators and estate owners pieced together extensive estates, the state lost hundreds of thousands of hectares of public land on which it had intended to settle foreign immigrants.

Cornelio Saavedra, the architect of the military conquest of the frontier, was appalled by the land speculation and social instability that followed in the wake of military advances south.[42] He linked the appropriation of indigenous land to fraud and speculation and to the state's loss of control of the frontier territory and its wealth. In 1861, for example, Saavedra complained about "an evil that we have seen for many years, resulting from the fraud that occurs in the alienation of indigenous lands and that has produced chaos related to the legitimacy of land rights. It may well be that the indio who sells land deceives the buyer, pretending to be the owner. It may well be that the buyer, abusing the ignorance of the indio, induces him to err. The end result . . . is a tangle of interminable litigation, a confusion of rights."[43] He characterized the frontier's property regime as "dispossessed indios, defrauded Spaniards, the province deprived of commerce and agriculture, and the confusion of properties."[44] Saavedra further observed that "there is not a piece of that territory . . . that has not been sold, mortgaged, ceded freely or willed, and other deceitful dissimulations," concluding that "the majority of the [property] contracts are fraudulent."[45]

By the late eighteenth century, some Mapuches had already sold land around Concepción, Arauco Province, and by 1860, according to the anthropologist José Bengoa, almost all of the land between the Bío Bío and Malleco rivers was occupied and a number of large haciendas had been formed, often by high-level political officials and military leaders.[46] As Tomás Guevara observed in his early history of the Mapuche, one of the main methods of colonization employed by military leaders on the frontier was to purchase land for trifles from Mapuche lonkos. Military officers then sold the haciendas and turned a tidy profit.[47] In part, this appropriation of Mapuche land, as in the case of illegal rentals and sharecropping agreements, resulted from contrasting understandings of property and territory. Mapuches did not fence their land or define clear boundaries. In addition, Mapuche understandings of property clashed with those set down in the colonization laws. Mapuches often moved from community to community in search of fertile land or ecological resources, pasture for their livestock, piñones and other forest products, salt, and shellfish. A lonko might move with his entire family or extended kinship line, or an individual might enter other communities and enjoy usufruct rights to land based on reciprocal labor exchanges. The very impermanence of Mapuche settlement, the necessity of mobility to both survival and the accumulation of wealth, primarily in the form of livestock, meant that property rights held a different meaning. When Mapuches sold or rented land to Chileans, they probably understood the transaction to be a granting of usufruct rights—a loan or rental or, perhaps, part of an exchange, a type of sharecropping. In the 1860s, for example, the botanist Claudio Gay observed that "a cacique who leaves his land to live with another tribe does not lose his terrain, if others come to establish themselves on it, the community members who remain give it to them but only as a loan. . . . All the land purchased by the Spaniards has been sold by indios who have no rights to it."[48]

Land auctions exacerbated the problem of the fraudulent formation of large estates in the frontier territory. In 1873, the government initiated auctions to distribute the "empty" land of the frontier, particularly in Malleco Province, south of Arauco and Bío Bío. Despite legal strictures, which limited the amount of land that could be purchased at auction to 400 hectares, land speculators and wealthy Chileans, often frontier military and political authorities or merchants, pieced together large estates employing a number of illicit methods. At times, they used proxies to purchase plots. In other cases, speculators purchased lots and then resold them, allowing landowners to form large haciendas. In still other cases, despite the legal proscriptions, landowners simply purchased a number of lots at auctions, building extensive estates out of their 400-hectare parcels.[49] Many new landowners cultivated the good land, logged native forests, and resold the land at elevated prices. Few paid fully for the lots. A report by the Land Measurement Office observed that, following the first land auctions in 1873, "Those who purchased land often committed abuses, extending the borders with the goal of logging the forests and profiting from the first few harvests, while only making the first payment."[50] The high rates of delinquent payments reflected the fact that for most, this was a short-term investment made profitable by the quick wealth to be earned from logging the native forests or burning the forests to clear land and raise its value while producing bonanza harvests fertilized with the ashes of forest fires.[51]

Both the land auction law of 1873–74 and foreign colonization laws were written with the understanding that settlement of the frontier could not and should not be undertaken by Chilean rural laborers or campesino smallholders. They simply lacked the skills, knowledge, and capital to make the frontier productive and profitable. Only foreign settlers and those with enough resources to make a down payment on a large plot of land at auction had sufficient cultural and material resources to promote the frontier's economic development. Land auctions and schemes to promote foreign immigration transferred enormous stretches of land once occupied by campesinos to regional elites, as well as to Santiago's upper crust. As the report of the Congressional Colonization Commission narrated, "Many of the first occupants [of the frontier] were soldiers who returned victorious from the north and whose presence forced the Araucanian to surrender." In addition, landless peasants from central Chile followed the military line southward, "quickly filling the land with small agriculturalists in what were later the provinces of Bío-Bío, Arauco, Malleco, and Cautín." With land auctions and land grants to foreign settlers, this large campesino population lost its lands, leading to "the depopulation of that part of the country."[52]

A case that typified the dispossession of these smallholder ocupantes took place southeast of the city of Osorno. In 1894, surveyors measured 59,200 hectares of public land in Cancura to be sold at public auction. Like most of the land auctioned off in the south, however, part of it was occupied and had been cleared and planted by more than two hundred ocupantes. In 1895, following the auction, these campesinos petitioned the government: "with the auction, we are left without even a tiny piece of land to cultivate the crops most indispensable to the subsistence of our families, since the land cultivated with the sweat of our brows has passed into the hands of wealthy landowners. We have repeatedly petitioned the *Jefes* [political bosses] of this province for provisional titles to the public land we have cultivated, but they have never listened to us."[53] The intendant of Llanquihue responded to this petition by noting that "colonization is not for national citizens but for contracted foreigners." The campesinos' petition was denied, and many remained on the land as the new landowners' inquilinos or emigrated to Argentina.[54]

A similar and notorious case on the Pellahuén *fundo* (estate) in Imperial and Cañete erupted in 1913 in a bloody battle between colonos and ocupantes on one side and carabineros on the other. Pellahuén had been purchased in 1889 by Jervasio Alarcón from Estéban Freire without any defined boundaries other than rivers and estuaries, which allowed Alarcón to expand his recently purchased property by a number of hectares. A report by the Land Measurement Office in 1913 used Pellahuén to illustrate the many cases of fraudulent titles, wills, and proofs of sale that it had found in the frontier. It described how a "señor" (Freire) had a deed to land from 1884, years after the prohibition of sales of indigenous land, based on a sale by an indigenous "cacique" that had been registered illegitimately by a local notary. Strikingly, Freire had no idea where the land was, since, "as in all the deeds, this one could not have vaguer boundaries. They were more or less thus: on the north a swamp forest, on the east an estuary without name, on the south a fallen tree, and on the west a burned oak." The title to the land called Pellahuén was confirmed finally by the Chilean Supreme Court. However, it was difficult to locate the boundaries in the "ocean of mountains and forests." The report stated, "In one place the boundaries coincided . . . but the fundo was too small . . . That could not be it. In another place the exact boundaries appeared, but there were no valuable forests. . . . Finally the fundo ended up with 20,000 hectares of excellent land and woods."[55]

When an administrative change shifted one of the sub-delegations the estate lay on from Cañete to Imperial, the re-registration of the property expanded its boundaries to include a number of places occupied by about 1,500 ocupantes, many of whom worked in the region's forests and saw the wood they cut, as well as their harvests, lost to Alarcón, the nominally new owner of the land. In 1905, Alarcón described the estate as consisting of 61,000 hectares, of which 30,000 hectares were suitable for sheep and cattle. The remaining 31,000 hectares were covered with mountain forests in the Nahuelbuta coastal cordillera.[56] In 1913, Alarcón rented Pellahuén to Manuel A. Rios, who threatened to take the land and harvests of the ocupantes and colonos on the estate. When the campesinos resisted, Alarcón called on his brother, Matias Alarcón, the governor of Imperial, to provide carabineros to evict them, leading to a bloody conflict that was described by the Conservative Party Deputy Emilio Claro as "a true battle between an armed force and an enormous group of men, women, and children."

Claro stood up in Congress to denounce the fact that "some colonos nacionales who had been born and lived in those places that they had made
fertile with their sweat, clearing the forest and making the land suitable for
cultivation, irrigated the land with their blood."[57] Many of these laborers
had organized a union in 1906 and demanded distribution of the estate.[58]
For his action on behalf of the putative landowner, his brother, the governor was rebuked by the region's intendant, as well as by the regional and
national press and deputies such as Claro, but little was done. The carabineros remained stationed in Repolcura, in effect a police force in the service
of Alarcón and Rios.[59]

The social upheaval on the frontier led the Congressional Colonization Commission to conclude that, before new concessions of public land
could be made, it was "indispensable . . . to clarify and protect the land that
belongs to the state: to know with scientific certainty what land the state
owns."[60] As early as 1899, the General Inspectorate of Land and Colonization had attributed widespread land fraud on the frontier to its own limited administrative reach: "unfortunately, the enormous amount of work
of these offices, the considerable distances in the frontier region, the lack
of will of a number of judges, and government officials' absolute lack of resources to defend the public interest . . . have constituted an accumulation
of disadvantages for the defense of the public interest that have given rise to
numerous legal adventures that, unfortunately, have been accepted by the
lower courts, depriving the state of enormous extensions of its best land."
The state's weak presence on the frontier "emboldened . . . the adventurers
and exploiters of forests and public lands to attempt whatever absurd and
unjust lawsuit to assert ownership of whatever public land they want."[61]

A report by the Land Measurement Office concluded that the majority of land titles registered with the Property Registry in the frontier were
the product of fraud. Even worse, rampant speculation had given birth to
enormous and largely idle latifundia: "those who, with incomparable self-
ishness, believe themselves to be the owners of these great extensions of
land, [and] want to keep it without working it themselves or permitting
others to work it." These landowners, availing themselves of false titles,
also dislodged and expelled colono smallholders who had cleared and prepared land for planting: "it is impossible to erase from memory the desolate
image of the expulsion of the colono and his family. They burn his house,
destroy his crop, and raze everything." The colonos, the report observed,
suffered under the burden of constant uncertainty, since any day someone with a false title could evict them from the land they had cleared and
cultivated.[62]

For example, a government official in charge of frontier colonies noted
in 1911 that hundreds of colono and ocupante smallholders in the Depart-

ment of Villarrica lived under the threat of expulsion by private parties who attempted to appropriate their land. According to the official, "land thieves" relied on the fact that state officials enjoyed only limited authority in the region and would not oppose the registration of their titles in the Conservador de Bienes y Raíces, "even when they had no material possession of the land in the titles." The colonization official continued, "As these titles do not describe the surface area of the land except for vague and imprecise boundaries, they can include the extensions desired in sales to third parties."[63] The Land Measurement Office report concluded that the expansion of the large estate onto land cleared by colonos and ocupantes had led to "disputes that at times degenerate into bloody conflicts. . . . The class struggle in this zone has reached its highest point."[64]

Colonization Concessions

As a result of its failure to attract European immigrants to southern Chile, the Chilean state handed enormous land concessions in 1896 and then between 1900 and 1906 to private companies charged with settling immigrants on southern soil, in essence subcontracting the business of bringing settlers to the frontier territories. The logic driving concessions to colonization companies, rather than individual colonos, was shaped partly by the ecological realities of the southern forests. State officials reasoned that individual settlers could do little on their own to clear land and bring it under cultivation. Only private enterprises with access to capital could do the job of transforming the dense tangle of frontier vegetation into fallow land planted with grasses and cereals. A report by the colonization inspectorate issued in 1910 argued that "the impenetrable forests of our region place obstacles almost impossible to overcome in the way of individual effort; if a colono who receives a piece of land covered with forest has to clear it slowly since he has no resources other than his own labor, making it productive only after a number of years, it will be a very difficult task to populate this region." Not only could colonization companies clear the forest more efficiently than individual settlers; they could also take advantage of the forests by installing sawmills and developing the forestry industry: "a company that possesses large amounts of capital, that begins by exploiting the forest with its sawmills and establishes other industries, from the first moment will establish a nucleus of population that will grow as the industries organized by the company progress."[65]

Based on the argument that only large companies with access to capital could populate and make productive the forests of the southern frontier, the state granted staggeringly large stretches of land to colonization concessions contracted to bring European immigrants to Chile. The first contract, between the state and Carlos Colson in 1896, provided a model of things to come. Colson agreed to bring five thousand immigrants to Chile. Three thousand would be settled on 300,000 hectares in Cautín or Valdivia, 1,500 would be settled on 225,000 hectares in Llanquihue, and 1,500 would be settled on 100,000 hectares on Chiloé.[66] Similarly, in 1905, Luis Silva Rivas received rights to a vast empire of 200,000 hectares in Llaima with the agreement that he would settle thirty immigrant families over the next three years and introduce more families on whatever land remained after the first three years of his contract.[67] Also in Cautín Province, the Budi colonization concession received rights to 75,000 hectares to settle three hundred families from the Canary Islands, and the Nueva Italia concession received the rights to 31,000 hectares to settle one hundred Italian immigrants.[68] In Valdivia, the Sociedad Colonizadora e Industrial de Valdivia was granted a concession to settle European immigrants on 22,575 hectares.[69] During the first decade of the twentieth century, similar contracts disposed of extensive tracts of frontier territory throughout the provinces of Llanquihue, Valdivia, and Cautín.[70]

As with earlier programs to subsidize immigration to southern Chile, colonization concessions failed to populate the frontier with industrious European settlers. The land inspector noted in his report of 1908 that colonization companies had acquired "huge expanses of territory," but settled fewer than 5 percent of the number of immigrants promised, violating their contracts with the state.[71] The concessions often ran up against the obstacle of local elites and land speculators. One official from the land inspectorate argued that local land speculators had engaged in fraud to steal land from the colonization companies. According to his account, in the case of the Nueva Italia concession, speculators who had had their eyes on large stands of araucaria and coigüe forest in the Nahuelbuta cordillera had been impeded by Mapuche reducciones, which enjoyed legal protection and had staked claims to valuable stands of araucaria pine. They then turned to the lands conceded to Nueva Italia to arrange the settlement of the Mapuches there in order to free up the Mapuche-owned forests.[72] In 1906, the land inspector reported that Nueva Italia had many enemies; the most intransigent were those who occupied the concession's land and distributed it to their inquilinos: "the shysters, so abundant in the neighboring towns, took advantage of the anger that the foreign presence produced among the Chileans who aspired to own the land that their exploiters offered them and initiated many lawsuits."[73] They had used their inquilinos to demonstrate occupation and, therefore, property rights.

In 1896 and 1898, the Chilean government passed two laws designed to repatriate thirty thousand Chileans who had left southern Chile for Argentina as colonos nacionales. The 1896 colonization law provided for the settlement of Chilean workers from Argentina as "colonos nacionales" with the right to 80 hectares for each male family head and up to 40 hectares for each son older than sixteen in the provinces of Cautín, Malleco, and Valdivia. The law of 1898 established requirements for these colonos: that they know how to read and write (something not required of foreign colonos), that they not have been convicted of a crime, and that they be "fathers of families." These restrictions effectively discriminated against many Mapuche and non-Mapuche ocupantes who were illiterate. Rather than redressing the extraordinary injustice that had led to the formation of large estates and the expulsion of rural laborers, the laws served as a tool for local elites to claim public land by claiming the status of colono nacional. As in the case of Nueva Italia, landowners and merchants in the frontier frequently established land claims within the properties granted to colonization companies or on public land by installing inquilinos to clear land and cultivate crops, thus demonstrating their rights of material occupation. In 1903, the land inspectorate reported that the national colonization laws had backfired, because "numerous private parties who believe they exercise some right conceded by this law have occupied public lands asserting their status of colono nacional. This has produced serious obstacles, since they not only prevent the ordered plan for the distribution of land and subvert the actions of the authorities, but they also impede the populating of the zone, which is in the interests of the country and create ... prejudicial antagonism between Chileans and foreigners."[74]

In 1902, the land inspectorate reported that with the national colonization laws it had received a torrent of requests for land: "all the industrialists and office workers (*empleados*) in the towns on the frontier have presented petitions, forgetting that the fundamental requisite is to cultivate personally the land. They understand this to mean introducing inquilinos and workers."[75] In 1908, the land inspectorate observed that "there are numerous colonos who do not live on their plots of land."[76] In the Freire colony in Cautín, a government report noted, "many colonos do not reside on their land, leaving it in charge of inquilinos, who later become the owners themselves, asserting their rights [based on occupation and "improvements"], which provokes conflicts and court cases."[77] Indeed, over the following decades, inquilinos throughout southern Chile invoked rights of material possession to claim the land they worked and the status of colono.

The colonization companies also encountered an impediment to establishing dominion over hundreds of thousands of hectares of southern forest land in the numerous campesinos who already occupied land auctioned or leased by the state. Colonization contracts required the companies to respect the property rights of Mapuche communities, ocupantes, and colonos nacionales. In many cases, colonization companies paid both Mapuche and non-Mapuche campesinos for the "improvements" they had made clearing forest and planting crops or pasture and then left them on the land as their arrendatarios or inquilinos, arranging matters extrajudicially. As in the case of wealthy colonos and their inquilinos, this method of resolving land conflicts would later backfire when campesinos demanded rights to the property based on their long-standing occupation and their labor in clearing and cultivating land they considered public.[78]

A classic example of conflicts between large properties and campesino ocupantes took place on the Budi concession. Budi's foreign settlers encountered trouble in the "large number of ocupantes who had no title and who declared themselves to be the tenacious enemies of the company, as well as the *indios* who, lacking any order and arrangement, occupy large extensions of land." Not all of these "ocupantes" were independent small-holders. A significant number were inquilinos and *medieros* (sharecroppers) employed by local landowners who claimed thousands of hectares, although they lived in towns in the region.[79] The Budi concession worked tirelessly to expel both Mapuche and non-Mapuche ocupantes within land granted to the concession, despite its contractual obligation to respect their property rights. The presence of campesinos within the vast area granted to the Budi concession, their refusal to vacate the land, and a spate of lawsuits initiated by local elites who also claimed these "public lands" led to a decree in 1907 that granted the company 7,932 hectares for eighty-eight foreign colonos it had brought to Chile. The company ceded to the state 6,025 hectares to settle a large number of Mapuches and eighty-one non-Mapuche ocupantes, who occupied land within the concession. However, the same decree also granted the company absolute dominion over an additional 42,078 hectares of valuable forest.[80]

Despite this generous allotment, Budi did not respect the stipulation in the contract that required it to acknowledge the rights of campesinos to small plots of land they occupied within the boundaries of the concession. Over the course of the following decades, the company used force to dislodge ocupantes from their lands. In 1912, for example, a group of local merchants in the towns of Nueva Imperial, Carahue, and Puerto Saavedra petitioned the government to intervene on behalf of five hundred "ocupantes nacionales" who hoped to receive titles as colonos. Budi had instigated hundreds of court cases to have the ocupantes expelled, and in many cases had been able to expropriate their harvests and their land, even where the ocupantes had worked the land for as long as thirty years. Budi's legal efforts to evict the ocupantes, as well as numerous Mapuches, and gain control of their harvests and land posed a major threat to commerce in the region. As the petition noted, "The success or failure of commerce depends directly on the aspiring colonos and indígenas" in Budi. Notably, while the settlement with the government in 1907 required the surveying and remeasurement of the concession's borders and the land allotments of Mapuches and ocupantes, five years later this work had yet to be completed, and the Budi company had taken advantage of this lapse by going to court to have the campesinos evicted. Meanwhile, Budi established itself as one of the country's leading lumber producers, logging the forests around Lake Budi and in the foothills of the coastal cordillera.[81]

Like Budi, many colonization concessions simply refused to recognize the property rights of Mapuche reducciones and non-Mapuche ocupantes and colonos. In 1906, for example, Manuel Calfuala, representing a number of Mapuche communities in Rancahue, near Panguipulli, petitioned the president for support in a conflict with the Queule colonization concession over land in the Rancahue valley, which, he stated, the communities had occupied "from time immemorial" and which, he also stated, his "grandparents and great-grandparents had always worked and occupied." The Queule company, he complained, had expropriated most of the communities' land, including a cleared plain, cleared mountain forest, and land planted with crops. The company also fenced in the communities' land and installed three sawmills and storage facilities for wood cut by the colonos it settled within the concession. That, according to the petition, was when the problems began: "they felled trees in the midst of our crops, they threw their wood on our wheat, they pulled up our plants, killed our livestock with firearms . . . , and harassed us as much as they could." The land inspectorate upheld the communities' claims and ratified their denunciations of the Queule company's abuses.[82]

After only a few years, the state began to attempt to reclaim the hundreds of thousands of hectares granted to colonization companies because of their abysmal failure when it came to settling European colonists. Recuperating the land proved to be far harder than leasing or granting it, however, and the state quickly became entangled in countless lawsuits with the owners of colonization companies that lasted for years. In 1909, for example, the state attempted to cancel the contract signed with Luis Silva Rivas, who had managed to settle only seventeen Europeans on his many thousands of hectares. In addition, these seventeen families soon abandoned their land and were replaced by other Europeans who were already residents in Chile. Silva Rivas did introduce eighteen other European immigrants on the land conceded to him by the state, but they also were already residing in Chile, having been contracted by the state colonization agency. The government's inability to assert its authority was made clear when it failed to cancel the contract because of a number of lawsuits Silva Rivas had initiated against it for failing to fulfill its end of the contract. The presence of numerous campesino "usurpers" within his Llaima concession, he claimed, had made it impossible to settle foreign colonos.[83] In 1909, a compromise was reached. Silva Rivas received definitive title to 26,625 hectares in Llaima. From this land, 2,210 hectares were distributed to the thirty-five European colono families introduced by Silva Rivas. In addition, Silva Rivas agreed to allow the settlement of ocupantes who already resided within the boundaries of the concession.[84] In the end, however, of the nearly one hundred ocupantes on the concession, only twenty-eight received the status of colono nacional, including three families repatriated from Argentina, settled on 2,165 hectares. In 1912, Silva Rivas was granted 9,971 hectares more for introducing ten European colonos. In 1918, his allotment was extended to 40,000 hectares in a resolution of the various lawsuits he had brought against the Chilean state.[85] While his legal battles with the state wound their way through court, Silva Rivas engaged in a profitable business logging Cunco's raulí forests. His successes in logging led him to head Chile's timber industry trade association.

In a similar case, the Sociedad Lanin in Lonquimay, organized by Juan Staeding to bring German immigrants to Chile, was purchased by Francisco Puelma Tupper, whose father, Francisco Puelma Castillo, had been a minister of war and a senator from the Radical Party, as well as a major landowner in the region. The Puelma family held five properties that belonged to the San Ignacio Pemehue hacienda, with well over 100,000 hectares, as a long-term lease in Lonquimay. The colonization contract, which gave the family another huge chunk of land to add to their empire in Lonquimay, required Puelma Tupper to settle Chilean colonos on the land acquired by the Sociedad Lanin. However, as a government report noted, by 1910 he had failed to bring even one colono to the property, which was occupied by numerous ocupantes, many since the 1890s, before the concession was even granted. It appeared that, like the Nueva Italia, Llaima, and Budi concessions, Puelma Tupper had failed to adhere to a decree from 1908 that required him to recognize the rights of established ocupantes. Puelma Tupper followed Silva Rivas's legal arguments, claiming that his colonization project had been impeded by "the many usurpers of public lands, who without any right, have installed themselves on the concession's land."[86] Rather than respect the property rights of the ocupantes and colonos, many of whom had returned to Chile from Argentina, he sought their eviction in local courts.[87]

Colonization concessions like those belonging to Puelma Tupper, Luis Silva Rivas, and the Budi and the Nueva Italia companies brought few European settlers to the south, but formed extensive properties, often expelling ocupantes and members of Mapuche communities from land they had worked for years or converting them into inquilinos on their own land. Even in cases where the contracts of the private companies were canceled because of their failure to introduce foreign colonos, the concessions managed to hold on to their lands through lawsuits against the state in local courts. In the case of the Tirúa i Quidico concessionaires, for example, even though their contract of 1906 had been canceled, they continued in 1908 "to enjoy use of the land on which they have established livestock, sawmills for the exploitation of the forests, imposing rents on the inhabitants of the region as if they were their inquilinos and harassing the indigenous as if they were the true owners of the soil."[88] The land inspectorate noted that many of the colonization companies had converted themselves into public companies with a number of stockholders and a lot of capital, "increasing the probability of their succeeding in their cases against the state."[89] Worse, the concessions used the money made from exploiting public forests to finance their "unjust" legal cases against the state.[90]

Squatters and Settlers

Campesinos often challenged the authority of both the state and large landowners employing colonization laws to stake their claims to land on the frontier. Government reports frequently complained that campesinos occupied public land illegally, disregarding the authority of land and colonization officials. In addition, as we have seen, colonization concessions and large estates were often unable to exercise their dominion over their land because it was occupied by Mapuche and non-Mapuche agriculturalists who claimed it in terms of either the indigenous settlement or the colonization laws. They were supported in their resistance to dispossession by a growing nationalist and populist movement that rejected foreign colonization as the central method of frontier settlement, as well as the monopoly held by haciendas. Social reformist parties such as the Democratic Party proposed "Chileanizing" the frontier by settling landless laborers, many of whom had migrated across the cordillera to Argentina, on public land.[91] By the first decade of the twentieth century, even as the Chilean state granted huge swaths of land to colonization companies, the nationalist sentiment that the frontier had to be made Chilean by distributing land to Chilean laborers had begun to make its mark on colonization law, encoding both a series of property rights claimed by landless laborers and a central contradiction with other colonization policies based on land auctions and colonization concessions.

As early as 1899, the General Inspectorate of Land and Colonization, while supporting foreign colonization, noted the importance of returning the "thirty thousand Chileans who populate the Argentine provinces next to the cordillera de los Andes, . . . depriving their country of the valuable resource of their labor and their recognized hardiness for the most arduous work."[92] In 1907, the Chilean consul in Neuquén, Argentina, had reported that "the Chileans keep leaving their country in large numbers and now not only do they occupy the valleys of the cordillera, they have extended throughout these pampas that they work with admirable tenacity, obtaining plentiful profits, principally in the cultivation of maize and wheat." He argued that to stem the exodus of Chileans who could be put to work promoting the economic development of the frontier, the government should begin promoting colonization with Chileans rather than foreign immigrants.[93] And in 1912, the Congressional Colonization Commission noted with concern that colonization policies had led to the expulsion of tens of thousands of Chilean laborers from their land and their flight either into the mountains or to Argentina. The state increasingly promoted the repatriation and settlement of Chileans from Argentina who, according to the commission, had "immigrated and made fertile with their labor foreign land because they did not even obtain for themselves a small part of the large extensions granted to concessions based on the promise, generally not fulfilled, of populating with foreigners what, in reality, had been depopulated by Chileans."[94]

Colonization officials contended that settling Chilean laborers on small plots would bring social stability to a region wracked by chronic land conflicts.[95] Agustín Edwards, the minister of foreign relations and colonization, argued in 1903 that the implementation of national colonization laws would establish social order and discipline, rooted in the male-headed family and fixed settlement, on the frontier: "the acquisition of a piece of land to cultivate by the sons of our country . . . gives naturally and logically ideas of order [and] respect for authority and established institutions, [and it] creates the stable settlement of the family in a determined place."[96] In 1908, finally recognizing the existence in the frontier of a significant population of "national" peasants, a new law to distribute land to the "ocupantes" of public lands was passed. Fathers of families, as well as widows, were conceded 40 hectares, and 20 hectares for each male child older than twelve if they had been in material possession of their land before 1 January 1905.[97]

The laws on national colonization of 1896 and 1898 had provoked a wave of land occupations by poor peasants or ocupantes. A land inspectorate report in 1903 noted that "a problem that must be resolved as soon as possible [is that of] all the petitioners to be colonos nacionales in Temuco who believe themselves to be authorized to invade and occupy, using force, whatever lands they wish, without taking into account whether they are occupied by others, have been sold at public auction, belong to the indigenous, or are designated for foreign colonos." According to the report, aspiring colonos impeded the functioning of the offices of the General Inspectorate of Land and Colonization throughout the frontier by illegally invading land: "they prevent, with complaints to the ministry and death threats for the [land] office personnel, our following the duties that the regulations require of us, creating disorder and lack of respect for the office's orders and increasing greatly the demands of those who aspire to be national colonos who today demand their settlement wherever they wish."[98]

The more sweeping national colonization law of 1908, which placed fewer restrictions on potential settlers (removing the literacy requirement, for example), provoked further unrest throughout the frontier. In 1913, Luis Urrutia Ibáñez, who had served as general inspector of land and colonization between 1909 and 1912, published a report in the bulletin of the Forest Department on the "anarchical" situation of property in southern Chile, echoing the land inspectorate's report of 1903. He noted that constant conflicts over land had produced "violence, fraud, and even crime." But he disputed the claims that there was a class struggle in the south in which the large landowner "oppressed and evicted the weak." To the contrary, he argued, "Here the weak, as a general rule, are stronger, owing to the fact that the aspiring colonos, who are thousands and all resolute, audacious adventurers, are organized, in fact, in resistance societies. . . . We have even seen cases of hundreds of colonos organized militarily who have disarmed the police, imprisoned government officials and prevented the implementation of judicial orders." Urrutia Ibáñez looked at the violent social convulsions in the south from the perspective of estate owners who faced continual invasions from poor campesinos. The peasant usurpers, he wrote, understood estate land to be public and therefore theirs to colonize: "the struggle that exists between the aspiring colonos and the large landowners is due to the fact that the former do not respect the property titles of the latter and attempt to colonize the entire region as if it were vacant."[99] Unlike in central Chile, *hacendados* (owners of haciendas) enjoyed little security in their landed possessions. They faced, Urrutia Ibáñez contended, the constant threat of appropriation by the landless poor, who viewed their estates as public property.

Both rural estates' failure to impose their hegemony over local society and the failure of state colonization officials to protect the borders of Mapuches' reducciones and settle colonos or ocupantes on small plots resulted from the state's limited administrative presence on the frontier. As the Congressional Colonization Commission noted, the failure to grant definitive titles to ocupantes was caused by the "reduced personnel" of the state's colonization agencies who were unable to engage in the close "vigilance" of would-be colonos or confirm that they had fulfilled the legal requirements that entitled them to land. The commission's report argued for the need to name more "administrators of colonies, surveyors, and various judicial agents" on the frontier.[100]

The members of the commission were surprised by a steady stream of Mapuche and mestizo peasant petitioners who viewed the commission as a court with the power to decide land claims. The tour of the south turned into an extended hearing, as commissioners traveled through the provinces receiving hundreds of petitions, many of which it found to be justified. As a result, the commission's report concluded that colonos had not received their land despite fulfilling legal requisites and were persecuted by "unjust thieves of public lands." The report also noted that colonization concessions were expelling ocupantes, without legal right, from within the boundaries of their property. But the commission was also frustrated to find that many ocupantes continued "to install themselves where they should not," an echo of Urrutia Ibáñez's complaints about the hegemony of the weak over the strong. The commission, however, turned Urrutia Ibáñez's description of frontier social relations on its head. State authorities, it wrote in its report, "in certain cases appear, supporting the arbitrary acts of the strong rather than the rights of the weak." This had led to a state of social unrest and an enormous number of legal battles that were undermining the central government's efforts to impose its presence and establish social order on the frontier.[101]

Similar laments were heard from the intendants of southern provinces during the first two decades of the twentieth century. The intendant of Llanquihue observed that the problem of widespread criminality in the province—especially the most common crime, cattle rustling—was due to uncharted forest land and uncertain property boundaries: "there are no precise boundaries between one fundo and another, and the roads and paths border forests and even extend into territories that are barely known. That is why it is easy for the residents of the countryside to steal animals, often compelled by necessity."[102] Similarly, the intendant of Arauco noted that "cattle rustling is very common owing to the lack of police. . . . [T]he theft of animals is very frequent. . . . [L]arge extensions of forests and few police are the basic causes."[103] And the governor of Cañete reported that, "as in the entire south of the country, property here is very badly constituted, with large extensions of state land in the power of private individuals. I see it as of the utmost importance that the government take measures . . . to recover the land that belongs to the state."[104]

The lack of a stable property regime underpinned by a precise mapping of territory and property boundaries constituted a significant obstacle to the economic development of the frontier. The "usurpers" of state land and of the land of indigenous communities did not, according to the Congressional Colonization Commission, make their land productive because the lack of clean titles rendered their legal possession precarious. They encountered difficulties selling fraudulently acquired land and obtaining credit.

And the situation was only worsened as the railroad extended its reach into remote corners of the frontier and land values soared. Hundreds of cases of land bound up in litigation—between private parties and between individuals and the state—contributed to the frontier's economic stagnation. "The ordered possession of property, the mapping of state, individual, and indigenous property," the commission's report argued, would open the door for "commercial transactions." And with "the rule of law established in those rich regions would come subdivisions and sales; credit will be born and industry will progress, with an increase in public and private wealth."[105]

Similarly, in 1913, the governor of Cañete argued that in the frontier territory the accumulation of vast extensions of land with fraudulent and vague titles caused the "subtraction of these properties from all commerce. Credit does not place confidence in them, and there is no serious capital that dares to exploit them."[106] In 1918, the land inspectorate argued that the frontier's isolation had prevented government officials from establishing a solid inventory of the nation's abundant forest resources and developing the forestry industry, "which was one of the state's most certain resources for the future."[107] As late as 1927, the Ministry of Development reported that "the economic life of the south was impeded" because "the greater part of the land was out of production since it was the subject of litigation that did not allow credit institutions to lend the support necessary for the progress of the zone." The "profound error" of entrusting settlement of the frontier to colonization companies had led to the "formation of the latifundium and the stagnation of agricultural production, because only one hand cannot be assigned the exploitation of thousands of hectares."[108]

Initially, the state had promoted burning native forest to open land for foreign settlement. Clearing forests, fencing land, and planting crops represented the most important means of demonstrating material occupation and obtaining land titles for settler colonos, large landowners, and Mapuche communities. In addition, national colonization laws after 1896 required colonos to fence their land and cultivate at least half of it. This implied clearing forest to plant cereal crops and forage for livestock. Demonstrating "improvements" in the land by fencing, clearing, and cultivating plots was the key to establishing property rights for poor campesinos and Mapuches alike. Yet by the end of the nineteenth century, deforestation and its attendant ecological woes, from flooding to drought and soil erosion, had begun to undermine the state's efforts to develop the frontier's agricultural economy. Forestry constituted both one of the central economic activities in the southern territories and an important source of public revenue. However, a commercially viable forestry industry required both state regulation of deforestation and an ordered property regime. Turbulent land and labor relations, unchartered property boundaries, land fraud, and the weak presence of the state all undermined the transformation of the southern forests into an engine of economic growth. By the first decade of the twentieth century, colonization and forestry officials had begun to link the state's loss of public land to growing ecological crises in the southern frontier. To halt the rampant destruction of the forests and exploit a resource that could rival the nitrates of the northern Atacama Desert, government officials reasoned, would require the state to impose its authority forcefully over the social and ecological anarchy that defined frontier territory during the early twentieth century.

Natural Disorder
Ecological Crisis, the State, and the
Origins of Modern Forestry

In 1911, Lorenzo Anadón traveled through Chile's southern provinces with the Congressional Colonization Commission. He later wrote a long letter about the state of the southern forests that was included in the commission's final report. Anadón described with "true pain . . . the fires, euphemistically called *roces,* that will bring about the extinction of the marvelous vegetation. For days from Lota to the south, the air was constantly smoky, becoming bothersome to breathe when the flames approached the railroad line." He noted that the French geologist Pedro Pissis's *Physical Geography of Chile* (1875) described how "from the Strait of Magellan to the 34th Parallel, the arborescent vegetation forms thick forests that, in certain points, occupy the entire width of Chile, as is the case between the 42nd and 39th parallels. . . . Remembering this, I found it difficult to believe that in such a brief time there was not a site in this large region immune [from fires]. . . . Pissis is ancient history." For Anadón, the massive destruction of the forests by fire was a tragic waste of a valuable natural resource: "this spectacle was equally strange for the traveler who witnessed the destruction of the same forests whose excellent woods were piled in the railroad stations we passed."[1] He placed the blame on the roces used to clear forests for logging, agriculture, and pasture. But he also noted among southern landowners an intense animosity toward the forest. Landowners incinerated forests not only out of desire to make land available for cultivation. They also viewed the forest as an enemy. He contrasted this attitude with the approach of the "indios," who "conserve eternally the admirable jungles on their land."[2]

In his letter to the commission, Anadón laid out a number of arguments made with increasing frequency after the turn of the century by proponents of forest conservation and silviculture. First, he contended that the destruction of the southern forests had wrought a change of climate prejudicial to agriculture throughout the region; while landowners cleared land for crops or herds of livestock, they actually damaged their own interests by producing widespread soil erosion. Anadón described how deforestation in the mountains had produced winter flooding in the valleys, summer droughts, and the desiccation of watersheds throughout southern provinces such as Malleco and Bío Bío. He urged Congress to pass laws that would ensure the conservation of forests on hillsides and mountain slopes. Second, Anadón enthusiastically advocated reforesting already cleared land with North American pine and eucalyptus. He cited the examples of the Lota and Curanilahue coal companies near the city of Concepción, which had already cultivated 25 million Monterey pine trees and 12 million eucalyptus trees.

Anadón's letter accompanied a report from the head of the newly organized Forest Department, the German-born forester Federico Albert, who also argued for regulating forest exploitation and promoting plantations of exotic species of trees.[3] Together, these documents established a blueprint for Chilean land and forestry policy on the frontier. Both Anadón and Albert linked the implementation of scientific forest management to the assertion of state authority over frontier territory. Their calls for state regulation of forests and forestry were part and parcel of the effort by the Congressional Colonization Commission to establish the administrative authority of the central government in Santiago over the property regime in the south following decades of unregulated colonization. As Anadón and Albert underlined, the anarchy of southern property relations had led to the loss of large extensions of native forest to large estates, land speculators, and peasant squatters, all of whom incinerated a resource that could generate considerable revenues for the state.

Wheat, Livestock, and Soil Erosion

Vicente Pérez Rosales, a government agent in charge of settling German colonists in Valdivia and Llanquihue during the 1850s, set the pattern for burning forests in southern Chile to clear land for colonization. For Pérez Rosales, the dense Valdivian rain forests were an impediment to attracting foreign settlers to Chile. In addition, the incessant winter rain made cultivating cereal crops difficult. Reducing rainfall by destroying the forests, Pérez Rosales reasoned, would improve agriculture and aid European colonists: "Valdivia's climate will be less unfortunate for the agriculturalist when the hand of man clears its forests and forcibly directs with canals the flow of its rivers which invade everything. The change in climate is already notable in the places where land has been cleared and placed under cultivation."[4] Pérez Rosales dealt the first major blow to the forests on an early reconnaissance trip. He wrote, "As I traveled I offered Pichi-Juan [an indigenous guide] thirty days' wages, which at that time meant thirty pesos, to burn the forests between Chanchán and the mountains." These first great forest fires burned for three months. "That frightful blaze, which neither the greenness of the trees nor their ever dark and dank bases nor the almost daily torrential rains could contain, had carried on its work of devastation for three months," he wrote. "The resulting smoke . . . shrouded the sun." The fires cleared a stretch of forests 15 by 5 leagues (roughly 90 square miles), making available to the German settlers level and fertile soil "of the best quality."[5]

Around the towns of Valdivia and Osorno, settlers followed Pérez Rosales's example. In a description of his travels to Valdivia, the German botanist Rudolph Philippi noted that his brother Bernard, who had been contracted before Pérez Rosales to oversee German immigration to Chile, had drawn a map of the region in 1849 in which the zones east from Osorno to the peaks of the Andes cordillera and west to Reloncaví Bay were blank spaces he described as "forests that are impenetrable because of *quila* [wild bamboo]." By the following year, this forest "had disappeared because of an enormous fire."[6] A decade after Pérez Rosales's first great inferno, settlers incinerated a stretch of alerce forest between Puerto Montt and Puerto Varas 125 kilometers long and 3 kilometers wide. While the goal had been to clear the forest for crops and pasture, the fire left, with its ashes, the enormous trunks of the alerces, rendering the thin soil unsuitable for agriculture.[7] Fifty years later, the gigantic trunks, most more than a meter wide, and invasive underbrush and scrub continued to cover the area once dominated by the coastal alerce forest, posing an impediment to agriculture. Anadón painted a devastating picture of the incinerated alerce forests as he followed the road leading from Puerto Montt to Puerto Varas: "on both sides of the road remain numerous alerce trunks, of trees that were colossal, burned a long time ago and that popular lore refers to as 'the cemetery.'" Blackberry bushes, bamboo, and weeds had claimed land cleared by the forest fires.[8]

Settlers set fires to forests driven by the twin goals of reducing rainfall and clearing land for agriculture. But they also employed fire to combat the ubiquitous wild bamboo, which opportunistically invaded soils cleared of their forest cover. As Philippi observed, in the Valdivian forest and the temperate beech forests farther north, the bamboos posed such an obstacle to human settlement and agriculture that campesinos who had crossed the Bío Bío River into the frontier region south of Concepción "fear the flourishing of the quila, surrounding it with all kinds of superstitions, since they blame it for bringing all sorts of illnesses." Once forest was cleared for planting, then settlers had to embark on an unceasing battle against the invasive quilas and coligües.[9] They turned repeatedly to fire to clear land of the bamboo or simply abandoned already cleared land to set fire to new stretches of forest, at times leaving their cattle to feed on the bamboo's nutritious leaves. Mapuche groups may have used fire to manage forests and pasture, as well as to clear limited stretches of land for crops, but colonization sparked a burning of the southern temperate forests that was unprecedented in scale.

In setting fire to southern forests, settlers reproduced landowners' practices in central Chile. A wheat boom in the mid-nineteenth century, sparked by exploding demand in Australia and the US West Coast, led to the final destruction of the patches of forest that remained on central Chile's large estates, especially those on hillsides and in the foothills of the coastal and Andes cordilleras. As a study from 1904 described, "As wheat exports increased, landowners made painstaking efforts to clear their soil of the trees that covered it; only a few trees were free of the destruction from the Maule River north." Until the mid-nineteenth century, landowners had let their cattle roam the forest stands on hillsides to find forage in the underbrush. With the growth in wheat cultivation, however, "the hills, above all, were denuded; the vegetation and humus were washed away by the rain; and the streams that before descended through their ravines dried out."[10]

In 1903, the intendant of the south-central Maule Province described the ecological disasters caused by the expansion of cereal cultivation and the destruction of the region's forests: "the native forests that were abundant before have been destroyed by fire and unregulated exploitation of their wood. As is natural, the springs or watersheds have desiccated to the point where many rivers in the summer dry up. . . . [R]ain in the fall and spring is almost unknown now." As he described, "The torrential rain of winter [fell on] mountainsides cleared of vegetation without penetrating the soil," provoking soil erosion, sedimentation in rivers, and flooding in the valleys. "The disappearance of the ancient [Maule] forests," he concluded, "has resulted in the erosion of land under cultivation, the rapid accumulation of wood in the estuaries and rivers, the flooding of lowlands, and dry years, some droughts that last six months from November to May."[11]

As early as 1873, a congressional commission charged with elaborating regulations on clearing forests had found that, "in a relatively short period of time, in fewer than fifty years, the appearance of the country, its climate, its fertility, and its health have experienced a lamentable transformation. If not desert, then something very similar to a desert, has invaded entire provinces. Cultivable land has become scarcer and scarcer. . . . The rain has come late; less rain is falling, and it is badly distributed. The rivers have lost slowly but visibly their flow of water. The watersheds are disappearing."[12] The commission elaborated a Forest Law that imposed new limits on hacienda owners, prohibiting them from destroying forests that protected wa-

tersheds and authorized the president to ban clearing forests on hillsides
and in the mountains. Clearing land with fire north of the Bío Bío River
was prohibited. South of the Bío Bío, in the frontier region, permission
from the regional governor was required. Public forests would be leased,
and their logging would be overseen by provincial intendants. The law also
established the new office of Inspector General of Forests.[13]

Despite these regulations, by 1900 settlers in the south had followed the
example of estate owners farther north, ignoring the Forest Law of 1873 and
setting fire to most of the forests in the frontier's valleys and plains, as well
as to significant stretches in the foothills of the coastal and Andes cordilleras. The ashes from these roces supplied exceptional fertilizer that helped
southern soil produce extraordinary wheat harvests and reignite Chile's
stagnant agricultural economy. In the southern provinces of Arauco, Bío
Bío, Malleco, Cautín, Valdivia, and Llanquihue, wheat production more
than tripled between 1879 and 1917. The province with the most significant
growth in cereal production was Malleco, where wheat yields quadrupled
during this period, making Malleco Chile's most important producer of
cereals, followed by Bío Bío and Valdivia.[14]

Landowners initially encountered difficulties cultivating wheat because
of southern Chile's inclement climate. Heavy rainfall, which irrigated the
dense native forests, placed a significant obstacle in the way of extending
agriculture from central Chile across the Bío Bío. As George McBride observed in 1936, "The south . . . has too much rainfall. An excess of precipita-
tion makes possible the extensive forests which protected the Araucanian
in his resistance to white intrusion and impeded the cultivation of the land
from which the Indian had been driven. Frequent summer rains made difficult the growing of wheat and barley, without which the men of Mediter-
ranean descent were handicapped in their efforts to colonize these southern lands."[15] Landowners in the frontier's new provinces thus adhered to
the blueprint laid out by Pérez Rosales, setting fire to the southern forests
to clear land and reduce rainfall, guided by the theory that eliminating forest would manufacture a climate more amenable to agriculture.

Landowners also developed arrangements to intensify the exploitation
of their estates' soil and earn money from rents. As *El Agricultor*, the bulletin of the southern agriculturalists' association, noted, southern landowners often rented plots on their properties to arrendatarios or medieros.
These rural workers had little investment in preserving the fertility of the
estates' soil through crop rotation. They also lacked the capital to invest in
fertilizer or new technology that might increase crop yields. Instead, they
wrung what they could out of the soil, relying on burning forest to produce
healthy harvests, and then moved on. "We know of many estates," *El Agricultor* noted, where, "because of the absence of many landowners from
their fundos, the land is worked by arrendatarios or medieros who are in-

terested only in extracting the maximum production from the soil, which will soon be exhausted in a proportion much greater than when it is in the power of the owners who rotate crops and rest the soil." *El Agricultor* concluded that because of this system of working estates, "Harvests decline in our country in open view, and soil erosion is every day more acute."[16] A study of southern agriculture found that most land had been dedicated to monocultural cereal production and lacked the type of crop rotation that might preserve soil fertility. This was exacerbated by the system of renting out land in which the arrendatario was concerned only with "obtaining the maximum earnings with the minimum sacrifice, uninterested in maintaining the valuable capital represented by the soil."[17] Landowners, often in possession of land with dubious borders or questionable titles, had little incentive to plan for the long term. Instead, they produced bonanza cereal harvests with minimal labor and capital inputs over the short term or extracted rents from their laborers and the forests. Insecure land titles and property rights on the frontier induced landowners to look to inexpensive labor and burning forests as a way to turn quick profits.

An agronomist's study of haciendas in the region around Temuco in 1910 made landowners' reliance on the double exploitation of forest soil and inexpensive labor clear. In this zone, ecological necessity drove the expansion of the hacienda. Large estates required access to extensive stretches of forest; soil exhaustion drove estate owners to burn forests and then profit quickly from the ensuing wheat harvests. Burned-over land was then turned over to native grasses, shrubs, and bamboo, which provided pasture for livestock. The agronomist concluded that "these large landowners show very little concern for improving their systems of cultivation since they still obtain magnificent harvests from clearing land with fire, livestock, and exploiting the forests."[18] He found that "for ten years, more or less, a significant change in the climate has become noticeable in [the Temuco] region. Every year, the rain is lessening [and] the air is less humid." He attributed this change to "the cutting down of the forests," noting that "every year, great extensions of forests have been destroyed. . . . Wheat is cultivated every year on the land cleared with . . . fires in virgin forests or second-growth forests. . . . They fell the shrubs and trees, no matter their diameter . . . , leaving the trunks [and] many fallen trees that have not burned." Then, he noted, landowners used fire again to burn invasive underbrush, bramble, and bamboo. "This system of cultivation can only exhaust the soil," the agronomist concluded, "because [the agricultural] value of the soil is very poor." Around Temuco, landowners tended to invest very little capital in their estates and left in charge administrators who sought to increase production quickly and cheaply by burning forest to bring new land under cultivation.[19]

Burning the forests shaped labor relations on Temuco's estates. Inquilinos and medieros stayed for only short periods because the land they worked quickly lost its productivity. Instead of remaining for long periods, even for generations, as occurred in central Chile, these workers sought contracts on other properties with stretches of forest that had not yet been cleared, because planting after roces was "the only way to actually produce satisfactory harvests."[20] An article published in the nationally circulated *Pacífico Magazine* noted that southern landowners provided stands of native forest to their inquilinos as a *regalía,* or as part of labor arrangements that often took the place of wages or was part of workers' remuneration. Prudencio Tardío, the author of the article, interviewed a landowner near the town of Victoria, in Malleco Province, who lamented the destruction of the once abundant forest, which had stretched from Victoria to Concepción, over the course of one generation; the landowner noted the cleared hillsides eroded by rainfall and the general "desolation [of the territory] cleared of all vegetation." He blamed the general belief that the soil could be made productive only by burning the forest. Relying on forest fires to fertilize southern soil had shaped the institution of inquilinaje, he said, because "we couldn't have inquilinos unless we gave them annual rights to new roces of 15–20 *cuadras* for their crops, since the following year, their harvests were only average, and in the third year, they were bad."[21] As in Cautín, in Malleco, providing inquilinos with forest to burn was a key component of labor agreements and contributed to producing an impermanent population of inquilinos and medieros who moved from estate to estate.

Large estates monopolized the best fertile land in the central plains, pushing Mapuche communities and non-Mapuche colono and ocupante smallholders onto more marginal soil, often in the mountains or foothills of the cordilleras. Soil erosion contributed to this process. By the first decade of the twentieth century advancing soil exhaustion had already begun to force campesino smallholders to sell their land to neighboring estates. As the agronomist's study of Temuco reported, "A certain unease has become noticeable among the people because exhaustion of all of the agricultural land has become apparent." This was particularly true for smallholders in the region: "the discontent of the small producers who see their harvests decrease owing to the terrible fertility of the soil is every day more general. Many of them have barely been able to harvest the seeds they need to cultivate after two, three, or four years, and they sell their land to neighboring large landowners who dedicate it to pasturing livestock." Soil exhaustion thus propelled the formation of ever larger estates and an increasingly land-starved campesino population.[22]

Smallholders' difficult situation was most acute among Mapuche communities, who composed the majority of Cautín's peasant producers. By the early twentieth century, wheat production, accompanied by the rearing of small herds of livestock, constituted the cornerstone of many Mapuche communities' agricultural economies. In regions with dense Mapuche populations, reducciones contributed a significant share to regional wheat production, cultivating their small extensions of land intensively. As the sub-inspector of land and colonization for Pitrufquén observed in 1905, "Today the indíjenas are no longer nomads. They acquired the habits of sedentary life a long time ago. They cultivate and harvest during the appropriate seasons, employing oxen and plows, and some even own threshing machines that they have purchased with their own money." He noted that "the majority of the commerce in the small villages of the frontier . . . is moved by the indíjenas. During harvest season, it is enough to see the considerable number of carts loaded with wheat that they take to the great warehouses of Temuco and Imperial to deduce that they are not as indolent as they are painted."[23] *El Diario Austral* similarly editorialized in 1916 that Mapuche communities constituted the motor of the frontier's agricultural economy: "the small pieces of land to which the indios have been reduced produce more than have the total harvests of cereals [on estates] in these wealthy departments. If we ask in every warehouse where their cereal comes from, they will answer unanimously that three-fifths of the wheat and oats comes from the indígenas."[24]

However, despite the productivity of their intensively cultivated small plots, as early as the 1910s many Mapuche communities—especially those whose agricultural labors focused on cultivating cereals in the zones around Temuco and Imperial—began to confront exhausted soil on their reducciones. Even as they supplied the bulk of cereals to local markets, Mapuche communities became increasingly vulnerable to ecological calamity. Settlement on reducciones removed the buffer of mobility and migration between ecological niches and exposed the newly organized communities to increased risk as droughts, changing patterns of rainfall, and advancing soil erosion began to appear throughout southern Chile. Most Mapuche communities lacked the capital to invest in fertilizer or in the extensions of land required for systems of crop rotation that could preserve the eroded soil. In addition, the use of plows and oxen, particularly among communities near Temuco and Imperial, intensified exploitation of the soil. Fields that might have recovered while they lay fallow were now plowed and cultivated every year. Cattle, more intensively grouped together within the confines of Mapuches' small plots and no longer roaming a variety of microecological niches, trampled and compacted soil, preventing both the regeneration of vegetation and the absorption of water. To open up fertile soil for crops and pasture, Mapuches had to clear their stands of forest, leading to the ecological costs associated with deforestation. As intensive cereal cultivation replaced Mapuches' once extensive economies, soil exhaustion eroded their communities' ability to maintain themselves.[25]

In 1912, for example, a report by the land and colonization inspectorate noted that because of "terrible harvests that produced nothing," Mapuche communities in Cautín "suffered the severity of hunger so that hundreds of them went out to feed themselves at sunrise on wild herbs and strawberries in the fields, only returning at night to sleep in their huts." Similarly, in 1914, a letter to the intendant of Cautín from the Mapuche organization Sociedad Caupolicán, based in Temuco, described a crisis of starvation among reducciones in Cautín. The directors of the Sociedad Caupolicán requested from the government "a subsidy in food or money for a majority of the Araucanian race that is today dying of hunger. . . . In fact, the society has received a number of complaints that various indigenous people in different reducciones have died because their means of subsistence have been exhausted."[26] These Mapuche communities' encounter with starvation was a relatively new experience. Travelers to southern Chile in the nineteenth century had uniformly commented on the Mapuches' prosperity and health, as well as on the abundance of crops and of forest and maritime products that sustained them.

By the early twentieth century, on both large properties and the small plots occupied by Mapuches or ocupante and colono smallholders, wheat cultivation had begun to create serious ecological problems. The agronomist Roberto Opazo noted in 1920, for example, that "clearing the forest has changed the climate of the frontier as the agricultural region of the center has advanced to the south. We have hotter summers without rainfall, which puts us in a position, in terms of summer cultivation, that is equal to the central region." In Malleco and Cautín, Opazo wrote, "The first method of occupation was to clear land for which they used fire that caused incalculable damage. . . . Since the soil was thin, it was soon exhausted by wheat cultivation, and it was necessary to move on elsewhere to repeat the same cultivation, leaving the first terrain abandoned to a state of complete exhaustion."[27]

Reports of the devastating impact of deforestation on southern agriculture were frequent during the first two decades of the twentieth century. In 1913, Tardío interviewed a colono in Malleco who described the widespread destruction of the forests by fire and the resulting changes in the province's soil and climate: "what a calamity! . . . The rain washed away the ashes from the fires and carried away the humus, leaving us with impoverished soil, eroding the hillsides, and inundating the rivers [with soil], producing the immense dunes we now have along the coast. The springs that once ran through every ravine have dried up." The colono echoed the reports of government land officials, noting that "it is undeniable that the

2.1 Roce, Bahia Cayatue, Lago Todos los Santos, First Half of the 20th Century (Courtesy of Museo Histórico Nacional)

climate has changed enormously. The winters and the summers are much drier, the variations between hot and cold weather are greater, and today we have to even begin to think of canals for irrigation."[28] That same year, the Radical Party's deputy for Valparaíso, Alfredo Frigolett, took the floor of Congress to describe a recent trip to the frontier, where he had observed "the ruinous decadence of agricultural production in the Araucanía. Because of the lack of rainfall, the harvests generally have been lost."[29]

As early as 1900, southern provinces such as Malleco and Bío Bío had begun to see diminishing wheat harvests due to soil erosion. During the first years of settlement in Malleco, wheat yields were 55 quintals per hectare. By 1900, the yields had fallen precipitously, to 15.9 quintals per hectare.[30] Federico Albert, charged with developing government forest regulations, noted that where soil was cleared for planting cereals "the layer of humus . . . is converted into ashes that are soon washed away by the rain, which explains the rapid deterioration of much of the soil cleared by fire that during the first year offer abundant harvests; the second year, barely normal harvests; and the third year, terrible harvests." Albert wrote that clearing land with fire (see figure 2.1) had caused "lack of rainfall in springtime, the extraordinary droughts of summertime, and the lack of water sufficient for irrigation," which were the subject of "constant complaints" by agriculturalists.[31]

Introducing large livestock herds placed additional pressure on southern forests and exacerbated ecological problems, because landowners, both small and large, looked to hillsides and mountain slopes covered with forest to pasture their herds, particularly as soil exhaustion began to claim agricultural land in the valleys and plains. After the first extraordinary harvests of wheat, yields began to decline by the end of the nineteenth century, and landowners looked for alternatives to cereals. They often placed cattle on already overworked soil, replacing cereal cultivation with pasture. Diversifying production by introducing cattle and sheep herds on estate land in combination with cereals was also desirable because of economic pressure. During the last decades of the nineteenth century, wheat prices continued to decline steadily, while meat prices soared due to expanding urban demand.[32] By 1917, southern Chile had become the nation's largest livestock-producing region, aided by a protective tariff on meat imports from Argentina.[33]

While Valdivia and Llanquihue provinces became central locations for raising livestock, estate owners farther north, in Malleco and Cautín, also introduced cattle on land once devoted to cereals, as well as on hillsides and mountain slopes that were not suited to agriculture. As they burned forests on hillsides and in the foothills of the cordilleras to plant pasture, they produced new ecological crises. Albert noted, "This is the great error of the majority of agriculturalists who believe they can produce more pasture on hillsides by completely burning and destroying the trees and bushes. Even if they do, indeed, obtain good results during the first year, this does not happen in the following years. As the years pass, the surface soil with the layer rich in humus is lost, . . . [with] the soil deteriorating significantly." Similarly, in 1920 Opazo described how southern landowners had cleared the slopes of hills and mountainsides of forest to plant pasture. As a consequence, he lamented, "The climate has changed notably." Summers were drier and hotter, and winter rain was torrential, a recipe for soil erosion on cleared slopes. Estate owners' large herds of livestock exacerbated soil erosion by preventing the regeneration of vegetation and compacting the soil, preventing the absorption of water and nutrients. The destruction of mountain forests and the introduction of livestock herds on land once covered by forests provoked erosion on hillsides and flooding in streams and estuaries. In addition, landowners placed herds of livestock on land that had already been degraded by wheat cultivation in the valleys.[34]

Many Mapuche communities continued their own cross-border livestock trade, maintaining nineteenth-century patterns of trade and transhumance. One government report noted, "Despite the truly prohibitive tariff imposed on cattle imported from Argentina, the indíjenas imported this year almost 1 million pesos worth of animals. . . . And they did this by a true economic miracle: they took things of little value or rented out their services to obtain them. The married men traded wool cloth that their women and children had produced during the winter for the livestock, and the single men worked as caretakers of the livestock on the Argentine estancias and received their salaries in a certain quantity of heifers or female calves that they later brought to Chile."[35] Similarly, a land inspectorate report noted that "in every indigenous family, there are two or three individuals at least who are in [Argentina]. . . . I have seen that they take the woven wool cloth that women produce during the winter to Argentina and obtain a good price in animals: they can sell a good shawl for two oxen of decent quality; for a 'llama,' or riding blanket, they acquire a heifer." In addition, he noted, single men also traveled to Argentina, where they often stayed for two or three years, laboring as field hands on cattle-ranching estancias "in exchange for animals that they later introduce into [Chile]." He concluded, "The indios, who have extended the trade in animals pretty far with [Argentina], thus constitute an economic factor of relative importance for [Chile]."[36] Yet another government report observed, "Today there are fewer animals among the indígenas than there were before."[37] Mapuche communities pastured cattle on their reducciones, but their limited allotments constrained the size of their herds, making migration into the mountain valleys of the cordillera or to Argentina essential to their survival.

As in central Chile, livestock rearing was a significant cause of deforestation in the south. By the early 1920s, a study by foresters with the US Forest Service found that clearing forests with fire had devastated 9 million of Chile's estimated 38.9 million acres of forest land.[38] As the study put it, "The practice of burning over forest land to make pasturage has prevailed for centuries in the northern and central provinces, and is the principal cause for the almost total disappearance of the original forests of those regions. In the regions where clearing land for agricultural use is still going on [the south], it is a common practice to burn the slashing broadcast and let the fire run until it stops. This, of course, gradually eats into the remaining green timber and results in the devastation of large areas not yet needed for farming." There was one notable exception: "forest fires burn over large areas every year except in the limited region inhabited by the native Araucanians, who are very careful to keep fires from the [araucaria] pine forests on which they depend during certain seasons for much of their food."[39] Given Mapuche communities' radically reduced access to land, the overworked and overgrazed soil on the reducciones, and limited herds, mountain forests constituted an invaluable resource. The members of reducciones could collect piñones and take their small herds of cattle to pasture in the remaining forests and valleys high up in the Andes and coastal cordilleras. Many Mapuche communities continued to regard forests as a resource essential to survival and worked to protect forests from fire.

Frontier Forests, the Timber Industry, and Forestry Science

Along with wheat and livestock, lumber emerged as a major product from southern Chile in the late nineteenth century. The lumber business followed the frontier as it advanced to the Malleco River and then pushed farther south. By the early twentieth century, the extension of the railroad on a longitudinal path through southern Chile brought distant stretches of native forest, containing valuable wood such as roble, raulí, tepa, coigüe, and araucaria pine, within closer reach and made timbering a profitable proposition. The railroad also produced increased demand for the wood it was bringing within the scope of the market. Alongside the coal and copper mining industries, the state railroad constituted the most important national market for wood, which it used for ties. Initially, most timber was transported on the transversal rivers, which crossed Chile from the eastern Andes cordillera to the Pacific. But by 1900, the state railroad had begun to reach out from its north-to-south course into the mountain chains on both sides of the central valley. As one study noted, "The principal business that was developed was the exploitation of wood that was logged and milled by hand and transported by river to Concepción. The first steam-driven sawmill was introduced near Angol in 1884, and the number of these mills increased gradually, following the path of the railroad south and then into the mountains."[40]

Numerous sawmills were working the mountain forests of Malleco by the 1920s. A report by the province's intendant noted that in the Department of Collipulli, fundos in the central valley were dedicated almost totally to cultivating cereals and raising livestock. Fundos in the Andes cordillera and the cordillera's foothills, however, logged the forests with sawmills driven by steam engines or hydraulic wheels.[41] In 1920, there were 650 small sawmills throughout southern Chile.[42] In the Cautín Province alone in 1928, 200 sawmills worked forests of raulí, roble, and lingue.[43] Opazo found that "the wood business" had created an extensive infrastructure of workshops and factories producing boxes, doors, window frames, moldings, and parquet in southern cities such as Concepción, Temuco, and Valdivia. Above all, logging lingue had become a very profitable business linked to the expansion of the cattle industry in the south during the late nineteenth century. Lingue bark was employed in tanning and exported north on the new rail lines, as well as used in tanneries in southern cities.

The lingue industry set a pattern for wasteful and illegal logging in public forests throughout the south. A study of Chile's forests conducted for the Ministerio de Industria i Obras Públicas (Ministry of Public Works) in 1904 observed that the extractors of bark from lingues rampantly violated the Forest Law of 1873 by felling and peeling the bark of lingues on public land without obtaining the required authorization from regional intendants or signing rental contracts with the state. Squatters on public land exploited "considerable extensions of public forests; they have established sawmills and with the most tranquil spirit sell the wood, competing with private landowners." The report noted that "the incentives offered by the purchasers of lingue bark used for tanning hides feeds the interests of the exploiters of public forests. They damage the trees, stripping the bark, leaving them abandoned, without making further use of them." This required land officials "to defend the public interest and impose fines for stripping lingues in public forests."[44]

Often the business of extracting lingue bark from forests lay in the hands of an itinerant campesino population who squatted in the forests. Gustave Verniory, a Belgian engineer employed by a railroad enterprise in the frontier territory during the 1880s and 1890s, described "the numerous rotos who, to avoid conscription, hid in the forests and took advantage of their seclusion to accumulate lingue. When they could freely leave their hiding places, they offered it to the many merchant houses."[45] These small lingue producers were part of a floating population of landless laborers who worked the forests, extracting wood for charcoal, firewood, railroad ties, and tanneries. They held no legal title to the forests and were frequently denounced by government officials as squatters who violated both regulations on forests and the property rights of private landholders and the state.

A similar informal logging economy had developed in the southern alerce forests. The German traveler Paul Treutler noted in 1859 that, while agricultural production in Valdivia was at a "very low level," landowners and itinerant landless laborers had developed a dynamic logging economy in the province's forests. Commerce in alerce beams, posts, and boards was central to Valdivia's economy, and by the mid-nineteenth century it had made a considerable dent in the native forests. As Treutler described, "Many people are dedicated year-round to the search for alerce stands; they are paid handsomely according to the quantity of their finds." The lumber industry was so important to Valdivia's economy that the peones who worked in the forests composed a majority of the province's labor force and received their pay in boards. "Every Saturday you can see hundreds of men, women, and children leaving the forest carrying boards on their heads on the way to Valdivia to pay for goods in the stores," Treutler wrote. "All of the traders, such as butchers, bakers, and so on, had piles of boards on their patios, and when they accumulated an appreciable amount, they sold them to the wood warehouses."[46]

The railroad's expansion south through the central valley and into the Andes cordillera made logging an increasingly viable commercial activity. As one study noted, about half the cargo shipped on the state railroads con

sisted of wood; of all cargo shipped by rail, wood figured first in terms of
weight, quantity, value, and distance traveled.[47] The railroad itself became
a major market for wood extracted from the forests, providing an important source of revenue for many landless laborers, squatters, and Mapuche
communities who relied on selling ties for their livelihood. Verniory noted
that "numerous *leñadores* [woodcutters] worked in the mountain forests
extracting logs of roble pellín for ties."[48] Indeed, the railroad's voracious
appetite for roble was an important cause of deforestation in south-central
Chile. By 1914, Albert observed, the production of ties had destroyed roble
forests in a zone that stretched from Linares to Maule and farther south to
Concepción.[49]

Despite the Forest Law of 1873, concessionaires, colonos, ocupantes,
and estate laborers continued to burn forest with the support of—or, at
least, a blind eye from—regional officials. When Tardío asked landowners
whether they feared fines or imprisonment for violating the Forest Law, he
was told, "No, señor, we don't even think about it. There are even recommendations from the higher authorities for us to burn [the forest], since
they believe that forest fires will clear land that will be cultivated in the
future."[50] Regional officials had an interest in promoting the economic
growth of their territories that often conflicted with the forest regulations
handed down by authorities from the land and colonization inspectorate
offices. Surveyors, agronomists, and foresters frequently complained that
colonos, ocupantes, and aspiring colonos both occupied public land illegally and extracted wood from forests without authorization. In 1911, for
example, a government surveyor reported that he had found two small mobile sawmills working native forests illegally on public land in Guardaba:
"both mills work with the wood that the colonos, ocupantes, and aspiring
colonos deliver and that they exploit. In Purén, there are three more that
do the same."[51] Similarly, despite the state's efforts to restrict logging in
public forests, ocupantes and aspiring colonos continued to invade forests
and extract wood, often selling it to small mills. In 1907, a surveyor noted
with irritation that he frequently confronted squatters who refused to obey
his authority during his travels throughout the south and who informed
him that they were colonos or had received verbal permission from government authorities to occupy public land. "They do not follow my orders
to obey the colonization laws and refrain from exploiting public forests,"
he reported.[52]

In the Villarrica Forest Reserve, established with land reacquired by
the state after the termination of the Sociedad Lanin colonization concession, state officials faced constant conflicts with campesinos who sought
land. The forest reserve was undermined by a separate law dictating that

land restored to the state be used to repatriate peasants who had left for Argentina. The competing and contradictory logic of colonization and forest laws created a space for campesinos to claim land and forests within the reserve. By 1917, the Forest Department still had not established its control over the Villarrica reserve's forests. Non-Mapuche campesinos continued to occupy land and petition for titles as colonos, while Mapuche communities claimed land within the reserve that they had occupied for generations.[53]

The state's promotion of logging and clearing land with fire was clearest when it came to the colonies established with colonos nacionales, or "national settlers," many of whom were repatriated from Argentina. With fertile lowlands dominated by large estates, the land granted to colonies tended to be less suited to agriculture but frequently was covered with potentially valuable forests. Colonization officials looked at logging as a vital economic activity for the subsistence of the colonos settled in the new colonies. In general, while colonos' harvests were inadequate, they were able to make a living logging the native forests or renting their forests to small sawmills.[54] In 1906, for example, the Inspección Jeneral de Tierras i Colonización noted that thousands of Chileans had been resettled in five colonies in Cautín and Valdivia, where there was "vast opportunity for the exploitation of wood that is as adequate for construction as [it is] for the production of furniture."[55] A typical example was the Colonia Huichahue, whose annual report for 1906 stated that of the 20,000 hectares it had been granted, 6,000 hectares of forest land had been cleared by some two hundred colono families. The colony cultivated wheat, peas, potatoes, and vegetables. However, it also produced large quantities of milled wood; lingue sales constituted its major source of income.[56] Similarly, the Huñivales colony in Cautín was "the richest and most important in the province owing to the sawmills that are used to exploit raulí on a large scale."[57]

Large estates in regions covered with forests also made logging an important activity. A typical example was Felipe Smith's Puello estate. Smith owned a number of estates in the Cunco region of Cautín Province, which he logged, mostly for raulí. He pieced together Puello by purchasing lots from land speculators who had acquired them at public auctions in Santiago. Two of Puello's lots, composing about 700 hectares, originally had been purchased in 1893 by Manuel Bunster, a member of the frontier's famous Bunster clan. While the auctions were intended to provide land to colonos who would clear their plots and turn them to agriculture, as José Bengoa argues, Bunster's interest was clearly speculative. Ten years later, Smith purchased both lots from Bunster, as well as a contiguous 500-hectare lot that also had been auctioned in 1893. Three years later, he

2.2 *Logging Camp in Pucón (Courtesy of Museo Histórico Nacional)*

2.3 *Campesino Transporting Logs with his Oxen, c. 1930 (Courtesy of Museo Histórico Nacional)*

purchased two additional adjacent lots from Bunster, thus piecing together a fairly sizeable property covered with valuable forests. Smith proceeded to work his newly organized estate in the cheapest and most profitable way possible. Puello's forests constituted its most valuable resource, and Smith installed a sawmill, logging roble and raulí. An agronomist's report found that timbering was Puello's major activity, but after forests had been felled, Smith pastured cattle on the underbrush, bamboo, and brambles that invaded the cleared land. This prevented any regeneration of the forest. By 1941, the degraded nature of the estate's forests and soil prompted the agronomist to note the pressing need to reforest with Monterey pine.[58]

Colonization companies also engaged in a booming timber business. Indeed, while most colonization companies failed to settle foreign immigrants on their extensive land, they were able to turn quick profits by logging the forests granted to them by the state. In Llanquihue in 1905, for example, a report to the land inspectorate noted that "the concessions of large extensions of public land in the valleys of the principal rivers . . . with the goal of introducing a certain number of European colonos has had as a logical consequence the formation of private stockholding companies with the end of bringing together large amounts of capital . . . to exploit the valuable forests."[59] In a typical case involving Llanquihue's valuable alerce forests, Laurencia de Solminihac, who had been granted rights to 55,000 hectares to settle immigrant colonos, sued the Chilean state for damages and requested compensation in the form of legal rights to establish logging operations in stands of old-growth alerce. She had won the colonization concession in 1905 but succeeded in settling only thirteen immigrant families, all of whom had abandoned their plots almost immediately. A land inspectorate report used Solminihac's concession as an example of the "numerous concessionaires who have not fulfilled any of the obligations in their contracts with the government" and who "exploit the land they receive to be colonized without introducing colonos."[60] Like Luis Silva Rivas and other southern colonization concessionaires, Solminihac claimed that her enterprise had failed because of squatters on the land who opposed the settlement of the foreign colonists. In the end, the government granted her 6,500 hectares to introduce nine new settler families. In response, Solminihac sued the state and petitioned for compensation in the form of logging leases. As in many cases, the primary purpose of the colonization concession appeared to be to gain access to valuable stands of timber. In the end, the government rejected Solminihac's petition, noting that her primary activity had been logging, not introducing European settlers.[61]

In a similar case in 1904, the Chilean state granted Roberto Schmidt an enormous tract of land just south of the Toltén River in exchange for bringing European immigrants to Chile. The concession was eventually canceled because, as the land inspectorate observed, Schmidt had devoted his energy to expelling a large number of ocupantes from the land instead of introducing immigrant settlers. He then distributed the land settled by the ocupantes, and even by some colonos who had legal land titles, to family members and local merchants, who logged the concession's valuable forests. Schmidt, the inspectorate reported, had reserved "extensions of land that ranged from 600 [hectares] to 1,000 hectares with magnificent forests for them; they have established engines and sawmills ... to produce handsome boards of laurel, lingue and pellín."[62] As an outraged surveyor employed by the land inspectorate argued, "In this region, they do not observe any of the rules governing the exploitation of forests."[63]

In a number of cases, however, concessions engaged in extensive logging operations that were both legal and encouraged by the colonization authorities. The Budi colonization company logged the forests around Lake Budi in Cautín, installing a number of sawmills on the land granted by the state.[64] Budi's logging operations were large enough to employ hundreds of workers.[65] As part of its agreement to introduce foreign settlers, Budi received from the state a number of sawmills, since milling wood was the company's major source of revenue.[66] Similarly, as we have seen, Luis Silva Rivas established a logging empire on his concession in Llaima, mining the region's valuable raulí forests, and the Nueva Italia concession logged the Nahuelbuta cordillera's stands of araucaria, coigüe, and lenga with state approval. In these cases, colonization officials held that clearing the forests for agriculture and logging were the only ways to generate revenue that could maintain the colonies of European settlers.

Despite the growth of commercial forestry on land occupied by squatters, settlers, and colonization companies, the potential of the forests had yet to be truly exploited. In the 1920s, loggers still employed rudimentary production methods that led to tremendous waste. Their primary instrument was fire, used to clear the dense tangle of underbrush, creepers, and bamboo that covered most forest floors. As Opazo observed of the frontier, "In the central valley, the woods have disappeared, and mountain forests that remain are exploited for firewood and charcoal. In winter, the thinnest trees are felled, underbrush and quila, which is a bamboo species of brush indigenous to our jungle, are cut. Only the large trees are left standing. This roce constitutes a forest carpet which dries in the spring and is burned in the summer, producing a colossal fire that leaves a layer of ashes on the soil and blackened almost without vegetation the robles, lingues, laurels, raulís, and other trees of exploitable wood." This wasteful form of clearing forests and timbering was driven largely by economics: efficient forest exploitation, "more in harmony with the principles of forestry economy," simply was not within the means of most southern landowners or consistent with their primary interest in agriculture.[67]

A number of factors made logging a secondary activity on many estates and impeded the modernization of the industry. Most important was the limited extension of railroads.[68] Albert reported, for example, that "in many regions landowners do not possess the means to transport wood to

the market. . . . Given the impossibility of assuming the costs [of building roads] themselves, we have wasted a great deal of forest wealth in the
center of the provinces of Arauco, Bío-Bío, Malleco, Cautín, Valdivia and
Llanquihue." Merchant houses' monopoly of markets also made logging
and milling a costly business for landowners, who could find more certain
profits in planting forage and cereals. Loggers and mill owners, like many
agricultural producers, suffered from a perennial lack of capital and relied
on advances from purchasers. Merchant houses extended these advances
and then used their hold over markets to dictate the prices of wood. Indebted and with few options for selling wood, loggers were forced to accept low prices. The threat of completely losing a shipment of wood to rot,
fungus, and worms added to merchants' leverage over timberers and drove
down prices.[69]

In addition, markets for potentially valuable varieties of Chilean wood
were limited by competition with inexpensive imports from North America. For the most part, it was easier and cheaper for the urban and mining
centers of central and northern Chile to import wood from the US West
Coast than to purchase it from southern Chile.[70] In 1913, the Union of
Frontier Wood Producers, whose directors included Silva Rivas, met in
the city of Temuco to put together a program of reforms to promote the
development of the forestry industry. Their two main concerns were with
high freight rates on the railroads and competition with imported Douglas
fir. As Guillermo Bañados, the Democratic Party's deputy from Valparaíso,
put it, "The industry is today in agony and decay because of the lack of
protection from the state." He proposed that congress pass new tariffs on
Douglas fir "that is in ruinous competition with national woods."[71]

The woeful state of the forestry industry was described vividly by
Héctor Anguita, the Radical Party's deputy for Temuco, Imperial, and
Llaima, who reported on his conversations with hacendados from Cautín.
"A majority," he noted, "are resolved to begin clearing their forests with
fire extensively." The problem was that managing the woods for logging
was not profitable: "today, the hacendados who work forestry estates have
to overcome great difficulties in transporting their wood to the stations
because of the lack of roads and bridges." In addition, when the product
finally arrives in urban markets, he said, "There is no interest in acquiring national wood because of the competition from foreign wood. Today,
these hacendados sell their wood for tiny profits and, in many cases, lose
a large amount of money to earn something to pay off the loans that are
coming due." Thus, wood-producing haciendas "have taken the painful decision to burn forests and destroy a large part of their own fortune and the
country's wealth."[72]

For many southern landowners, logging continued to be a secondary activity carried out only during summer months, an adjunct to clearing land for wheat or forage.[73] Inquilinos provided cheap labor to log forests on large estates during off-seasons, when their labor was not required for agriculture. In addition, landowners often contracted with small, portable mills to log their forests (see figure 2.4). While the forests belonged to estates, the milling operations often were carried out by hundreds of small mills that migrated from forest stand to forest stand throughout southern Chile. The level of technology was fairly low, with almost all of these sawmills using circular saws that produced a great deal of waste. In addition, landowners engaged in logging commonly used fire to clear underbrush. Typically, these fires spread and took a great deal of valuable forest with them. A US Forest Service report on the Chilean forestry industry from 1946 noted, "Often the area to be cleared was burned over prior to tree felling, usually after the underbrush had been cut, piled and dried at the close of the harvest season, or, alternatively, in the spring or following summer." These fires often took up to 80 percent of the original stand of trees adequate for milling and spread to other adjacent timber stands. The report concluded, "This and other waste in logging and milling was looked upon with almost complete indifference, for the forest had little or no real value in the eyes of the landowner and its ultimate destruction and disposal were regarded as a desirable step in clearing the land for agricultural use." This system of logging was inefficient and destroyed large swathes of forest. However, it was also inexpensive and required little capital, an advantage for southern landowners who were interested in turning a quick profit.[74]

The very nature of Chile's forests proved to be an obstacle to the development of a modern forestry industry. A report by Mark Jefferson for the American Geographical Society in 1921, for example, observed that the heterogeneous nature of Chile's southern forests and the relative lack of unified stands of conifers had impeded the development of the forestry industry. Jefferson found an almost mirror-like similarity between the temperate rain forests of the Northern Hemisphere and the Southern Hemisphere. But, he argued, "Chilean forests are of very moderate value, while those of Washington are unrivaled."[75] Similarly, in 1922, Lilian E. Elliott traveled to southern Chile, where he observed that "an impressive picture is created by the density and extent of the southern forests of Chile, among the last of the great primeval tree-covered areas in the world."[76] However, he contended, "Two-thirds of this quantity must be left out of consideration as regards opportunity for organized commercial effort such as paper-pulp making. Lack of large 'social' woods, and thin or patchy distribution is, of course, a bar to industrial effort on a great scale."[77] Elliott

2.4 Saw Mill in the Countryside (Courtesy of Museo Histórico Nacional)

underlined the problems logging enterprises faced in the sparse population of conifers mixed in dense forests overrun by underbrush: "a great deal has been said concerning the suitability of the South Chilean forests for making paper pulp, but up to the middle of 1921 no manufacture has been commenced. Expert opinion has proposed new plantations of eucalyptus etc., owing to the non-social character of Chilean timberlands. Were the Chilean conifers more closely grouped the problem might have been solved long ago." Landowners interested in extracting a profit from their properties by logging confronted another problem in the age of the forests: many of the trees in forest stands were hundreds of years old and no longer suitable for lumber.[78]

The general view, shared by Chilean landowners and loggers, that Chile's native forests held minimal commercial value was a major factor in the massive deforestation that afflicted southern Chile during the colonization of the frontier. As forestry experts underlined, the application of modern forestry techniques to manage the native forest and ensure the reproduction of its most valuable species, while clearing invasive second-growth trees, underbrush, creepers, and bamboo that held little value—or, in Elliott's words, to cultivate "social woods"—was within the economic reach of neither large landowners, who continued to rely on merchant houses for credit and to look to agriculture and ranching to make the most profitable use of their land, nor land-starved smallholders and squatters, who depended on exploiting the forests for their subsistence.

The State and the Forests

By the first decade of the twentieth century, the ecological crises provoked by the destruction of southern Chile's forests led to a reconsideration of the state's colonization policies, as well as calls to regulate the exploitation of the forests. In 1904, the Ministry of Public Works commissioned a study of Chile's forests by Victorino Rojas Magallanes to establish a new law "to regulate cutting and burning forests and to foment tree plantations."[79] Rojas introduced his study with an appeal to forestry science, noting that "the disorderly felling of our forests, without following the principles of silviculture that constitutes today in advanced nations one of the most important branches of agronomy, has brought deplorable consequences."[80] He concluded, "Our agricultural future is closely tied to the repopulation of the forests. Forgetting the principles of forestry economy has brought the consequences that disturb our agriculture, principally in the central valley: flooding during winter, prolonged droughts in summer, abrupt changes in temperature that bring the terrible frosts of spring." Rojas viewed the unregulated destruction of the "ancient forests" and lack of a "rational" system of forestry as leading to the waste of valuable woods that could earn high prices in foreign markets.[81]

Rojas proposed a series of concrete regulatory measures, including restrictions on cutting forests and the creation of public forest reserves managed by state foresters and guarded by a corps of forest guards: "only the state will be able to make use of the forest reserves. . . . State leases of public reserves . . . must be according to rules of silviculture." In addition, Rojas argued that the state needed to both provide incentives to landowners to cultivate tree plantations of exotic species such as pine and eucalyptus and take charge of reforestation itself. Landowners required subsidies and protective policies because "it takes twenty or more years for trees to develop, during which period capital does not earn any interest. . . . This mission corresponds to the state, the municipalities, or the large capitalists." Rojas ended his treatise on forests and forestry by reemphasizing that "the work of repopulating forests must be undertaken by the state."[82]

Rojas was not alone in demanding an active government role in regulating logging and promoting forestation. Since the 1870s, the Sociedad Nacional de Agricultura (SNA) had been an active proponent of both restricting the destruction of forests and cultivating tree plantations. It played a major role in distributing plants and seeds, primarily of eucalyptus, cultivated in its botanical gardens to landowners to form tree plantations. In addition, estate owners who were active in the SNA began tree nurseries and plantations to take advantage of growing markets for wood in deforested central and northern Chile. Salvador Izquierdo organized a tree nursery in 1888 that, one study claimed, rivaled any in Europe and the United States, and he formed the Sociedad de Plantaciones with the goal of supplying Santiago with fuel and firewood. Around Concepción there were a number of pine plantations by 1904, including one owned by Pedro del Río at the mouth of the Bío Bío; one by Claudio Vicuña in Bucalemu; and one by the Cousiño family in Lota, as well as the hacienda Los Ríos in Lebu. These plantations were conveniently located close to one of Chile's most important ports, as well as the national railroad network.[83]

Yet despite the success of this private initiative in organizing plantations on already cleared and eroded sandy soil near the coast, landowners militated for state support. In 1898, the SNA lobbied for state subsidies for landowners who planted trees, the naming of state forest inspectors, and ten-year property-tax exemptions for land planted with trees.[84] In fact, the SNA's active promotion of tree plantations was taken up early by the Chilean state. In 1889, the Liberal Party government of José Manuel Balmaceda contracted Federico Albert to head a new Water and Forest Department.[85] Albert's mission was to combat the advance of soil erosion and dunes in agricultural areas. In 1890, he organized the first state-run tree nursery in Linares. In 1899, he was appointed by a government commission to study the effects of the invasion of dunes in the littoral stretching from Aconcagua to Arauco. The following year, he was commissioned to cultivate tree plantations in Chanco, in Maule Province, to prevent the invasion of dunes. Albert then became the director of the Zoological and Botanical Experimental Stations under the Ministry of Industry and Public Works. His office oversaw the formation of tree plantations, supplying landowners with trees at low cost. In 1910, Albert helped organize and became the first director of the Ministry of Land and Colonization's Department of Water and Forests.[86]

Albert articulated an early conservationism defined by his commitment to the sustainable management of Chile's forests and tree plantations in the interests of the state and nation. With his colleague in the Forest Department, the agronomist Ernesto Maldonado, Albert sought to employ forestry science to manage forests sustainably and combat the spread of dunes, soil erosion, and droughts wrought by unchecked deforestation. Rather than seek the preservation of forests, Albert and Maldonado labored to employ forestry science and the power of the state to engineer nature in the name of a greater social good, resolving ecological crises that threatened to undermine national economic growth and exhaust resources that they viewed as belonging to the nation. Like Gifford Pinchot in the United States, Albert saw regulated logging as a key conservationist strategy and understood forestry's major role as promoting the renewable resource of tree plantations of both native and exotic species. He drew on the same Prussian and French *dirigiste* tradition as Pinchot to argue that state-directed forestation was essential to national strength.[87]

In his report on the state of the frontier's forests and the impact of colonization from 1911, Albert tied the protection of the forests to the protection of public land. He argued that the state should immediately issue a decree that made all land south of the Bío Bío River public and that required property owners to make a case within a defined period of time for their land titles. In addition, he contended, the state should survey all of the land that had already been auctioned or leased and determine whether it was suitable for agriculture or forestry. If it was suitable for forestry, all clearing with fire should be prohibited, and other forms of exploitation should be subjected to special regulation. He called on the state to develop a system for classifying and protecting forested land, especially through the formation of state-owned forest reserves in all of the provinces of the frontier. For Albert, the calamity of colonization was the unrestricted transfer of public forests to private landowners. He explicitly linked the reassertion of state control of public land and colonization to forest management. The loss of public land in the south, he held, had led inexorably to the disappearance of the forests.[88] Albert noted in 1913, "Today we see the last vestiges of public land that remained in Arauco, Malleco, Cautín, Valdivia, and Llanquihue disappear. Together with the change of state-owned property into private property has come the destruction of the forests without measure or moderation." Landowners looked only to short-term gains and ignored the future of their soil, to say nothing of community interests such as the conservation of watersheds.[89] Albert argued that the state should put an end to its distribution of forest land for colonization, "since the state should receive a growing and permanent income from that forest wealth that still remains."[90]

Albert made the well-known arguments that maintaining and managing forests was indispensable to agriculture because forests prevented erosion and flooding and moderated fluctuations in temperature. But forests were also valuable because they were the basis for the logging, milling, construction, and furniture industries. Albert underlined especially the importance of "the production of cellulose for paper."[91] He was alarmed by the virtual disappearance of roble pellín in the south-central provinces such as Maule and the invasion of less valuable second-growth species, bushes, and bamboo, which prevented the reproduction of logged-over roble forests. Albert argued that following logging, "the hand of man" should intervene to direct the development of second-growth forests by leaving in every hectare a certain number of robles and cypresses. He called attention to the "rapid disappearance of raulí farther south," which were also replaced by invasive maqui, coligues (also referred to as colihues), lingues and avellanos. In Valdivia, Llanquihue, and Chiloé, Albert noted the fast disappearance of cedars and alerces "owing to the rapid exploitation of the large trees and the slow growth of the new plants that are often consumed by animals and flames used to clear land for pasture."[92]

Albert advocated cultivating plantations of species such as maritime pine, Monterey pine, Douglas fir, álamo (poplar), and eucalyptus, as well as managing second-growth native forests to prevent the spread of invasive species: "the forests, like wheat, need to be cultivated and cut during the period of the harvest to prevent the loss of their products and the invasion of weeds and underbrush such as the maqui, the quila, and so on, which have no value."[93] The old-growth forests posed a problem for Albert. "We have to consider the imperfections of the virgin forest themselves," he wrote. "There has never been cultivation of trees here, and as a result, we see standing out from the multitude some of gigantic size, not all healthy, since most are hundreds, even thousands, of years old, decrepit, rotten inside, [and] about to fall."[94] According to Albert, often only a small portion of the forest held commercial value. Rather than preserve these forests, Albert proposed, they should be logged and scientifically managed to direct their regrowth.[95]

Like Rojas, Albert proposed strict regulations on clearing land with fire, enforced by an active forest guard. He also advocated transferring forests to the state: "it is necessary ... to reserve for the state the unproductive and productive forest land that is still in its power, since private industry cannot be expected to dedicate itself to the long-term cultivation of [industrial forests].... At the same time that the state establishes reserves on its property, it should find a way to acquire other land ... to benefit the collectivity and maintain the species that are of greatest importance." Albert viewed taking charge of forestation as the state's responsibility. The state should provide a model for landowners of how to manage native forests and cultivate tree plantations in its public forest reserves, as this was an initiative that landowners, with a few notable exceptions, had failed to take up. The state should also play a major role, Albert believed, in providing incentives and subsidies to landowners to promote cultivated forests of native and exotic species.[96]

Albert's proposals for scientific forest management were largely embraced by both landowners' associations and state officials in charge of the colonization of the frontier. His arguments about the detrimental impact of deforestation on agriculture carried weight with landowners who had already experienced flooding, summer droughts, and soil erosion. The southern landowners' trade association became an early convert to tree plantations.[97] The association emphasized that it neither endorsed infringing the rights of private property nor criticized landowners for clearing forest to plant crops and pasture. However, there had been "abuse in the exploitation of forests, in which the right to property should have been efficiently limited in certain cases." Plantations could return value to depleted soil and create revenue for both landowners and the state: "if the procedures developed in countries with advanced forestry were followed," the argument went, landowners could earn "many times the sum invested . . . a secure income every year." Best of all, tree plantations were relatively inexpensive because they required "paying only a few workers."[98]

The newspaper *El Diario Austral* of Temuco published numerous articles in favor of forest management, including an interview in 1916 with a German immigrant wood producer who advocated cultivating tree plantations that would serve as "a continual source of prime material" for the paper and pulp, as well as wood, industries. "In Germany," he argued, "cultivated forests are exploited. . . . The repopulation of the forests in this country is an urgent necessity."[99] Similarly, in 1911, *El Sur* of Concepción editorialized that southern Chile's economic future lay in commercial forestry and the production of paper and pulp: "today the world suffers a hunger for paper. . . . Chile is a country that could respond to this demand, and instead we dedicate ourselves to destroying forests and reducing our millionaire riches to ashes." The paper observed that in many zones in the frontier, "rain has disappeared; the climate is worse." Certain zones of the southern provinces had been turned into "a sterile desert" that had caused the population "to flee the region."[100]

The appeal of scientific forest management was made clear in an article published in *El Sur* that was written by Carlos Vicente Risopatrón, a former Conservative Party deputy, a leading figure in the SNA, and a landowner from Malleco Province. Risopatrón began by describing how the roce had completed the conquest of the frontier by eliminating the "sterile and useless" rain forest, subduing the "bellicose indio," and extending agriculture and ranching in Malleco. He argued that, rather than regulate roces, the state should promote "the substitution of wild, uncivilized, sterile forest with the civilized and productive forest. . . . This is what we want for Chilean soil, that there are many trees harmoniously planted, that there are many useful and profitable woods." Risopatrón invoked the example of forestry in France, where, he argued, almost all of the forests were plantations of exotic species, which had replaced native forests long since destroyed. "All of these forests are homogeneous," he wrote, "and those that are not perfectly ordered in their distribution in the soil do not have the multitude of vegetation and useless plants that make the forests in Chile inaccessible, nor are they mixed with useless trees." The model for Chile, Risopatrón contended, should be newly formed plantations on the haciendas belonging to the coal-mining companies in Concepción and Arauco. In the coal-mining regions, he noted, the complete elimination of the native forest had turned agricultural land to dust. But plantations of pine and eucalyptus now covered the entire region and had restored value to the land while stimulating industrial development.[101]

In fact, the Lota coal-mining company, owned by the Cousiño family, had pioneered pine and eucalyptus plantations in Concepción. During the late nineteenth century, Lota had purchased more than 60,000 hectares of land covered with forests in the coastal cordillera, from the Bío Bío River to Curanilahue, to supply its mining operations. As Lota's tunnels extended farther under the ocean floor, its demand for wood grew, even as its native forests disappeared, as one study noted, "since nobody cared for the second-growth forest that was frequently lost to immense fires." In this context, Lota contracted with the German forester Konrad Peters in 1902 to cultivate tree plantations to replace its vanishing native forests. Peters successfully developed plantations of eucalyptus to supply timber for the coal mines. He also cultivated 800 hectares of Monterey pine. Although North American pine grew rapidly, its wood was not resilient enough for the mines, and Peters and Lota abandoned their efforts to cultivate pine plantations between 1920 and 1935, when other uses for pine emerged in the construction, paper, and pulp industries.[102]

In 1911, President Ramón Barros Luco sent a new forest law to Congress that reproduced Albert's forestry proposals. After providing a lengthy description of the disastrous impact of deforestation and a review of the "extremely important role played by forests in the economies of nations," the president proposed that a law promoting "the replanting and conservation of the forests" was in "the public interest." Barros Luco underlined the central role of the state in developing tree plantations south of the Bío Bío: "the development of forests in the province of Bío-Bío and southward where the state still possesses large extensions of forests can be realized in such a way that the action of the state alone guarantees the needs of the public interest of those provinces." Chile, Barros Luco argued, "must dedicate itself to cultivating forests. . . . It is the duty of the public powers to foment the planting and conservation of forests. . . . As it is not always possible to demand that a private party plant trees on land whose forest repopulation is immediately necessary, the project gives the state the power to proceed in determined cases with the expropriation of the lands that are indispensable with that end, when they are not ceded or sold."[103]

Barros Luco's proposed forest law laid out a program for the state to cultivate tree plantations managed by the new Inspectorate of Forests, Fishing, and Hunting. The law also established the creation of a special fund in the National Treasury for cultivating plantations derived from money from fines levied for illegal forest clearing and game and fish poaching, as well as the revenues generated by leasing logging rights in public forests. In addition, the law provided a series of financial incentives for private landowners to conserve native forests and cultivate tree plantations on their property. The law gave the president a series of regulatory powers, including the right to declare certain land to be of public use and thus the property of the state, as well as the right to limit clearing forests with fire and logging.

The forest law floundered in Congress, yet the government expanded its executive powers to regulate forests in a series of decrees drawing on the Forest Law of 1873 that allowed it to assert authority over land and forests in southern Chile. Beginning in 1907, Chilean governments began to reclaim land from concessions, leases, and colonization companies and to form public forest reserves and national parks where forest exploitation could be regulated. In 1907, for example, the Ministry of Industry and Public Works established a forest reserve in Malleco because "of the manifest necessity of conserving for the state the public forests in the territory being colonized." The logging rights in these forests had been leased to the Sociedad Fredes y Padilla between 1900 and 1910.[104] Similarly, after a definitive title was granted to the Lanin colonization company, which by 1912 had become the Sociedad Ganadera Lanin, the land remaining from the original concession not included in the final title was recuperated by the state for a forest reserve in Villarrica. The state also organized forest reserves on the Tirúa River, Arauco Province, and in Malalcahuello, in Malleco Province. In general, this was land that had been auctioned or leased but whose owners had failed to make payments or property that had been granted to concessionaires who had never fulfilled their colonization contracts.[105] In keeping with Albert's conservationist thinking, the goal of these reserves was not to preserve pristine wilderness but to establish sustainable logging in the native forests by leasing logging rights and imposing the oversight of the Forest Department.[106]

Despite the activities of the Forest Department, the state's presence remained weak on the frontier. Many state officials pointed to the monopolization of land by fraudulently formed large estates as the single greatest impediment to the development of modern forest management and commercial forestry. The rational exploitation and management of the forests, accompanied by reforestation, as advocated by Albert and others, required a long-term vision that most southern landowners did not possess. In 1911, for example, the Sociedad de Fomento Fabríl (Industrial Development Society) published a blueprint for the establishment of a pulp and paper plant in Llanquihue Province. The report noted that the major obstacle to building the forestry and cellulose industries in a province rich in forests was that "a majority of the large estates suffer from a lack of clean property titles." In addition, large land concessions, which were granted contracts for limited periods (generally twenty years), occupied almost all of the valuable land and forests in the province. The concession contracts almost guaranteed that the concessionaires would seek to extract the maximum profit in the shortest time, forsaking any rational management and regeneration of the forests they felled.[107] Similarly, Opazo noted that the system of land auctions introduced during the 1870s had led to the organization of "large fundos that remain abandoned, awaiting the rise in property values to do profitable business without any investment." Landowners had little incentive to invest in modern forestry practices since speculation seemed the easiest means of turning a profit. In addition, landowners with shaky titles had little confidence in investments that could take decades to pay off. In 1920, Opazo recommended the imposition of a special tax on property that was not placed in production, as well as a law that would "allow the state to expropriate [such property] to auction it off in plots whose size [would] be appropriate for efficient production."[108]

Arguments in favor of subdividing large estates and reasserting state ownership of public forests in the frontier were met with a countervailing narrative that ascribed deforestation to the activities of poor campesinos. A number of early proponents of forestry science and forest conservation cast campesinos, not large estates, as the major obstacle to rational forest management. Luis Urrutia Ibáñez, who as the general inspector of land and colonization had denounced the tyranny of campesinos over large landholders, argued in the Forest Department bulletin that distributing public forests to colono smallholders was a grave error, since most lacked capital and needed to make their plots produce immediately: "to subsist and install himself on the land, the colono has to do away with the forest. . . . [C]learing with fire is the last effective means at the disposal of the colono to make his plot productive through cultivating pasture and cereals that flourish after burning the forest." In addition, colonos were "uneducated, lacking the foresight and the will to look past the immediate profit produced by clearing with fire to the profit one hundred times greater offered later by the firewood and lumber of the same forest." Instead, the state should only grant land covered with forests to "large and wealthy landowners who are guided by calculation, that is the foresight of business . . . [and who] can reserve the forests for future profit." In Malleco and Cautín, he contended, not unreasonably, the colonies of smallholders repatriated from Argentina had destroyed the forest completely in only a few years. Similarly, he argued that ocupantes in Loncoche and Villarrica were "committing the crime of felling lingue and ulmo, very useful trees, with the sole object of extracting the bark in order to sell it in the tanneries." The immediate destruction of the forests could be prevented by giving land with forests "to the capitalists whose interest and foresight counsels them to reserve the forests for the future."[109]

Urrutia Ibáñez's arguments increasingly shaped forest policy. Rojas, Albert, Anadón, Opazo, and other advocates of forestry science were able to push the state to promote pine plantations and impose new restrictions on forest exploitation. They were less successful in their efforts to exercise state governance over frontier land appropriated by large estates. In the end, forestation with exotic species represented a compromise. Scientific forestry served the interests of the state by producing a territory defined by well-organized tree plantations that would stimulate economic growth and provide a source of revenue for the National Treasury. At the same time, as Risopatrón had advocated, forestation allowed landowners to continue destroying native forests through roces and timbering. Plantations would both mitigate the negative environmental effects of deforestation and replenish forest resources. Rather than limit the activities of landowners, the state could now replace devastated native forests with modern monocultural plantations.

At the root of this compromise between state projects to modernize forestry production and estate owners' campaigns to protect their property rights was the use of forestry science to promote planting Monterey pine in place of the sustainable logging and management of native forests. While providing incentives for cultivating plantations on privately held land, the state gradually reasserted its control over the native forests that remained following the first decades of colonization, organizing them in forest reserves and national parks managed by trained foresters. As many landowners turned to planting pine on southern estates in coastal areas of Arauco, Malleco, and Concepción, the state began to establish oversight of native forests on public land, regulating the unchecked deforestation that had defined the first decades of colonization. The state continued to lease public forests to logging concerns in reserves but now imposed the authority of the Forest Department to supervise logging and reforestation with Monterey pine.

Written out of the compromise represented by forestation with Monterey pine were the poor campesino smallholders denounced by Urrutia Ibáñez, who found their access to the southern forests increasingly restricted. During the first decades of the twentieth century, land and colonization officials put forward new laws on national colonization and forestry that extended the institutional apparatuses of the state into the frontier. For poor campesinos, colonization laws held the promise of land ownership and reform of the unequal system of property ownership established during the initial years of settlement. However, state efforts to rationalize forest exploitation and promote commercial logging imposed restrictions on campesinos' access to forest resources by defining them as incapable of modern forestry practices. During the 1920s, campesinos' increasingly vociferous demands for land would expose the contradictory social and environmental logics of state colonization policy and challenge the authority of both large properties and the state on the frontier.

3

Forest Commons and Peasant Protest
on the Frontier, 1920s and 1930s

In 1934, campesinos on the Ránquil estate in Lonquimay rose up in rebellion. Four years earlier, workers in the region had organized a union, the Sindicato Agrícola Lonquimay, which led the uprising with the support of Chile's Communist Party and the communist-led national labor federation, the Federación Obrera de Chile (Workers' Federation of Chile; FOCH). An army of hundreds of starving and ill-clothed estate laborers, settlers, and squatters sacked the company stores (*pulperías*) belonging to the region's large estates. Armed with revolvers, rifles, and sticks, they also attacked pickets of carabineros and the big houses of the local landowners. On the Guayalí estate they assaulted the manager and carabineros who had been stationed on the property to enforce the landowner's authority. They also invaded the Lolco estate, burning the houses. The movement of more than six hundred men and women swept through the region, attacking the Mulchén Forest Reserve, killing a forest guard, and gaining adherents among inquilinos on the Mariposa, El Morro, and El Aguila estates.[1] On the El Rosario estate outside the city of Victoria, insurgent campesinos induced the inquilinos to rebel against "those who own the land."[2] On the Amargos estate, workers in the sawmills joined the rebellion "incited by communist emissaries."[3] The principal targets of the uprising were estates that belonged to two of the southern frontier region's most powerful landowning families, the Bunsters and Puelmas, which each controlled tens of thousands of hectares of mountain forests and meadows.

For more than a week, the peasant rebels controlled a significant area of Lonquimay, aided by heavy snows that closed the zone to the outside world and impeded the transportation of carabineros to Lonquimay's mountain cordillera. After ten days, the carabineros were finally able to quell the uprising, aided by two paramilitary organizations, the Milicia Republicana (Republican Militia) and the Guardia Blanca (White Guard).[4] A large number of rebels were taken prisoner, and hundreds more fled into the inaccessible mountain border region with Argentina or to Argentina. The prisoners were bound and marched toward the city of Temuco. Many never arrived. In effect, they became Chile's first *detenidos-desaparecidos* (disappeared people) in one of the worst massacres in Chilean history.[5]

Strikingly little has been written on what, until the agrarian reform of 1967–73, was Chile's largest and most violent rural uprising. In general, historians have tended to treat the Ránquil rebellion as unique and unrepresentative, an aberration in Chile's otherwise exceptionally peaceful and stable twentieth-century political history. Nonetheless, a close analysis of the uprising offers an extraordinary glimpse of land and labor relations throughout southern Chile. The Ránquil rebellion was far from an unwelcome interruption in an otherwise peaceful history of rural land and labor relations. Rather, during the first decades of the twentieth century throughout southern Chile—from Llanquihue and Valdivia to Cautín, Malleco, Bío Bío, and Arauco farther north—rural laborers, squatters, and settlers contested the authority of landowners over labor and asserted their rights to land on estates. They employed colonization laws to question the legitimacy of landowners' property rights and to establish their own rights as colonos to land they defined as public. Decades before the land invasions, or *tomas,* that shook rural Chile during the agrarian reform, landowners and the Chilean state confronted a major social upheaval that threatened the pillars of southern agriculture.

A close reading of the Ránquil rebellion allows a better understanding of the ecological processes that shaped campesinos' relationship to the state on the frontier. Conflicts over land were not just conflicts over the neatly drawn geometric shapes contained in surveyors' maps. Campesinos and estate owners fought over forests, water rights, and pasture rather than hectares. Struggles over land were necessarily struggles over ecological niches in the mountains and forests of the frontier, and were defined by competing understandings of the uses of natural resources. Throughout the 1920s and 1930s, as both landowners and the state increasingly looked to forestry as a new source of wealth, conflicts over land and labor were shaped by contests over access to native forests. In the case of the Ránquil rebellion, peasants fought for land, but they fought for access to specific kinds of land—native forests where they could collect the seeds of the araucaria pine and fuel wood, and fertile mountain valley grasslands (*invernadas*) where they could pasture their livestock when harsh winters made higher regions in the mountains inhospitable.

The state's contradictory colonization and forestry policies exacerbated rural laborers' conflicts with landowners. As Chilean governments began to establish key social reforms, including labor and social legislation, as well as the reform of colonization laws, they also sought to modernize forest management and exploitation. The new reforms, which created a legal framework for colonization with landless peasants, contradicted the state's regulation of forest exploitation and promotion of scientific methods of forest management. Modern forestry and forest laws served as instruments for extending state control over frontier territory, braking rampant land fraud and the unchecked clearing of forests with fire. They also served, however, as another wedge separating rural laborers from their means of subsistence by enclosing forests and mountain meadows in reserves and submitting them to the management of trained foresters. It is no coincidence that of the handful of people killed by the Ránquil rebels, one was a forest guard.

Inquilinaje in Southern Chile

For most of Chile's late colonial and modern history, inquilinaje shaped rural labor and property relations. Large estates (fundos and haciendas) enjoyed a monopoly of arable land, progressively expelling Mapuche and mestizo peasants over the course of the nineteenth century and twentieth century. The estates relied on a large population of landless peasants to fill their labor requirements. According to the estate and the agreement or contract established with the inquilino, landowners provided seeds, tools, animals, land for crops, pasture, and occasionally even wages to their workers in exchange for their labor. As we have seen, in the south these *regalías* also often included access to stands of native forest. Many inquilinaje contracts called for inquilinos' entire families to provide labor to estates. As late as the 1940s, for example, labor contracts in Osorno required the children of inquilinos to work exclusively on the same haciendas as their parents. In other regions, labor contracts granted landowners the right to search the homes of inquilinos on their estates. Inquilinos were frequently barred from leaving estates, and passageways into and out of estates were closely guarded.[6]

At times, landowners also established institutions that bolstered their paternalist control of their tenant laborers: chapels, rudimentary health clinics, and pulperías, at which inquilinos usually built up large debts, adding an element of debt peonage to tenant farming arrangements. This powerful paternalist control of inquilinos buttressed the landowners' political power. Until electoral reforms in 1958 established the secret ballot and penalties for fraud, adult male laborers provided valuable votes during elections for the favored candidates of their *patrones* (employers). Landowners consecrated the paternalist dimension of their political and social power by holding fiestas, barbecues, and rodeos for their workers during political campaigns and elections.

In Lonquimay, the enormous Puelma and Bunster estates were largely devoted to raising livestock, although their owners also installed sawmills to take advantage of the abundant forests of raulí, roble, and araucaria pine in the Andes cordillera. In his memoir of life in the region during the 1920s and 1930s, Harry Fahrenkrog Reinhold, an administrator of the Puelma estates, recalled that each inquilino was permitted to pasture his cattle on the Puelmas' extensive land and was allotted a small plot (2–3 cuadras) to cultivate crops. Once a year, the estates' absentee owners arrived to oversee the two-week rodeos in which the livestock were corralled and separated into groups for fattening, slaughter, and breeding. While lower mountain valleys were employed for invernada winter pasture, the high-altitude valleys supplied veranadas: cattle *mallines* (wetland meadows), *coironales* (thatch grass on mountain hillsides), and wild bamboo, as well as "forests with all kinds of nutritious plants, such as the *liuto* [*Alstroemeria,* or Peruvian lily], the *lechuguilla* [small lettuces], etc. In the month when the piñones fell, the animals ate them, acquiring an exceptional fattening."[7] Raising cattle and exploiting the forest were the pillars of survival in the Alto Bío Bío. The seeds of the cones of the araucaria pine gave the local campesino population an important alternative to both fodder and wheat. Landowners provided their laborers with access not only to small plots to cultivate crops, but to different ecological niches, veranadas and invernadas, meadows and forests, in the mountains where they could collect forest products and pasture livestock.

The uprising of inquilinos in Alto Bío Bío in 1934 shocked many observers who were convinced of the paternalist beneficence of the region's landowners. Indeed, the first reports of the rebellion claimed that the rebels were not inquilinos but squatter ocupantes and aspiring colonos. Once the presence of inquilinos was established, the explanation for the revolt then shifted to the presence of outside agitators who had duped purportedly credulous inquilinos. This refusal to recognize the role of inquilinos in the revolt stemmed from a necessary denial of the labor system's violence and instability in the south. On the frontier, land speculators and landowners employed inquilinaje to establish legal rights to large tracts of unoccupied land. They hired inquilinos and inquilinos-medieros, as well as arrendatarios, to clear and cultivate public land, which they then claimed as their property. Sharecroppers and peasant renters were more common on large estates than inquilinos, although they were often called "inquilinos" and signed inquilinaje contracts. Estate owners frequently lived in towns or cities and left their property in the charge of managers (*mayordomos*). Inquilinos had little contact with their patrones. As George McBride noted in 1936, "Where laborers live on large farms they bear little of the traditional relation to the patron that exists in central Chile."[8] In Lonquimay, the Bunsters and Puelmas left their estates to mayordomos and visited only a few times a year.[9]

The attenuated relationship between estate laborers and landowners was further undermined by workers' mobility. Southern Chile's inquilino laborers tended to be far less permanent than their counterparts in central Chile, where families might work for generations on a single estate. As McBride observed, in the frontier region "there is no such fixed population as that characterizing the agricultural regions of the center. There are few inquilinos proper in the south. Much of the labor is secured only as needed, particularly at wheat harvest. It is supplied largely by migrant harvest hands."[10] Mark Jefferson wrote in 1921, in his study of colonization in Chile for the American Geographical Society, "The inquilinos do not suffice to gather the harvest. There is always lack of hands for that. So the landlords have always clamored for help, but only two months in the year. The sons of inquilinos may drift to the city . . . , but there are few factories, and there is little need for hands. Many go to the Argentine, both to the southern region and near Mendoza. The lot of the landless Chilean is very hard. He cannot find work in a country that has always encouraged and aided immigration."[11]

Even on fundos where more stable labor arrangements prevailed, inquilinos suffered the often despotic and, at times, violent authority of landowners. Landowners frequently refused to allow their resident laborers to leave their fundos without permission, in effect making them virtual prisoners of the estates. Rural workers, as in the case of the Puelmas' properties, were obliged to purchase their food and basic goods at the pulperías, where they often fell heavily into debt to landowners or to stores that operated as concessions on the estates. In many cases, they were paid in tokens that could be redeemed only at the pulpería. Because of the restricted presence of the state in the region, rural workers were isolated, with little access to the institutions that might have offered them some protection, such as courts and labor inspectorates, public schools and hospitals, and even churches. Frequently local officials, judges, and the carabineros operated as landowners' partners in establishing discipline on the fundos. In some cases, local officials installed pickets of carabineros on estates to provide security and enforce discipline.[12]

Through the 1950s, inquilinaje constituted a major weapon in landowners' arsenal for appropriating the land of indigenous reducciones and mestizo campesinos, as well as public land that putatively belonged to the state. Estate owners, anxious to gain control of pasture or forest land, would enter into inquilinaje contracts with their poor neighbors whose small plots provided insufficient land for their animals and crops. In these contracts, often signed with thumbprints, the borders of the estates were ingeniously expanded to include the small plots in the description of the inquilinos' duties. Landowners then proffered the contracts as evidence of the estates' legal ownership of land that had once belonged to their laborers, who were then permitted to occupy their old plots as inquilinos. Landowners acquired at once cleared land and an inexpensive labor force to work it.

The inquilinaje contract had a powerful legal effect. Peasants who had once enjoyed rights as indígenas, ocupantes, or colonos suddenly became inquilinos with no rights at all. Inquilinos' status as secondary citizens, subjects not of the state but of their patrones, meant that they qualified neither as indigenous members of reducciones nor as Chilean colonos. Landowners frequently sought to trick Mapuches and colonos by getting them to sign inquilinaje contracts and then taking over their land. In December 1925, for example, Palmera Maldonado de Mansilla sent a letter to the Colonization Section of the Ministry of Agriculture asking for protection against the encroachments of her wealthy neighbor, one Señor Descouvieres, who had built a house and begun to pasture his animals on her land. Maldonado petitioned for the status of colono since she had occupied public land in Carelmanu, Llanquihue, for sixteen years and introduced significant "improvements." Maldonado's petition was rejected. As a widow of a colono, she could have inherited her husband's rights to land, but Maldonado and her twenty-one-year-old son had signed an inquilinaje contract with Descouvieres in which they agreed to work on his land, tend his cattle, and build fences. In exchange, they were to be paid 600 pesos a year (along with separate wages for working in the dairy) and to be granted two or three cows for milking, land for crops, and the rights to pasture twenty of their cows on the land they relinquished. By signing the contract, Maldonado lost all rights to the land she and her family had worked for years. She no longer qualified to be a colono, and Descouvieres acquired both her labor and her land.[13]

In another case, in 1935 the president of a rural laborers' union, the Sindicato Profesional Agrícola de Pichi-Ropulli, which represented a large number of ocupante smallholders in the Department of La Unión, wrote to the Ministry of Land and Colonization to protest abuses committed by Temistocles Conejeros. Conejeros had installed a sawmill on the ocupantes' land, destroyed their houses, and prevented them from planting crops. The ocupantes had cultivated wheat and extracted wood from the local public forests with permission from the Dirección General de Tierras y Colonización (General Directorate of Land and Colonization) since 1932. According to a report by the directorate dated 1935, however, a number of the ocupantes were also employed as inquilinos by the neighboring estate of Carlos Foitzick, where they logged his forests. In this case, government officials argued that their signing of contracts as inquilinos meant that they had forfeited the legal status of ocupantes with rights to be settled as colonos on public land.[14]

In 1939, the land directorate reported on a number of anomalies in the colonization of the southern frontier. Most serious was landowners' use of inquilinaje to appropriate not only land occupied by ocupantes but also land held by Mapuche reducciones. In commenting on the colonization law's requirement of material possession as a prerequisite for a legal title, the directorate's memo noted, "It is not enough . . . to have contracts of inquilinaje with ocupantes who do not know how to sign their names and who by use of their thumbprints authorize a labor contract whose content and consequence they do not know and who are then evicted maliciously." The directorate noted that peasant ocupantes were frequently enticed to sign inquilinaje contracts that they did not understand via promises of a full resolution of their land claims in exchange for small sums of money. The claims were, in fact, resolved, but in favor of the landowners who, inquilinaje contracts in hand, proceeded to evict their new laborers from their small plots.[15]

Because of the landowners' manipulation of inquilinaje to steal peasants' land, the labor institution did not have the same stability on the frontier that it enjoyed in Chile's central valley. Inquilinos were, for the most part, former smallholders (ocupantes or colonos) or members of Mapuche reducciones whose memories of dispossession were quite fresh. A petition in 1929 to the governor of Victoria by the Sindicato Agrícola Lonquimay protesting evictions of ocupantes, colonos, and inquilinos from land claimed by large landowners reflected the open wound of the earlier land appropriations that had sent thousands of campesinos across the cordillera to Argentina during the first decades of the century: "we don't want to repeat here the massacres that are registered in our legendary history. We do not want once again to experience the horrible acts that were carried out on this land against our ancestors, fires, violent expulsions and long caravans of refugees crossing the Chilean-Argentine Andes. These memories live in the memories of today's children of this land."[16] By the 1920s, campesinos' sense that they been expelled unjustly from public land, or that they enjoyed rights to public land based on their material occupation and labor, was reinforced by new political movements for social reform, including state policies designed to redress the abuses that had occurred during the process of colonization.

The Inquilinos' Uprising

The election of Chilean President Arturo Alessandri in 1920 provided a political opening for an explosion of rural unrest in the south. Alessandri's populist political campaigns, and a military coup in 1925 by mid-level officers committed to social reform, created new political spaces for urban and rural workers. The "social" laws of 1924–25 provided the framework for Chile's first labor relations system by legalizing unions and the right to strike and institutionalizing state regulation of working conditions and collective bargaining. In 1927, Carlos Ibáñez del Campo, one of the officers who had supported the military coup of 1925 and Alessandri's program of social reforms, and who had served as Alessandri's minister of the interior, ascended to the presidency after a fraudulent election that garnered the victor 92 percent of the vote. Ibáñez called out the armed forces a number of times to repress workers' strikes. He also established a secret police force to infiltrate opposition political groups, especially unions and the Communist Party; fixed elections; and acquired dictatorial powers. At the same time, however, he engineered a number of key reforms, including the creation of a labor code in 1931 that brought together the 1924–25 laws, a new Forest Law (1931), and the Southern Property Law (1927), which was designed to resolve the conflicts over land and titles on the frontier once and for all. The Southern Property Law sought to create a stable regime of property out of the infinite lawsuits and violent conflicts on the frontier by establishing the ownership rights of those in material possession of land as a fundamental legal principal. It demanded that those who occupied land provide titles and maps in petitions for definitive titles within a defined period to ensure that property claims "did not damage the interests of the state." Beyond resolving decades of struggles over southern land, the law established the authority of the state in the frontier region by enabling it to "establish what land belongs to it and what land belongs to private parties, and to form a cadastral map on a solid and certain basis."[17]

The law was designed to regularize southern Chile's property regime in response to "a movement of opinion generated by the landowners . . . who were alarmed because credit institutions did not extend their benefits to properties whose titles were considered vitiated by prohibitive regulations . . . and because the courts did not fully recognize the legality of the titles to the immense majority of land located [in the south]."[18] The law also permitted the state to distribute land occupied and cultivated before 1 January 1921 to ocupantes who were "fathers of families." The goal was to increase production and establish a large population of smallholders, "thus combating pernicious social doctrines." The law facilitated the procedures for those who had purchased land at auctions, colonos nacionales, foreign colonos, and repatriated colonos from Argentina to make their land claims official. In addition, Ibáñez's government created the Caja de Colonización Agrícola (Agricultural Colonization Fund) to purchase and subdivide large estates and distribute the land to peasants, the first step on what would be a long road to agrarian reform in Chile. Ibáñez may have presided over an authoritarian regime, but it was a regime interested in technocratic reforms and modernization, as well as the extension of the state's administrative and coercive apparatuses into the ungoverned regions of the frontier.[19]

The political climate of the 1920s led to an eruption of conflicts over labor and land on the frontier. Rural workers, organized at times in unions and cooperatives, increasingly thwarted the power of landowners and began to claim the land they worked on estates. Throughout the 1920s and 1930s, inquilinos, medieros, arrendatarios, ocupantes, and colonos, as well as members of Mapuche reducciones, invaded, occupied, and squatted on large estates, contending that the land was public. Rural workers articulated arguments that would animate movements for agrarian reform during the 1960 and 1970s. They made the soil productive, contributing to national progress though their labor, while estate owners left their land idle. In addition, campesinos frequently cited landowners' rampant destruction of Chile's native forests to denounce their vandalism of national resources. They invoked the forests' status as public property to defend their own access to the woods' wealth, emphasizing that landowners had robbed the nation of southern Chile's forest resources.

On 4 November 1925, the Winckler brothers, owners of the large Cancha Rayada estate in Llanquihue, petitioned the Ministerio de Tierras y Colonización (Ministry of Land and Colonization) to request support in a conflict with their workers. The Wincklers had been granted a title to Cancha Rayada in 1872 and had established a productive estate dedicated to raising livestock and timbering. However, in September 1925 inquilinos and part-time workers employed by the Wincklers petitioned to establish a colony on Cancha Rayada. As the Land Ministry noted, "A number of these inquilinos, or *contratistas,* despite having signed inquilinaje contracts, have petitioned the government for the status of colonos nacionales." The rebellious inquilinos claimed that the land was public and that the Wincklers' titles were invalid.[20] At least some of the inquilinos and part-time workers were Mapuches who had ceded their land to Winkler and signed inquilinaje contracts as a condition for remaining on land that was legally defined as public, and that they had occupied for decades without titles.[21] In a similar case, on a number of estates around Osorno, inquilinos rose up against their patrones following the passage of the Southern Property Law. The Osorno office of the Ministry of Southern Property reported that "the inquilinos . . . oppose their patrón who they had respected and recognized until yesterday, calling themselves ocupantes of public land."[22]

Frequently the land invasions engineered by inquilinos reflected a social landscape of landless peasants, squatters, and estate laborers that included both Mapuches and Chileans, as non-Mapuches were often called. In 1926, for example, Luis Silva Rivas sent a telegram to the land directorate denouncing the fact that more than 2,000 hectares of his enormous Llaima concession in Molulco, a center of logging in Cunco's raulí and roble forests, had been occupied by campesinos organized in a cooperative. In the words of a government forester employed to manage forests and timber production in the region, the ocupantes were "armed and disposed to resist any effort to evict them." The conflict made its way to Congress, where Silva Rivas's supporters claimed that the ocupantes were inquilinos contracted by the concession and thus held no rights to public land. A government surveyor, clearly aware of the legal implications of inquilinaje contracts and sympathetic to the campesinos' cause, claimed that they were not inquilinos but ocupantes who could be legally settled as colonos. The government finally resolved to settle the ocupantes on the land and compensate Silva Rivas by granting him 2 hectares of land elsewhere for every hectare he lost to the ocupantes. A census of forty-five ocupante families, who were granted an average of 50 hectares apiece, included many Mapuche last names. The Mapuches were joined by a large number of Chilean ocupantes repatriated from Argentina who had organized a cooperative, the Sociedad Cooperativa de Concepción.[23]

Despite the resolution of the conflict, little had changed in the following year, and the campesinos had yet to receive land. In 1927, they wrote President Ibáñez to praise his government and the new Southern Property Law. They had heard that "the government intends to cancel all concessions of land based on the [colonization] Law of 1874 that have not fulfilled their contracts or that are not legally constituted." The letter denounced the enormous concession made to Silva Rivas, arguing that Silva Rivas had never fulfilled his end of the colonization contract and should forfeit his land. The campesinos employed a nationalist idiom to decry the concessionaire's "unjust actions" against "our country and its humble children" and his failure to live up to the stipulations in his contract with the state that he respect the land occupied by numerous Mapuche and non-Mapuche smallholders in Llaima. They were particularly angered by Silva Rivas's exploitation of native forests on which they relied for survival. "Three years ago, Silva Rivas installed sawmills in Cunco," they wrote, "and he is exploiting the raulí forests that are becoming rare in our country and extracting millions of álamo (poplar) logs, all because of the lack of will of the government, which should have prevented this extraction and appropriation of wealth belonging to the state." The letter noted that because of the government's negligence, Silva Rivas had devastated local forests: "today, if the government ordered that the sawmills be paralyzed, it would find much wood already milled and piled in the railroad station in Cunco,

... [I]t is illegal to exploit land that belongs to the government, province, and country."[24]

In the wake of the Southern Property Law, land invasions and labor conflicts shook Cautín Province. A report from the Ministerio de Fomento (Ministry of Development) in 1929 noted that throughout the province, "motivated by the Southern Property Law, a number of people have dedicated themselves today to introducing propaganda among the inquilinos on estates to convince them that said law concedes them property rights to the land they work."[25] In 1933, Felipe Santiago Smith, who like Luis Silva Rivas logged Cunco's raulí and roble forests, petitioned the intendant of Cautín for support from carabineros because some of the inquilinos on his Nueva Escocia estate "pretend that the estate is public and have rebelled against its administrator . . . in such a way that it is impossible to continue with agricultural production."[26] In another case, a landowner near Temuco sent a telegram to the intendant to ask for the help of carabineros because a strike and "insubordination" among his inquilinos had paralyzed his logging and milling operations.[27] On the Long-Long estate in Cautín, the landowner reported that a group of workers were preventing "the normal development of labor in the fields" and "maintaining an active campaign of agitation on the estate." These workers, he reported, "do not show the consideration and respect that they owe to the patrón."[28]

Rebellious inquilinos were joined by squatters who occupied land on estates that they claimed was public. In Cautín, the owner of the San Juan de Trovolhue estate wrote to the intendant, "Lately, with the implementation of the laws on the Constitution of Southern Property . . . my estate, along with a number of other estates in the region, has been invaded by a considerable number of alleged ocupantes and indígenas who are trying to take possession of my estate without any right."[29] The owners of the Las Ñiochas estate in Imperial sent a similar petition, denouncing a land invasion by "numerous groups of indígenas," as well as "some Chilean individuals."[30] The campesinos argued that the land was covered with extensive forests that the landowners intended to log and that were public property. On the Toltén ranching and logging estate in Cautín, the Ministry of the Interior reported in 1930 on the "violent occupations by individuals who claim rights to the land."[31]

Forests and Peasant Protest

A common thread running through almost all of the conflicts between rural laborers and estate owners in the south was access to both publicly and privately owned forests. Campesinos frequently invoked the state's interest in exercising governance over the frontier territory to argue that loggers were unpatriotic vandals who were destroying a national and public resource. Yet as the state cracked down on land fraud and logging in the frontier territory, enforcing new regulations on clearing forests with fire and establishing national parks and forest reserves, campesinos found themselves excluded from forests they depended on for survival. Worse, they discovered that landowners and loggers maintained their access to the forests by winning timbering leases, albeit with supervision by state foresters and surveyors. Campesinos' fury at the enclosing of native forests in parks and reserves was only intensified by the sight of logging companies going about their business, apparently unchecked, on public land. Foresters and surveyors charged with representing the state's interests and protecting a public resource appeared in the campesinos' eyes to be loggers' allies and bore the particular brunt of their anger.

State efforts to reform property relations and restore control of public land through the Southern Property Law were tied to efforts to regulate forestry and forests. In December 1925, a new Forest Law was passed to promote both the conservation of native forests and the "development of forest plantations, which are essential to Chile's climate, agricultural production, and public health."[32] The law provided financial incentives for private parties to plant forests on *"terrenos forestales"* (forest land) and established state-run nurseries to provide trees to landowners at subsidized prices.[33] The Forest Law also created the legal authority for the president to establish forest reserves and national parks on public land.[34] After the law went into effect in 1925, Chile established its first national park, the Benjamín Vicuña Mackenna National Park, on 71,600 hectares carved out of the Villarrica Forest Reserve. Unlike forest reserves, the purpose of national parks was to preserve the forests' "splendid beauty" and attract tourists "who want to experience the intense emotions produced by the contemplation of the marvels of nature."[35]

The Forest Law of 1925 codified a new version of conservation more akin to ideas of wilderness preservation that had emerged in the United States with John Muir than to the forestry science-influenced conservation of Gifford Pinchot and Federico Albert. This version of conservation implied protecting wilderness, usually in remote and inaccessible areas with few commercial possibilities, from burning and logging. The foresters who designed national parks sought to inspire nationalist sentiment by displaying Chile's natural wonders to the public. They also planned to make these mountain regions of the frontier engines of development by promoting tourism on the state railroad. Travel to the forests would have a number of benefits, ranging from making more accessible remote zones near the border with Argentina, stimulating economic development, and preserving quickly vanishing native forests, to promoting nationalist sentiment and the health and vigor of the growing urban population by exposing them to pristine wilderness.

Nonetheless, the Forest Law also established the conservationist vision of Federico Albert and his collaborator in the Forest Department, the agronomist Ernesto Maldonado, enshrining in law the legislation they had written in 1911. Along with wilderness preservation, the law laid out a blueprint for regulated commercial forestry. The problem of clearing forests with fire and clear-cutting was not primarily deforestation but, rather, the waste of a potential source of public revenue. The Forest Law created a system of leases to privately owned timber companies in state-administered forest reserves. Landowners and logging companies won contracts to exploit public forests but provided a significant portion of their earnings to the state. In addition, they were responsible for reforestation with exotic species. Timber production was to be carefully overseen by foresters from the Forest Department, who would guarantee the application of "scientific" and sustainable methods of forest management. The Forest Law also regulated clearing land with fire, the major source of deforestation in the south. The law required large landowners and smallholders alike to petition regional governments for permission to use fire to clear forests.

The new forest regulations spurred myriad conflicts among the frontier's rural laborers, the state, and landowners, in part because they flatly contradicted national colonization laws and the more recent Southern Property Law, which gave landless peasants a legal basis for occupying and claiming public frontier land. Campesinos were particularly incensed at the closing off of public land and forests. The forester in charge of the Villarrica Forest Reserve, for example, confronted an almost continual invasion by Mapuche and non-Mapuche campesinos, including many repatriated laborers from Argentina who returned to the Chilean side of the cordillera looking for small plots of land. He reported in 1929 that "people without any authorization or right" were continually trying to take possession of land and cultivate crops.[36] The forester was at a disadvantage in patrolling the forests because he had few resources at his disposal. Another government report noted that invasions of the Villarrica reserve by peasants, among them "many indígenas who have been settled in other parts," had been possible because the forester had only two forest guards in his employ.[37]

Three years later, the invasions of the Villarrica reserve had only intensified. The forester reported, "The advance of so many people who are claiming public land, most of them audacious with bad pasts, is becoming unsustainable. . . . [T]hese people ignore the law and illegally occupy land, often armed, without respecting even private property with already constituted titles. . . . With reference to the reserve . . . it impossible to sustain the avalanche of so many people who, in groups of ten, twenty, thirty or more individuals arrive and take possession of the land ... executing acts of true banditry because they respect nobody and go about armed." A group of forty people had settled in the Pocolpen valley in the reserve, high up in the cordillera, on land whose only value was its raulí forests, which the forester had succeeded in conserving "with a great deal of work." In addition, a large number of campesinos, many repatriated from Argentina, had invaded the forests on an estate within the reserve that belonged to Percy Compton in Pucón. The local carabineros reported, "They are convinced that nobody has the authority to throw them off the land that is public." The squatters had an "absolute lack of respect for authority" and were completely "disobedient" and "rebellious." The Quezada y Hermanos company, that had leased land in the Villarrica reserve presumably for logging and to pasture livestock, also faced an invasion by more than twenty landless peasants.[38]

The Villarrica reserve's forester was helpless to combat the wave of land invasions because he had no police force to back him up and could not even impose the fines called for in the Forest Law. The forester also confronted the "members of diverse colonization cooperatives" who had "violently" invaded the neighboring Vicuña Mackenna National Park.[39] He reported that in the Villarrica reserve, a large number of Mapuches from the reducción of Manuel Huaquivir had invaded an estate, evicted the property's "servants" (probably inquilinos), stolen their tools, and built a hut with wood from the forest. To top it off, they had treated the reserve's forester "insolently" and informed him that they had no intention of obeying his authority.[40] In this they resembled the repatriated landless peasants who "received" the forester with "insults and abuse, declaring that they had no reason to obey or respect me and that they did not recognize my authority over the reserve's land."[41]

The obvious contradictions of state colonization and forest policy sparked considerable outrage among landless campesinos, squatters, and settlers who witnessed the massive destruction of native forests on public land leased to logging companies. Both parks and reserves appeared to peasants who aspired to be settled as colonos as the illegitimate enclosure of public forests to which they enjoyed use rights. Foresters earned campesinos' ire by renting forest land to large landowners or logging contractors. In 1928, for example, squatters in the Villarrica Forest Reserve petitioned the governor of Villarrica to launch an investigation into the activities of the forester, who, in violation of the law, they maintained, had been leasing forests they occupied to timber companies.[42] In another typical case, in 1932, Valdivia's Deputy Prudencio Garrido, of the Ibáñez-identified Republican Confederation for Civic Action, denounced the evictions of ocupantes from public land by large landowners throughout the south. Garrido asserted that many expulsions were the work of "logging industrialists who have done nothing but exploit public land, especially the so-called Vicuña Mackenna forest reserves [*sic*]; these reserves have been denied to the poor, and yet they have not been denied to the rich who extract wood from the forests without paying any taxes." At the Ministry of Land and Colonization, he contended, any wealthy landowner or logging enterprise that paid for or leased public land was given preference over poor peasants when it came to distributing land titles.[43]

Garrido included in the congressional record a petition from campesino smallholders in Tracura, Cunco. The peasants denounced the theft of public forests by Francisco Croxatto, an Italian immigrant, and his son Atilo, who, they contended, had stolen 10,000 hectares of public land in Cunco. In 1921, the elder Croxatto had won a lease to log raulí forests in the mountains while agreeing to pay taxes of 40 centavos for each *pulgada* of wood he extracted.[44] Croxatto had felled more than 500,000 pulgadas, they argued, but paid nothing in taxes to the state, violating his contract. In addition, he had closed the only public road in the region, preventing the colonos and ocupantes from moving back and forth across the border. Over the years Atilo Croxatto had terrorized the region's peasants by evicting them from their houses. He employed the contract of 1921 to log the raulí forests to establish his authority, "which is absurd and illegal and a moral aberration." A group of aspiring colonos sent a similar petition denouncing the fact that land set aside for colonization had been taken over by large landowners in Cunco who were logging the public forests in the cordillera near Lake Colico "as if they were their own," in open violation "of the sacred interests of the nation" and the Forest Law. Worst of all, government foresters responsible for administering the reserves had turned a blind eye to this theft of the nation's forest wealth. The aspiring colonos included a number of documents that demonstrated that the landowners, Francisco Pérez and Felipe Smith, had engaged in illegal logging operations in the forest reserves. They included sworn testimony from many of the loggers' workers.[45]

It may well be that the foresters had been suborned or looked favorably on loggers and local landowners, allowing them access to the reserve's forests. But a number of government officials complained that they had little power to control the unchecked logging in the reserve. In 1930, for example, the sub-delegate of Cunco reported that, without the support of carabineros, there was very little he could do to prevent the use of fire to clear land for logging in public forests "that today represent the wealth of the nation in this southern region." He had battled to implement the Forest Law of 1925 and exercised oversight of the destruction of the forests, requiring permits for clearing with fire and fining violators of the law. Nonetheless, he wrote to the intendant of Cautín, private parties continued to destroy "large swaths of public forest." The region's residents continued to use fire to log and clear land without permits from the government and made it clear that they understood "that the sub-delegate and carabineros have an absolute lack of authority."[46]

It is difficult to know what effect the implementation of the new Forest Law and commercial logging had in Alto Bío Bío, where the Ránquil rebellion broke out. But we have some clues. In 1930, *El Comercio,* published by Froilán Rivera, one of the original organizers of the Sindicato Agrícola Lonquimay, denounced abuses in the Malalcahuello forest reserves in Lonquimay. A man named Anselmo Lagos had won a lease to the reserve's forests under the pretext of establishing a cellulose plant in the region. Instead, *El Comercio* reported, he devoted himself to logging araucaria pine forests and pasturing livestock in the reserve. In addition, he had made money by renting the forests to third parties. One of these subcontractors, Ramón Arostegui, "became a destructive plague destroying the araucaria pine with axes and fire, without taking into consideration that the abundant population of these trees constitutes a real public treasure, because these trees with their valuable fruit feed the Chilean population that lives in the cordillera. . . . [T]he people who most depend on this fruit are the indígenas, who use it in a thousand ways, making it their daily bread."[47] The paper denounced the "merciless destruction of a public treasure that is the property of the state." The logging was viewed as particularly pernicious because the state forester had ejected a large number of campesinos from the reserve. In 1931, twenty-five outraged colonos denounced the "intransigence" of the forester in charge of the reserve who, empowered by the Forest Law, had prevented them from cultivating crops and pasturing livestock on land they occupied in the cordillera.[48]

The conflict between the campesinos and the forester over access to the reserve's native forests became an issue for the Sindicato Agrícola Lonquimay. In 1930, Juan Segundo Leiva Tapia wrote a report to the union's directors informing them of negotiations he had held with the Ministry of Southern Property. Along with the expropriation and colonization of the Ránquil, Rahué, Guayalí, and Chilpaco estates, the union requested the suspension of settlement in the Alto Bío Bío Forest Reserve, where the destruction of the araucaria forests by logging threatened a cornerstone of many campesinos' subsistence. The union also petitioned for the colonization of the Malalcahuello reserve with the region's campesinos, hoping to wrest control of the reserve's araucaria forests from loggers such as Arostegui.[49] In early 1933, the governor of Victoria wrote to the intendant of Cautín to inform him of denunciations of the forester in charge of the Alto Bío Bío reserve by ocupantes who pastured livestock there.[50] The forester apparently had denied them access to the reserve's forests. In this context, the Ránquil rebels' invasion of forest reserves and their assault on the forester make sense as part of a broader emergent conflict among campesinos, landowners, and the state for control of southern forests.

The Ránquil Rebellion

In 1928, inquilinos who had worked on estates that belonged to the Puelma family for at least two generations, colonos, and ocupantes, many repatriated from Argentina at the beginning of the century, organized the Sindicato Agrícola Lonquimay and petitioned the government for the expropriation of land on the Puelma family's estates. The union's major goal was the colonization of public land in the cordillera that had been usurped by large landowners.[51] In 1929, the governor of Victoria reported that Juan Segundo Leiva Tapia had disseminated "subversive propaganda among the inquilinos of the Rahué, Villacura, Lolco, Chilpaco, and Ránquil estates making the inquilinos believe that the lands they occupy are public and that they have the right to occupy them, the result of which is that they do not obey their patrones."[52]

Leiva Tapia had studied the Southern Property Law and understood that the Puelma family did not have legal titles to its estates. The estates claimed by the Puelmas dated back to the nineteenth century, when the land was granted to José María Rodríguez (in 1863). Because the law that guaranteed Mapuches rights to the land they occupied did not include this region, Rodríguez was able to claim dominion of a vast "vacant territory," albeit one populated by many Mapuches-Pehuenches. Part of the land was then sold to Manuel Bulnes Pinto, an army general and son of a former president, in 1873, and another part was sold to the commander of the southern army, Cornelio Saavedra, in 1874. They, in turn, sold to Minister of War Francisco Puelma Castillo (in 1881 and 1879, respectively) in the ongoing carousel of land speculation among leaders of the southern military campaigns against the Mapuche. As the minister of war in 1881, Puelma had helped to negotiate a key border treaty with Argentina. He claimed all of the land in the Lonquimay valley that fell to Chile following the treaty. The Chilean government then pressured Puelma to provide land in his dominion to settle some of the thousands of Chilean peasants who had migrated to Argentina. While some of Lonquimay's valleys were given to the state to use for repatriation and colonization, Puelma retained a thirty-year lease to five large estates: Chilpaco, at 19,739 hectares; Rahué, at 28,000 hectares; Ránquil, at 37,625 hectares; Lolco, at 41,791 hectares; and Vilcura, at 51,607 hectares. In 1901, the Puelma heirs received titles to the estates in Lonquimay in an agreement struck by Agustín Baeza, the general inspector of colonization, and Eliodoro Yáñez, the legal representative of the Puelma estate who was then serving as the minister of land and colonization. Nobody seems to have mentioned any apparent conflict of interest. Nor did anyone seem to notice that, as with so many of the frontier's estates, the boundaries established in the new titles in 1901 actually differed from the boundaries defined in the original titles by including extensions of public land that Puelma Castillo had rented from the state beginning in 1889.[53]

The Sindicato Agrícola Lonquimay's first order of business was to petition the government for the expropriation and distribution of land on the Puelma family's estates.[54] The union found support in the presidential palace, where Ibáñez intervened personally on its behalf. Ibáñez met with Leiva Tapia and endorsed the union and the goal of colonizing Alto Bío Bío's fundos with rural workers organized in cooperatives.[55] The president also personally intervened to have the local Welfare Office register the Sindicato Lonquimay; appointed Leiva Tapia, "an inspired and patriotic man," to the position of sub-delegate; and denounced the exploitative conditions on Lonquimay's estates.[56] The union's slogan, "Populate, Produce, and Civilize," struck a chord with Ibáñez, as did Leiva Tapia's project of establishing public schools in the region and making productive large stretches of unworked land in the hands of local landowners. The union would engineer the state's initial penetration of a region ruled in a feudal manner by Lonquimay's landed elite. Leiva Tapia and Ibáñez also found common ground in the project of cultural uplift of the region's "mestizo and aboriginal Chileans" and the colonization of a key frontier area that, properly settled, could be made prosperous and productive.[57]

On 27 March 1930, President Ibáñez issued a new decree naming a commission to study the question of the Puelma estates' borders and the rights of the many ocupantes located in the region. One of the commissioners was Leiva Tapia. After two months, the commission presented a report favorable to the Sindicato Agrícola Lonquimay, finding that the land titles of 1901 did not correspond to the original titles of 1863 and, in fact, "deprived the state of enormous extensions of land." The commission recuperated for the state 30,000 hectares of land incorporated into the Puelma estate in 1901. But that still left 139,362 hectares on the five estates in the hands of Puelma's heirs and new owners to whom they had sold the Lolco property. The commission divided the Ránquil estate into two parts: 7,000 hectares for the Puelma heirs and 30,000 hectares for 130 families of ocupantes, most of them inquilinos. This victory was consolidated when the following year, another decree expropriated 4,000 hectares of the 22,860-hectare Guayalí estate that belonged to Martín Bunster Gómez, another major landowner in Lonquimay, to settle the many ocupantes in the region.[58]

Unfortunately for the campesinos, external events intervened. The world economic crisis of 1929 brought the Chilean economy to a halt, striking a death blow to the nitrate industry, which had driven export-led growth since the late nineteenth century. The Ibáñez regime fell in 1931 to be replaced by a series of short-lived governments. When Arturo Alessandri, president from 1920 to 1924 and part of 1925, finally returned to office to rule for a six-year term in 1932, the Sindicato Agrícola Lonquimay no longer had an ally in the president. Alessandri began a concerted campaign to shore up the old social and economic order. He shed the populism that had characterized his first government, backing conservative economic policies with rigorous repression of workers' unions in cities and mining districts and of peasant organizations in the countryside.

Ibáñez's reform of the system of colonization and landownership in the south was quickly halted. The new administration sought to push back the challenge to the rule of the hacienda and put a lid on campesinos' militant movements for land redistribution. The Alessandri government was especially concerned with the wave of land invasions, occupations, and labor conflicts throughout Chile's southern provinces. The newly appointed governor of Villarrica denounced the "alarming situation" created by land conflicts, "abuses of private property," and "armed land usurpations" in the "Pucón-Toltén-Villarrica" region.[59] A report by Alessandri's interior minister noted the "serious problems produced by the acts of [land] invasion. . . . In the region of the south, agricultural labor is suffering disturbances. . . . These individuals violate the right to private property and disrupt the public order." The minister instructed the intendant to send carabineros to expel rebellious workers and squatters from southern estate land and to work with the Servicio de Investigaciones (the Chilean equivalent of the US Federal Bureau of Investigation) to investigate the "instigators" and "professional agitators" who provoked land "usurpations."[60] In Lonquimay, the governor of Victoria informed the minister of land and colonization that the Sindicato Agrícola Lonquimay had "incited the inquilinaje on the estates of the cordillera to disobey their patrones under the pretext that they have usurped public land. . . . They have led them to believe that . . . since they live on this land, they will have their rights recognized as occupants of public land and will have the option for free titles. Thus incited, the inquilinaje have begun to reject the rights of the patrones, refusing to obey them, which has brought the complete disorganization of agricultural labor in an extensive area."[61]

Lonquimay's landowners had initiated a counteroffensive following the fall of the Ibáñez regime even before Alessandri reassumed power. In early 1932, for example, Congressman Juan Pradenas of the Democratic Party read a petition from the Sindicato Agrícola Lonquimay that denounced the theft of land in the Nitrito valley that had been allotted to the peasants on the Ránquil and Guayalí estates.[62] The columns of *El Comercio* were filled with similar petitions from the union requesting state protection. As a sign of the times, Ramón Arostegui, having lost his lease under Ibáñez to log araucaria forests in the Malalcahuello reserve, returned to take up his timbering operations once again.[63] In 1933, Leiva Tapia was arrested and sent into internal exile for six months on the islands of Guaytecas and Chiloé because of his membership in the Workers' Federation of Chile and his attendance, as a FOCH delegate, at the international conference of the Communist Confederación Sindical Latino Americana in Montevideo.[64]

The distribution of land to the ocupantes was stalled.[65] At particular issue were 8,000 hectares of invernadas that the commission charged with dividing Ránquil in 1930 had allotted to the members of the Sindicato Agrícola Lonquimay and that, because of their high quality, the Puelmas coveted. Similarly, while Ibáñez's director of land and colonization, Ernesto Maldonado, had promised 4,000 hectares of invernada belonging to Guayalí to the ocupantes during the division of the estate, Bunster, taking advantage of the favorable new political order with Alessandri in office, as well as his ties with local officials, had "ordered" the surveyors who had made the final measurements to designate a high-altitude craggy piece of land called Llanquén in their maps of the new colony.[66]

The Puelmas and the Bunsters, with the support of the local police, began to expel ocupantes and colonos from Ránquil.[67] On the Guayalí estate, the campesinos protested their settlement on Llanquén, refused to recognize the maps drawn by surveyors, and threatened to take up arms.[68] Despite numerous petitions for assistance, trips to Santiago, and meetings with congressmen by delegations from the Sindicato Agrícola Lonquimay, the expulsions of small groups of campesinos proceeded apace during the first years of Alessandri's government.[69] A petition from the union asked the regional intendancy to step in to prevent the violent abuses committed by Gonzalo Bunster Gómez, who, accompanied by an armed guard, had killed pigs and destroyed fences and crops that belonged to members of the Sindicato Agrícola Lonquimay who had settled on the 4,000 hectares within the Guayalí estate. Bunster had threatened the campesinos and tried to force them to abandon their houses. The union pointed out that Bunster employed the local judge as an administrator of his estates and provided subventions to the carabineros.[70]

In response to the escalation of landowners' violence, the local branch of the FOCH and the Sindicato Agrícola Lonquimay organized self-defense committees. In 1933, the FOCH declared that in Cura-Cautín it had organized a Consejo de Oficios Varios (Council of Various Occupations) with more than one hundred workers and campesinos. The Consejo de Oficios Varios in Lonquimay maintained "close contact with the Sindicato Agrario de Lonquimay [*sic*] that has numerous *consejos* [councils] of *Colonos* and *Ocupantes de Tierras* [land occupants]." By 1933, these consejos had "formed campesino guards to defend themselves from the usurpers of land . . . , the evil concessionaires Bunster and Puelma."[71] It is notable that the union regarded Bunster and Puelma as "concessionaires." The term implied what was common wisdom for the region's campesinos: that the Bunster and Puelma estates belonged to the public domain.

The Puelmas and the Bunsters did not confront the Sindicato Agrícola Lonquimay alone. They were joined by landowners throughout the frontier provinces who faced a wave of often violent confrontations with inquilinos, ocupantes, and colonos. In March 1933, for example, landowners held a large meeting in Angol to hear the report of a commission that had traveled to Santiago to discuss the situation of southern agriculture. The meeting and the delegation's trip to the capital were motivated by a deep sense of concern about the state of labor relations on the region's fundos. The landowners present at the meeting denounced "the usurpation of private property by inquilinos and medieros on the region's estates, who, instigated by a few professional agitators, refuse to recognize the authority and rights of the patrón with the false pretext that they are occupying public land. They violently resist accepting the conditions that they agreed to with their patrones."[72]

Like the Puelmas and the Bunsters, many large estate owners initiated a counteroffensive against rebellious colonos and ocupantes. In late 1932, the Budi company, now owned by the Alessandri family and one of the nation's most important timber companies, expelled members of Mapuche communities who had occupied land claimed by the enormous former concession. The company razed thirty huts, leaving five hundred people "indigent," as a petition for support to the Cautín intendancy put it.[73] In October 1932, the Toltén company began expelling Mapuche and non-Mapuche campesinos from land they claimed and that the company used for logging and pasturing livestock.[74] The company claimed that more than four hundred people had invaded five of its estates.[75] In January 1933, the Socialist Colonization Central, a leftist organization designed to promote colonization of the frontier with land-starved campesinos, sent a letter to the region's newspapers denouncing a wave of expulsions of ocupantes and colonos. According to the Central, large concessions and estates had burned the houses and taken the land of thousands of poor campesinos throughout the south. That month, a surveyor from the Ministry of Land and Colonization traveled to Temuco to look into a wave of evictions initiated by large estates, especially the Budi and Toltén companies.[76]

In 1934, the Socialist Party newspaper *La Opinión* published a letter from the Socialist Colonization Central about "the problem of southern land."[77] According to the Central, colonos and ocupantes were being ex-pelled from their land by the owners of large concessions, who enjoyed the support of regional authorities. By March 1934, the Central had worked to resettle two hundred ocupantes who had been thrown off their land and had their huts and harvests reduced to ashes by the Toltén company. The ocupantes had demanded that they be settled as colonos on the company's land, invoking the Southern Property Law and the rights of those who had cleared and made the land productive. They had originally invaded the es-tate in 1923 under the impression that the 15,000 hectares claimed by the company were public. An agreement was reached in which the Agricultural Colonization Fund agreed to purchase the land and settle the ocupantes.

However, with the fall of Ibáñez and restoration of Alessandri, the ocu-pantes were thrown off 13,000 hectares they had occupied.[78] They then or-ganized another movement to take back the estate, holding secret meetings on the run from the company's forest guards. As *La Opinión* reported, the ocupantes "had the goal of engaging the company's police force, which would have provoked a bloody struggle, since these are strategic mountain forest lands." The conflagration was averted when Juan de Dios Moraga, the president of the Socialist Colonization Central, intervened and mediated an arrangement between the Land Ministry and the ocupantes. Moraga spoke to a meeting of 350 ocupantes about the agreement. The entire speech was translated by "the indígena Segundo Colil Jiménez in the aboriginal lan-guage." The presence of Mapuche and non-Mapuche ocupantes also meant that the Indigenous Protectorate was brought in to help broker the deal.[79]

In Molulco, Llaima, the site of Silva Rivas's colonization concession and logging operations, a similar conflict between aspiring colonos and landowners broke out. In 1933, a large number of campesinos were thrown off land they believed to be public by a local landowner who claimed that their plots fell within the boundaries of his estate. The testimony collected by the local judge reflected the common condition of smallholders who occupied public land and petitioned to be granted titles as colonos. One campesino testified, "We are many colonos who entered this land called Molulco, on which my father had settled in 1909, and I began to occupy that land myself because my father died. We all knew that the land was public, and that is why we occupy it without knowing that is has been sold to the Comunidad Llaima [Silva Rivas's colonization company]. . . . Since we are many who occupy this land, we still haven't divided the *campo* [fields] and we are just simply living on it."[80]

The campesinos established a detailed description of the history of Silva Rivas's concession, including the concession's failure to settle foreign colonos in the Llaima valley, the reduction of the size of his concession to 40,000 hectares after 1911, and the original stipulation in the colonization contract that the Comunidad Llaima respect the rights of the campesinos who already lived within the boundaries of the enormous allotment. Despite restrictions on the colonization contract, Silva Rivas had astutely registered all the land granted to the Comunidad Llaima in the Conservador de Bienes y Raíces and had established titles to the land occupied by the campesinos. He then sold titled land in a classic method employed by the owners of colonization companies to turn a profit. Juan Bautista Hiriart purchased the land occupied by the ocupantes from Silva Rivas. He then installed a sawmill to exploit valuable native forests in the Llaima valley and had his workers expel the campesinos and burn their *ranchas* (shanties) and crops. As one colono testified, "Señor Hiriart has proceeded to set fire to the houses that the colonos of Lolulco [*sic*] reside in . . . , burning tools, fuel and all that the impoverished colonos possessed. . . . [F]ire was chosen to overcome the moral resistance of the colonos who refused to hand over the land they had worked for many years."[81] As the case wound its way through the judicial system in the early 1930s, the ocupantes introduced into evidence an article from *El Diario Austral* describing the "public alarm that exists because large landowners are setting fire to the houses of colonos in Villarrica, Pucón, Cunco and other places throughout the region."[82]

In the summer of 1934, a commission of surveyors traveled to Lonquimay to redraw the boundaries of the land granted to the ocupantes, inquilinos, and colonos. Their goal was to move the campesinos farther up the cordillera. According to government officials, including the land and colonization minister, the campesinos were not being expelled from Ránquil, only resettled, albeit with the assistance of carabineros.[83] The final straw came, however, when carabineros evicted sixty campesino families from Ránquil in April 1934. A month before, a force of twenty carabineros had expelled ocupantes on the Guayalí estate from the invernadas in the Nitrito valley.[84] Called to a union meeting, former inquilinos, ocupantes, and colonos organized the assault on the pulperías, big houses, and pickets of carabineros on the local estates.

Mapuche Reducciones and Peasant Mobilization

After the outbreak of the uprising, the regional and national press began to report that the rebellious members of the Sindicato Agrícola Lonquimay had been joined by members of neighboring Mapuche reducciones. *El Sur* reported on 2 July 1934, for example, that "one hundred indios have joined the revolt," including members of the reducciones Minas Queuco and Cauñicú, as well as "indígenas" who worked on the Guayalí estate as inquilinos.[85] And in Queco, one report declared that the "*indiada*" (group of Indians) in the region had risen up and formed a band of hundreds that sacked the estates in the region.[86] After the revolt had been crushed, *El Sur* reported that the lonko Ignacio Maripe (or Maripi) and members of the Ralco reducción had participated in the rebellion and suffered violent retribution, including torture and execution by carabineros.[87] This charge was repeated in Congress by Deputy Juan Prádenas of the Democratic Party.

The press accounts describing the rebellion as an indigenous uprising may reflect the hysterical exaggerations of local elites, as well as their view that all of the region's rural laborers, Mapuche and non-Mapuche, could be described as an "indiada," whatever their ethnic identity. The term contained the sense that Chile's rural poor were barbaric and backward. Yet despite the inflated rhetoric about the participation of indigenous reducciones, it appears that no reducción acted in the rebellion as a reducción. Rather, Mapuches, perhaps some of them laborers on the region's estates, participated as individuals. Of the sixty or so rebels detained after the rebellion, twelve were referred to as Mapuches in newspaper articles.

In Lonquimay, Mapuche reducciones had engaged in a long-standing conflict with local landowners. Most prominently, members of the Ralco reducción, which neighbored the Bunsters' Ralco and Guayalí estates, had been embroiled in an often violent struggle to retain land granted in their original título de merced. The land the indigenous community claimed on the Ralco estate just north of the Bío Bío River had been sold originally with the provision that the owner respect the many indigenous Mapuche-Pehuenches' and non-Mapuche ocupantes' rights of occupation. However, the owners marshaled titles based on fraudulent sales of plots on the estate by twenty-three of the reducción's 250 members in 1881 and had the police forcibly expel all of the Mapuches from Ralco on at least two occasions. The Ralco reducción, however, continued to return and occupy thousands of hectares it claimed in a permanent state of "*rebeldía*," as one Indian Court put it, during the first two decades of the twentieth century.[88]

In 1912, having been forcibly expelled from land they had occupied "since time immemorial," Benicio and Ignacio Maripe, lonkos of the Ralco community, along with the lonkos of the Queuco, Trapa-Trapa, Antonio Canío, Malla, and Callaqui communities, sent a number of petitions to the Land Ministry. They argued that they and their ancestors had been born and raised in the high-altitude valleys between the Bío Bío and Queuco rivers and the Andes cordillera, but that since 1900 they had been "systematically dispossessed" of the most fertile land for pasture and crops by "our wealthy neighbors who today occupy an elevated social position," so that they were "left with only sterile mountain land where our cattle have little pasture, and we have been forced to find our sustenance in the forests' piñones." The petition continued:

The dispossession has arrived at such an extreme, Your Excellency, that it is enough for a wealthy person to take any indigenous person to a notary, paying him an insignificant sum of money, so that they take our names and sell our territories and soon the land is taken from us. This is demonstrated in the numerous public documents that exist in Los Angeles, Angol, Temuco, etc., of sales made in this way and made in favor of the aforementioned gentlemen. On numerous occasions, we have presented petitions to be settled, but up to now we have achieved nothing. In recent years, we have traveled to the capital to present our demands, but very little—or, rather, nothing—has been done for us. In fact, when we return to our homes, we are persecuted and sometimes brutally mistreated by the agents of these aforementioned gentlemen.

As we have seen, in at least one case a government surveyor was unable to measure land to settle Lonquimay's Mapuche-Pehuenche communities because of opposition from large landowners who had purchased or leased pasture. It is also of note that loss of pasture to large estates had pushed communities such as Ralco into increased dependence on the mountains' araucaria forests, making them all the more zealous in their efforts to protect the forests from logging and fire.[89]

The lonkos' petitions were rejected because, as Land and Colonization Minister Luis Risopatrón argued, much of the land they demanded was titled to private landowners. Risopatrón also repeated landowners' arguments that the members of the Ralco community were Argentines, not Chileans, who had fled the Argentine army during the War of the Desert and taken refuge in Alto Bío Bío. As Argentines, they were not entitled to be settled on reducciones. That the Mapuche-Pehuenche communities of Alto Bío Bío and Lonquimay were nomads and not Chilean was a long-standing accusation made by landowners who had been granted or who leased pasture in the mountains and sought to expel the indigenous communities settled there.[90]

During the mid-1920s, the Ralco community continually petitioned to be settled on land claimed by the Bunster family and protested incursions into the land of the reducción. By the early 1930s, members of the reducción, along with Chilean ocupantes—numbering roughly 550 together—had occupied an 11,493-hectare invernada, which they claimed was public land, on the estate.[91] The Bunsters, in turn, waged a war of attrition against the ocupantes. In 1932, in a session of the House of Deputies, Prádenas requested that the government intervene on behalf of the hundreds of colonos and ocupantes who had been expelled from Guayalí. According to Prádenas, the conflict between Bunster and the local reducciones was "a sad and black history in which caciques have [been] disappeared, indígenas have been beaten, cornered in Alto Bío Bío on the frontier with Argentina, and had their land stolen."[92] A year before the rebellion, the Bunster family, with support from the courts and carabineros, evicted members of the Ralco reducción from land on the Ralco estate, as they had done to the aspiring colonos settled on Guayalí.[93]

Like many of the conflicts between large landowners and campesinos on the frontier during the 1920s and 1930s, the Ralco reducción's confrontation with the Bunsters included non-Mapuche ocupantes. While they articulated their rights to land in terms of their indigenous status and the legislation that governed indigenous affairs, they also defended the rights of poor nonindigenous campesinos to occupy public land. In fact, as government censuses reflected, both Mapuche-Pehuenches and non-Mapuche ocupantes occupied land on the Ralco estate and were engaged in conflicts with the Bunster family. Thus, members of the Ralco reducción petitioned the state to settle them on the land, but also defended the rights of neighboring ocupantes who were also squatting and who claimed the status of colonos. A few years after the Ránquil revolt, the Ralco reducción once again petitioned the Ministry of Land and Colonization for support in their continuing conflict with the Bunster family:

> We declare that we have occupied the land in Ralco since time
> immemorial, since our ancestors were the first to populate [the
> region].... Now we ask that the supreme government take interest
> in us, send us a surveyor to determine the borders of our reducción,
> respecting the rights of the poor Chileans who occupy the place on
> public land that neighbors our possession, excepting the representa-
> tives of the Bunster ... family who have always abused us and now
> are trying to take our land from us.[94]

The reducción defended its land in terms of its indigenous identity rooted in its occupation of the land from "time immemorial." But the ambiguous borders of the neighboring estates, whose land could be defined as public, also meant that members of the reducción claimed land rights in terms of the more general rights of ocupantes of public land. In this they joined hundreds of non-Mapuche campesinos in conflict with the Bunsters. Members of the Ralco reducción worked on the Bunsters' Guayalí estate with non-Mapuche campesinos, and like inquilinos throughout the south, they viewed estate land as public and rightfully theirs.

In southern Chile, Mapuches engaged in a repertoire of strategies to claim land on the frontier. Many left their reducciones, squeezed by so little land, and settled as ocupantes on public land or land they believed was public. Often this land had no title, ambiguous titles, or ill-defined boundaries. Often it lay on the borders of their reducciones. In these cases, Mapuches lived in close proximity to poor non-Mapuche ocupantes, who likewise occupied public land or land on the borders of estates whose titles and boundaries were often in question. These campesinos, squatters on land they believed to be public, were often thrown together in conflicts with large landowners who sought to expand their estates or claim public frontier land. Both Mapuches and non-Mapuches staked their claims in terms of the land's status as public and their rights to be considered colonos. In addition, they joined the broader movements of inquilinos and ocupantes to claim estate land as public and demand plots as colonos. Finally, for Mapuches who continued to live in reducciones, the original land grant, or título de merced, became a crucial weapon for defending land rights from usurpation. In these cases, Mapuches phrased their claims to land in terms of their residence in the reducciones.

What remains to be explained in the case of the Ránquil rebellion is why Lonquimay's campesinos, including some Mapuches from the region's reducciones, moved from legalistic strategies of appealing to state authorities, going to court, and invoking rights established in colonization and indigenous legislation to violent confrontation with landowners and the Chilean military. Mapuche-Pehuenches in Lonquimay, like Mapuches throughout the south, had employed the indigenous courts and protectorates to defend their claims to land occupied by estates. Similarly, the region's non-Mapuche campesinos had acted lawfully, building a union and petitioning for land under the Southern Property Law. They had even established a strong, if short-lived, alliance with Ibáñez during his presidency and had sent countless delegations and petitions to Santiago. What had changed was the state's response to their efforts to win access to land in Lonquimay's mountains. The Southern Property Law, along with the package of social reforms during the 1920s, raised expectations that the state would redistribute some estate land to the region's campesinos, settling them as colonos. Campesinos' hopes were dashed, however, when Alessandri returned to power. Juan Segundo Leiva Tapia, once a staunch supporter of Ibáñez, allied himself with the Communist Party and took up the banner of rural insurrection following his return from detention and internal exile under the Alessandri regime.

During the three decades that led up to the Ránquil rebellion, the Chilean state underwent a significant transformation. Beginning with the early "national colonization laws" of the first decade of the twentieth century and the parliamentary commission's study of colonization in 1911, and ending in the labor, land, and forest reforms implemented under the first Alessandri government and the Ibáñez government, the Chilean state asserted its authority—in fits and starts, to be sure—over the social, economic, and ecological life of the frontier. The plethora of laws and reforms during this period allowed the rural poor in southern Chile to view the state as an interlocutor and guarantor of their rights, above all to land. Yet reforms of colonization laws to expand settlement of the frontier with ocupante small-holders collided with reforms designed to modernize forestry production and conserve native forests. This contradiction incited considerable outrage among southern campesinos by raising expectations that they might be settled on public land while simultaneously restricting their access to key forest resources that were now organized in reserves and parks.

In addition, the reforms of this period reflected only one side of Chilean politics. On the other side stood the entrenched political and economic power of the traditional landowning class, with its resistance to reform and its fear of revolution. Intermittent periods of reform ended with repression; periods of political opening ended with authoritarian rule. The repression of the Lonquimay uprising, including the execution of Ignacio Maripe and imprisonment of other members of the Ralco community, reflected the Chilean state's periodic policy of meeting social unrest with violence. In the south, Alessandri's crackdown in the countryside put an end, momentarily, to campesinos' movements to take back land and forests from the frontier's large estates and logging concessions.

4

Changing Landscapes
Tree Plantations, Forestry, and State-Directed Development after 1930

In 1943, Adolfo Ibáñez wrote an article for *El Mercurio* describing a recent trip to his natal province of Concepción. He was astonished by the transformation of the countryside. "Speaking frankly," he wrote, "the province's old landscape had little about which to be enthusiastic. . . . [A]mong sand pits and almost sterile hills, between arid reddish ravines and . . . swampy plain[s], the cultivable agricultural land appeared, in reality, very scarce." The "modest and silent" process of reforestation over the previous decades, however, had covered the province with "forests and forests, millions of leafy trees of all ages." While Ibáñez had once known a barren and parched land, cleared of native forests, he now saw, "from the railroads and principal roads, every horizon adorned with the festive shade of eucalyptus and pine." Tree plantations had become one of the engines of Concepción's economy and had produced a new population of self-made men. Ibáñez built a mythical ideal of the frontier entrepreneur: the modest smallholder who made a fortune planting pine. "As a typical example, I can cite the case of a foresighted man who invested his modest savings in land that held little value," he wrote. "For a quarter of a century, he planted it with forests, and recently he has rejected an offer from an important industrial company worth millions of pesos for that property." For Ibáñez, the waves of green tree plantations represented more than a resuscitation of the province's dusty hills and valleys. They were inextricably linked to Concepción's industrial future, its complex of coal mines, textile plants, railroads, and a busy port. He celebrated the "enormous industrial value of the forests and unlimited use of wood in every aspect, not only for construction and furniture . . . , but also for plywood, compressed wood, [and] cellulose."[1]

Ibáñez's article reflected the extraordinary ecological changes in Chile's southern provinces—particularly in Ñuble, Concepción, Bío Bío, Arauco, and Malleco—during the 1930s and 1940s. At once description and prescription, the article served as a lesson to landowners throughout Chile:

the country's agricultural future lay in tree plantations. As agricultural production diminished yearly as a proportion of the nation's economy, and industrialization transformed Chile into a largely urban society, plantations would restore value to Chile's increasingly barren soil by combating erosion and linking the rural sector to developing industries. As a piece of propaganda, Ibáñez's article promoted the image of the frontier landholder as a classic small or medium-size farmer who was modern, rational, and productive, in implicit contrast with central Chile's large and still feudal estates. The article made little mention of the fact that these same landowners, like the Lota coal-mining company in the coastal cordillera in Concepción and Arauco, had overworked the frontier's land so intensely and cleared its forests so effectively that by the 1930s, Chile's former "breadbasket" confronted a protracted crisis of diminished yields and infertile soil. Nor in his celebration of the modest frontier planter did Ibáñez see fit to refer to the central role of the Chilean state in fomenting forestation and building the developing forestry industry. For Ibáñez, like many southern landowners, Concepción's forestry miracle was a result of individual initiative that had made Chile the site of Latin America's most extensive tree plantations.

The spread of pine during the 1930s and 1940s redrew southern Chile's natural and social landscapes. The nature of Monterey pine itself drove the reorganization of land and labor relations and state agrarian policy. Pine was unlike other agricultural commodities. It could be harvested only after twenty years; it required little labor and a great deal of technical forestry management and initial capital investment; and it restored value to overworked soil by staving off erosion. Pine plantations required landowners to have fairly extensive holdings, significant amounts of capital, and the ability to wait decades for profits. To meet these needs, a number of landholders formed public stockholding companies and entered into partnerships with Chile's social-security system, the major locus of capital in the country. Stockholders and future retirees could wait for their trees to mature and investments to pay off.

Landowners turned to pine to resolve a host of social and ecological problems. Having decimated the soil of their properties through labor arrangements such as sharecropping and renting out parcels to smallholders and through the burning of forests to plant wheat and pasture, they could now reclaim land that was producing diminishing yields and losing value. As rural unrest and demand for agrarian reform spread during the late 1920s and 1930s, the prospect of estates populated by trees rather than a restive resident labor force or *ocupante* squatters was increasingly appealing. Artificial pine forests allowed landowners to reclaim their soil from erosion and their property from rural laborers who claimed it as their own.

Finally, as the state began to implement the Southern Property Law and reacquire public land to colonize during the 1930s and 1940s, tree plantations allowed landowners to present themselves as modern, productive, and patriotic. Unlike the vilified owners of latifundia, these landowners were combating erosion, employing modern production techniques, and building a national resource that both revitalized agriculture and provided raw materials for industry.

While landowners in provinces such as Ñuble and Concepción played a major role in the turn to forestation with pine and commercial forestry during the 1940s and 1950s, these self-described "forestry pioneers," like their counterparts in the industrial sector, looked to the state for key subsidies and protections. They also looked to the state as a source of capital that would assist them in their efforts to transform deforested and eroded soil, as well as land covered in forest scrub and second-growth forest that held little commercial value, into ecologically and economically dynamic pine plantations. Indeed, the state's role in forestation had been initiated by Federico Albert during the late nineteenth century; after 1930 a series of Chilean governments built on Albert's prescriptions for state-directed forest management linked to the development of new industries and the diversification of the agricultural sector.

The State and Forestry Development

The period between the election of the populist President Arturo Alessandri in 1920 and the coalitions led by the Radical Party and initiated by the Popular Front coalition (1938–52), which included, in varying configurations, the Radical Party, the Socialist Party, the Liberal Party, and the Communist Party, produced a shift in state forest policy from a laissez-faire approach that encouraged burning and clearing native forests to state intervention to promote commercial logging and forestation, as well as the conservation of native forests on land held by the state. Like most of the social reforms passed in 1924 and 1925, Chile's Forest Law remained no more than a piece of paper, as critics often put it. While it finally made legal what Albert and other forestry advocates had long proposed, the new regulations on logging, clearing with fire, and forestation went largely unimplemented. The world crisis of 1929–30, however, sparked a major change in both the Chilean economy and Chile's forest policy. With Chile's mining export economy convulsed in a protracted crisis that would culminate in the closing of the northern nitrate mines and in agricultural production experiencing a steady decline in production and profitability, Chilean policy makers looked from the country's northern region to its southern frontier in search of a new motor of development. The southern provinces' forests, native and artificial, now appeared to be an untapped source of national wealth that might reignite another cycle of export-driven growth. In addition, as Chile embarked on a new path of domestic industrialization and economic diversification following the global recession, forests and forestry took on an increasingly prominent role in fantasies of national development. Forestry and forestation would rescue the rural economy by providing a new crop whose profitability lay as much in the way it restored value to overworked soil as it did in its harvests. In addition, forestry would help promote industrialization by providing raw materials for a series of industries, such as construction, wood, furniture, chemicals, and cellulose. Forests and forestry would be the signature of a modernized agriculture and rapidly industrializing nation.

In July 1931, in the midst of a recession that had caused massive unemployment and the dispersal of thousands of former mine workers from the north throughout the country, Ibáñez's government passed a new Forest Law as a "decree law." Like its predecessor in 1925, the law established tax exemptions and subsidies for tree plantations, regulated logging in native forests and the use of fire to clear land, created national parks and forest reserves, and established official commercial classifications for native Chilean woods. In addition, the law created a forest guard within the carabineros to enforce the forest code. The law sought to reassert state control of forest exploitation and promote commercial forestry as a means to reigniting Chile's stagnant export economy. A congressional commission on agriculture and colonization argued, "We must look for new industries that . . . can at least provide work for the great mass of the unemployed. . . . [I]f capital is invested in forming industries derived from the exploitation of wood, we will have taken a step toward solving many of our domestic problems."[2] Writing in 1936 for the American Geographical Society, George McBride echoed this commonly held assessment of the southern frontier's forestry potential: "in value of its resources the south is probably not a whit inferior to the nitrate regions of the north. Its importance should be more lasting by far."[3]

Forest management was tied to a broader strategy of development based on industrialization. In this case, however, the shift from export-oriented growth to import-substitution industrialization did not constitute a rupture. Forest-based industries were still extractive economies and would supply export commodities, as well as the raw materials for domestic industries. The authors of the Forest Law of 1931 saw in forest products a worthy replacement for nitrates. Chile would continue to mine its natural resources to promote development and growth, only trees, rather than nitrates, would provide jobs and revenue. In addition, despite the state's newly energetic role in promoting commercial forestry and imposing forest management, the industry and its raw materials would remain largely in the hands of private landowners. While state-owned forest reserves would contribute raw materials to the new industry, most of the frontier's native forests, tree plantations, sawmills, and paper and pulp factories would be private concerns. The irony of forestry development in Chile after 1930 lay in the central role played by the state in sponsoring an industry that, on the whole, was privately owned and whose profits were privately enjoyed. Both the second government of Arturo Alessandri (1932–38) and the Popular Front coalition led by the Radical Party and its successor center-left coalition governments (1938–52) made the development of commercial forestry a key element of agrarian and industrial policy, building on the forest laws of 1925 and 1931.

After 1931, state institutions began to invest heavily in pine plantations and to funnel assistance to landowners who cultivated tree plantations. During the 1930s, the Caja de Empleados Públicos (Public Employees Retirement Fund) purchased a number of estates throughout the frontier and planted them with pine.[4] In addition, the Caja de Seguro Obrero (Blue-Collar Workers' Retirement Fund) purchased the large Canteras hacienda in Bío Bío Province in 1934 on which it planted 10,000 hectares with pine. The plantation was designed to perform the dual function of providing pensions for retired and disabled workers and "increasing the forest wealth and contributing . . . to the improvement of the national economy."[5]

In 1937, the Ministerio de Previsión Social (Ministry of Social Welfare) and a new Comisión Maderera (Wood Commission) named by the Alessandri government began to plan how to employ the capital held by social-security institutions to develop tree plantations. *Cajas* (retirement funds administered by the state) would provide capital for reforestation and scientifically managed tree plantations would provide a secure future of annual returns for the cajas' retirees. The Wood Commission noted the "urgent need" for 500,000 hectares of Monterey pine plantations "to replace the Chilean forest species that are already exhausted." The commission argued that Monterey pine was ideal for this purpose because it grew much faster than native Chilean hardwoods and had a number of uses in "construction, carpentry, mine supports, wood boxes (for fruit etc.), the extraction of cellulose for the production of paper . . . and other valuable industrial products."[6]

In 1938, following the election of the Popular Front coalition, the Caja de Colonización Agrícola (Agricultural Colonization Fund) turned its attention to both the distribution of public land and the problem of deforestation. The fund reported grimly that reserves of raulí, laurel, and roble pellín were rapidly dwindling and could be found only in remote places far from the railroad and towns. The problem was both the use of fire to clear land and the lack of reforestation required for the "formation of the country's forestry capital." Nonetheless, the fund continued the policy of destroying forests to make room for agricultural crops and pasturage. In Chiloé, for example, it distributed 30,000 hectares, with 200 hectares per colono, with the requirement that the settlers clear the forest and plant pasture for livestock.[7] Yet while the fund viewed clearing native forests to promote agricultural development as imperative, it also began to promote reforestation, mostly to roll back erosion and protect watersheds in Chile's southern provinces. It noted that during the 1930s, the Forest Department had installed tree plantations in Llico, Chanco, Llolleo, and other locations in southern and central Chile to keep dunes from advancing on coastal agricultural land. In addition, the Lota coal company had cultivated the country's largest plantations, and the Public Employees Retirement Fund had begun to plant trees on its estates.[8]

The fund echoed Federico Albert's advocacy of state support to expand the scope of pine and eucalyptus plantations. Without state initiatives, it would be impossible to "form the forest capital that the country needs." Because of pine's long period of maturation, investments in plantations would take a long time to pay off; landowners were accustomed to seeing the results of their investments in a year or, at most, between three and five years in the case of vineyards and fruit trees. Pine might take two decades to reach harvestable age. In addition, agricultural producers worked with little capital, and "it is not possible materially to direct part of it to forest plantations." Reforestation could not be left to private initiative: "it is undoubtedly the state that is the first party that should take responsibility for this work, given our national idiosyncrasies and particular economic conditions. In second place [are] the semi-public institutions and companies [i.e., the social-security funds]."[9] By 1942, the Agricultural Colonization Fund had established eleven tree nurseries and twelve colonies dedicated to forestation. The goal was to replace logged roble forests with Monterey pine plantations.[10]

The state development agency, the Corporación de Fomento de la Producción (Economic Development Corporation; CORFO), organized by the Popular Front government in 1939 to promote industrialization and economic diversification, also worked to stimulate pine plantations and commercial forestry. Taking up where the Forest Law left off, CORFO provided a number of incentives to private landowners to replant their fundos with pine and eucalyptus. They included loans of 500 pesos (roughly $25) for each hectare planted, with a ceiling of 25,000 pesos ($1,250), at 5 percent interest over a five-year period. In addition, between 1939 and 1949, CORFO extended 2.924 million pesos (roughly $145,000) in loans and credits to forestry companies. The agency also supplied credits to be used for the installation of modern sawmills.[11] In 1946, *El Sur* of Concepción, a news-paper identified with the Radical Party, noted, "It is because of CORFO's labor and economic assistance that a number of forestry companies have recently been started in the center and south of the country."[12]

Through its Forestry Section, CORFO organized forestry companies with private landowners. The latter provided the land on their fundos while CORFO supplied capital to pay for the plantations and other costs. By 1942, four large mixed public-private forestry companies had been established through this kind of arrangement.[13] In addition, during the 1940s CORFO and the Agricultural Colonization Fund participated with private investors in the formation of the Sociedad Anónima Maderera del Sur, with CORFO providing two-thirds of the capital.[14] The company installed a plant for producing plywood near Lago Puyehue, east of the city of Osorno. The plant was surrounded by fundos dedicated to logging the cordillera's native forests and thus could count on a steady supply of lingue, olivillo, ulmo, laurel, tepa, and alerce.[15] In addition, CORFO put together the Consorcio Nacional de Productores de Madera (National Consortium of Wood Producers) to deal with the chronic lack of credit, transportation infrastructure, and marketing mechanisms suffered by southern wood producers. In 1947, CORFO, with the state railroad company and private investors, formed the Sociedad Impregnadora de Madera (Wood Preserving Company) to supply domestic and external markets for railroad ties, posts, and other treated wood. Finally, CORFO formed the Exportadora de Maderas de Chile (Chilean Wood Exporters Company) in 1941 to help promote Chilean wood exports.[16]

The agency provided credits to private companies that logged forests, produced plywood, distilled chemicals from woods, and provided materials for construction. It also organized cooperatives of wood producers to help small mills market their products (and overcome the monopolies held by merchants in towns). Small mills received additional "help [from CORFO] in grading and inspecting, and guidance and aid in the purchase of machinery and equipment."[17] The agency's National Committee on Wood planned the establishment of large drying plants and sawmills, conducted studies of foreign markets for Chilean wood, and established the basis for a system of classification of wood products and regulation of wood production that would improve the quality of the materials produced by the timber industry and facilitate the growth of exports of forest products. The agency also began to plan the installation of factories that would produce cellulose for export.[18] It helped to open new markets for Chilean wood in Argentina, Peru, Bolivia, and Uruguay, as well as in a number of European countries; between 1941 and 1949, the value of Chile's wood exports more than quadrupled.[19]

Private Landowners and the Expansion
of Pine Plantations in the 1940s

The state's push to develop forestry and forestation as an engine of Chile's economic and ecological recovery met was welcomed by many southern landowners. During the late 1930s and early 1940s, *El Sur* promoted pine plantations in a series of interviews with government officials and agronomists. In 1940, for example, the paper published an interview with the agronomist Adrian Molina titled "Forestation and Reforestation: The Way to Improve Our Soil and Create Sources of Wealth." With an audience of landowners in mind, Molina argued that tree plantations represented "easy, certain, and reproductive investments." And, he asserted, invoking a persuasive model, calculations made in the United States "have demonstrated that forest plantations [earn] a return on investment that is four times greater than that for the cultivation of cereals." The frontier provinces possessed soil that was ideal for forestation, since a large area was not suitable for agriculture because it was mountainous and rocky or eroded. Like other advocates of plantations, Molina underlined the climatological and public health benefits of plantations: "the hygienic and aesthetic value of forests is well known; together with purifying the air, they provide above all natural recreation for our spirit."[20]

Pine plantations were also promoted in *El Sur* by Alfredo Wolnitzky, an agronomist in the Agriculture Department at the Universidad de Concepción. Wolnitzky, like many of his peers in forestry science, looked with disfavor on Chile's native forest species, which, he argued, took hundreds of years to mature. Monterey pine, by contrast, grew so rapidly that it could be harvested profitably and expeditiously by private landowners. No landowner, he pointed out, was going to invest in species of trees that would benefit only future generations. Landowners could enjoy the fruits of pine, unlike native forests, in their own lifetimes. In addition, the crisis of soil erosion in the frontier provinces demanded an immediate solution that only pine could provide.[21]

El Sur's enthusiasm for pine plantations reflected the increasing interest of Concepción's landowners in forestation and forestry as an alternative to agriculture. The Lota coal-mining company had already developed plantations to replace the native forests it had logged and to provide timber and fuel for its coal mines. In addition, Chile's only paper producer, the Compañía Manufacturera de Papeles y Cartones (CMPC), also known as La Papelera, owned the large Los Pinares estate in Chiguayante, Concepción, from which "thousands of tons [of wood] leave the forest to be transformed into paper and byproducts of cellulose." In 1932, La Papelera founded its own cellulose plant and began producing pulp from araucaria pine and álamo (poplar) logged on its southern estates. The expansion of the company into a full-blown industrial forestry complex took off, however, when Jorge Alessandri became its president in 1938. Until the Second World War, the CMPC imported half of its pulp to produce paper from Finland and Sweden. The war cut off this supply of raw material, and the CMPC turned to southern pine plantations as an alternative to both the dwindling stock of native forests on its estates and foreign pulp. In 1940, it purchased Los Pinares, making it one of the country's largest producers of Monterey pine.[22]

An article published in *El Sur* in 1940 invoked La Papelera's forestry activities and plantations as a model. The paper noted that the company's exploitation of mountain forests by a labor force of two hundred men on Los Pinares would be "eternal," since the company was replacing the araucaria pine and other native species it logged in the coastal cordillera with plantations of Monterey pine that "in ten years more will constitute easy-to-exploit forests."[23] Similarly, in 1933, *El Campesino*, the publication of the landowners' association the Sociedad Nacional de Agricultura (National Agricultural Society), also called on landowners to follow the example of La Papelera; the company's growing consumption of pine trees to produce pulp guaranteed future markets and profits: "there is a major national need to stimulate plantations of Monterey pine. . . . Anyone who plants these trees can be certain that in ten to fifteen years, a commodity of great value will be produced that will be assured of markets, either in the Manufacturera [La Papalera] or in other companies that will have been established [to produce pulp] by that time."[24] Outside Concepción, then, led by the Lota coal-mining company and La Papelera, the process of replacing native forests with North American pine on privately owned estates had begun in earnest by the 1930s.

State aid for forestry and forestation sparked a boom in plantations throughout a number of southern provinces. The Forest Law called for subsidies of up to 200 pesos (roughly $25) per hectare planted with trees on land declared *forestal* (a property dedicated to forestry), with a maximum limit of 200,000 pesos a year. In addition, fundos with tree plantations were guaranteed thirty-year exemptions from property taxes. *El Agricultor*, the publication of the association of southern landowners, noted in 1943 that the thirty-year tax exemption for tree plantations included in the Forest Law of 1931 had stimulated a large number of owners of farms that were useless for any other cultivation to undertake the "patriotic task" of creating plantations and "transforming the decimated soils of the provinces of Concepción and Arauco into sources of wealth." *El Agricultor* also pointed out, "State institutions contributed effectively to the execution of this great work, installing nurseries of forestry trees in different locations, providing plants to people with limited resources, providing instruction about the best methods for performing the work."[25] In 1939, for example, Guillermo Otto requested that his plantations be certified by the general land directorate. Between 1926 and 1934, he had planted 595 hectares of his Coliumo fundo and Las Catalinas fundo, both in the Department of Tomé, Concepción Province. According to his calculations, Otto was owed 95,050 pesos ($4,750).[26] Similar petitions flooded the Forest Department beginning in the late 1930s. In 1945, Alberto Heck petitioned for tax exemptions on his Los Pinos fundo in Mulchén, which had been declared "forestal" in 1939. Heck had planted almost 300 hectares with Monterey pine.[27] That same year, Federico Duerr was granted tax exemptions for his Selva Negra fundo, on which he had planted 398 hectares with 795,000 pine trees.[28] Also in Mulchén, the Sociedad Forestal Hermann Corrsen received tax breaks because of the 375 hectares that it had planted on its San Luis fundo.[29]

A classic shift in landowners' practices following the new interest in forestation was a petition in 1932 from Schmitz and Company to the Ministry of Land and Colonization proposing the "agricultural and industrial exploitation" of 70,000 hectares on the Los Alpes and Los Quillayes fundos near San Fernando. The company used the language and standards of agronomy to phrase its petition. The haciendas lay on mountainous terrain covered by native forests of roble, coigüe, "and other species," and their productivity had been "deficient" because of the lack of capital and means of transportation required for profitable logging. Schmitz proposed to log the native forests to produce wood for construction, fuel, and firewood, and then replant the haciendas with Monterey pine, a species "that gives extraordinary results in this zone and could signify after ten or fifteen years an immense source of wealth for the country, as much as for its wood as for its byproduct . . . , cellulose." The company also wanted to purchase four neighboring fundos for its forestry activities.[30] As in this case, pine plantations often replaced native forests (usually second-growth or degraded forests) that had been logged for wood and charcoal. In 1938, for example, the owner of the San Cristobal fundo in Yumbel requested authorization from the General Directorate of Land and Colonization to fell a stand of quillays (soapbark trees) and litres to make charcoal. The primary goal, however, was "to transform the natural vegetation of this land to dedicate it to plantations of radiata pine."[31] Juan Moty, the owner of the Casas de Caliboro fundo in Los Angeles, Bío Bío Province, also petitioned for permission to fell a stand of quillay. He planned to replace the native vegetation with plantations of Monterey pine.[32]

During the 1930s, a number of landowners on the frontier also transformed their fundos into public forestry companies dedicated to the exploitation of pine plantations. These companies boasted about their modern methods of production directed by forestry science. They targeted especially the new urban population of white-collar workers by promising that investments in pine plantations would secure a stable rate of return and a prosperous retirement. By investing in *parcelas-bosques* (small plots of pine plantations), urban workers and growing middle-class sectors, particularly public employees, could earn a healthy return on investments whose profitability scientifically managed plantations would guarantee.[33] The plantations would constitute a shield against the chronic inflation of the 1930s and 1940s, which consumed the wage hikes and benefits blue-collar and white-collar workers had begun to earn with the Popular Front governments. As one company noted in its prospectus, "Money loses its value, the cost of living rises, and the small savings you may have been able to accumulate are rendered more and more insufficient. The human desire for a stable old age with a comfortable economic situation is every day more of a chimera."[34] Another business prospectus read, "No combination of savings and multiplication of capital exists today that can equal the profits produced by a plantation of [Monterey pines] that is technically well managed and on land in areas appropriate for forestation. It is unnecessary to say even a word about the future of pine plantations in this country. The theme is exhausted, and all of the studies done by [CORFO] . . . coincide in recommending the privileged conditions in Chile for this prime material."[35] The prospectus echoed the assessment of the director of the Agricultural Colonization Fund who argued, "The annual accumulation assured by the development of tree [plantations] has an economic value very superior . . . to what is offered by the insurance companies and other welfare institutions."[36]

Public forestry companies were designed to take advantage of the new forestry laws and state development programs. After 1939 the state began to promote jointly owned public-private forestry enterprises as a way to wrest revenue from Chile's forest resources. Pine plantations, as Albert and other proponents of forest management had argued, were an ideal source of raw materials for these new industries. And, of course, the parcelas-bosques companies would benefit from state support for reforestation, including tax exemptions and a subsidized supply of trees from state-run nurseries. In addition, new state-sponsored housing programs designed to meet the rising demand from rural-urban migrants and to create construction jobs for the armies of the unemployed after 1929 would also provide markets for domestically produced wood. Government studies showed that the existing urban population, mostly concentrated in Santiago, needed 150,000–300,000 new residences.[37] One report from 1946 noted that "a rising standard of living from industrialization and other activities under way [would] bring about a comparable increase in construction." In addition, the report argued, "Government efforts in the housing field (it is estimated that one-third of the structures erected over the past eight years have been built by the government) should tend to lower prices by direct competition and cheaper credits, thus also stimulating building."[38] Government programs to confront the pressing need for working-class housing created a potentially dynamic market for the southern timber industry.

While tree plantations were an attractive investment, few landowners had sufficient resources to purchase saplings from nurseries and then wait twenty years to turn a profit. The retirement funds for blue-collar workers (*obreros*), white-collar workers (*empleados*), and public employees constituted an important potential source of capital and credit. Indeed, beginning as early as 1934, forestry advocates pointed to Chile's social-security system—its cajas—as the one source of capital that could fuel forestation and the modernization of forestry production. In one case, the Comunidad Yrarrázaval Mac-Clure transformed three large fundos in Ñuble Province into subdivided *fundos forestales*. With the assistance of the Caja de Empleados Públicos y Periodistas and a forester hired by the fund, the company sold parcelas to five hundred white-collar public employees. The fund facilitated credit to the empleados so they could purchase parcelas-bosques on the fundos owned by the Comunidad Yrarrázaval Mac-Clure.[39] In a published interview, Leonicio Larraín described how, with his brother-in-law Alfredo Yrarrázaval, he had initiated a parcelas-bosques company in 1935: "the big problem was financing and finding people to acquire plantations. To meet these ends, we put in practice a system unknown in Chile: the parcela-bosque, which we offered to small investors who would pay small monthly quotas." Four thousand hectares was sold to the semi-fiscal funds: "the support of the Caja de Empleados Públicos y Periodistas was fundamental . . . since it stimulated these investments among its members and even modified its statutes so that every empleado could purchase up to five parcelas-bosques."[40] By 1950, the Comunidad Yrarrázaval Mac-Clure had planted 8,000 hectares with pine. In addition, it had founded the Sociedad Anónima Maderera Industrial "Pinos de Cholguan" in 1947 and the Sociedad Agrícola y Maderera Lago Laja, which was organized to exploit raulí, roble, and coigue in its forests on the 120,000-hectare Polcura hacienda in Ñuble Province. The company planned to replant the property, once cleared of its native forests, with Monterey pine.[41]

The Comunidad Yrarrázaval Mac-Clure has been widely heralded as a shining success story of Chile's pioneer forestry entrepreneurs. The company itself liked to refer to its "entrepreneurial and patriotic spirit" in taking up the task of reforestation.[42] Yet the community was formed with capital from Chile's social-security system, took advantage of important state subsidies and tax exemptions, and followed the lead of the

Blue-Collar Workers' Retirement Fund's Canteras hacienda in linking the cultivation of pine plantations to the capital investments of retirement funds. Even the southern landowners' association recognized that the entrepreneurial drive of landowners who planted pine owed something to the actions of the state. As *El Agricultor* put it in 1951, "The reforestation that is being accomplished in our country is one of the great progressive actions demonstrated by the owners of rural properties. . . . The state, in turn, has fomented this initiative through the tax exemptions on planted land for thirty years, also supplying plants from its nurseries to interested parties at infinitesimal prices."[43] A report in 1937 by a government commission named to study forestation noted that the formation of the Comunidad Yrarrázaval Mac-Clure was "based on diverse studies and information produced by the Undersecretariat of Commerce about the national and commercial importance of Monterey pine." In 1936, it had sold 256 "parcelas-bosques" planted with Monterey pine to 103 public employees, "who were able to acquire them thanks to the special credit facilities that were extended exclusively for this purpose . . . by the [Caja de Empleados Públicos y Periodistas]."[44]

In addition to the parcelas-bosques companies, landowners formed stockholding companies to purchase fundos and plant them with pine. In one case, the Sociedad Ganadera, Agrícola y Forestal Río Laja was formed to purchase the 7,000-hectare Colcheu hacienda in Concepción Province. The company proposed planting already farmed land, as well as cleared native forests, with álamos (poplars) to supply Concepción's match company and pine to supply raw material for paper production. A company prospectus stated, "Few have the large extensions of land adequate for [tree] plantations. . . . [P]rivate individuals have to wait fifteen or twenty years to obtain the maximum benefits. With the difficulty in acquiring credit [and] the high interest rates, very few are in a position to do this. . . . In the case of a stockholding company . . . , not only is it advisable, but it must be understood as a healthy policy of foresightedness to engage in this magnificent business that, from its first moment, will give value to the stocks, whose prices will rise progressively as the plantations grow."[45] The prospectus pointed out that a stockholding company was almost a prerequisite for planting pine because of the need for capital, large extensions of land, and the capacity to wait twenty years for profits. A stockholding company could do what the owners of small parcels could not: cultivate trees on a large scale, manage the plantations, and wait years for the initial investment to pay off. Another public company dedicated to forestation noted that its prospects were rosy because markets for pine trees were guaranteed by the state's promotion of new forestry industries, as well the construction of inexpensive public housing through the Caja de Habitación Popular (Low-Income Housing Fund).[46] Indeed, the security of investments in pine relied on domestic markets for paper and pulp and construction that were stimulated and protected by the state.

Other large companies followed suit. The Sociedad Anónima Colcura, originally established as part of the Lota coal-mining company, became an independent publicly owned company in 1947, with twenty-one fundos and 14,000 hectares of plantations. By the 1940s, Colcura had moved from simply supplying Lota's coal mines to exporting 25 percent of its production to Argentina, making it one of the country's largest exporters of forest products.[47] It was joined by Forestación Nacional, formed in 1946 by members of the Larraín family and dedicated to purchasing fundos and planting them with pine, as in the parcelas-bosques, with significant investments from white-collar workers facilitated by the Public Employees Retirement Fund.[48] Founded in 1893, Bosques y Industrias Madereras (BIMA) also turned to planting pine on land it purchased or leased from the state. It owned seventeen southern haciendas, as well as the Productora de Maderas Nahuelbuta wood company, with more than 50,000 hectares, during the 1940s. During that decade, the government required BIMA to reforest the public native forests, many of them araucaria forests in the Nahuelbuta cordillera, that it leased and logged. It also owned tree nurseries and employed forestry technicians to oversee plantations on land it had previously logged or that had already been cleared when it was purchased. Like other public companies, BIMA pointed to forestation with pine as security for investors: "we are happy to let our stock holders know that what is reforested today is a certain indicator of future earnings for the company."[49] Like other forestry companies during this period, BIMA received state support, not just in terms of leases of publicly owned native forests, but also via credits and subsidies.[50] In 1946, the company purchased large, modern sawmills and established a factory to produce pressed wood in "collaboration with the Corporación de Fomento and the Caja de la Habitación, which is very interested [in the project] since it is the most important consumer."[51]

The business of transforming fundos into public forestry companies followed a certain ecological logic. The fundos planted with pine were mostly located in provinces such as Arauco, Bío Bío, Concepción, Malleco, and Ñuble, where native forests had been devastated and where wheat cultivation had led to soil erosion and declining harvests. By the 1930s, *El Campesino* and *El Agricultor* had begun to analyze with alarm declining yields of wheat and to turn enthusiastically to Monterey pine as a crop with a future.[52] The very nature of pine as a commodity (as opposed to, say, sugar, wheat, or other agricultural commodities)—the facts that it restored value to eroded soil, took twenty years to harvest, and required significant capital investment—led to the formation of stockholding companies. In this sense, the ecological realities of deforestation, soil erosion, and Monterey pine itself shaped the organization of forestry production.

Of course, through ruthless overexploitation of the frontier's soil and labor and rampant destruction of its forests, it was the landowners themselves who had created the ecological conditions that made pine an attractive investment. By the 1940s, soil erosion in provinces that produced cereals on the frontier, particularly Malleco, had reached alarming proportions. A report in 1937 by an agronomist for the Agricultural Colonization Fund noted that deforestation and intensive cultivation had decimated the soil of Bío Bío and Malleco provinces: "on many estates . . . the soil covering agricultural terrain has become so impoverished that it is no longer profitable to plant because of the very low yields. . . . That explains why many of these estates are not worth even half of what they were before."[53] Another report by the fund argued, "In the provinces of Malleco and Cautín alone, the extension of forest land that should be declared in need of urgent reforestation is estimated at nearly 300,000 hectares."[54] Yet another study noted, "If years ago the region of Mulchén, Collipulli, and Traiguén was Chile's breadbasket, today the results obtained by planting are insignificant. Is this not enough of an example to demonstrate the serious consequences of the uncontrolled exploitation of the soil?"[55]

In the end, soil depletion in southern Chile was also a result of agricultural cultivation pushing up against its geographical limits. By the 1940s, all of the land suitable for agriculture had been placed under cultivation. What remained were the steep and rocky slopes of mountains, where logging and pasturing livestock were the dominant activities. As a report pointed out in 1946, "Most of the land suitable for crop production has already been cleared." Chile's "agrarian future" depended on the introduction of new techniques for the "intensive management of the rich farmland already in cultivation." There was no longer a need to clear forest for agriculture. Rather, Chile confronted the immediate challenge of managing the forests that remained and replanting forests to protect watersheds, streams, and rivers and ensure the continued productivity of the soil.[56]

Pine's suitability to southern soil made it an attractive alternative to agricultural crops. *Chile Maderero,* the bulletin published by the timber industry's trade association the Corporación Chilena de la Madera, for example, noted that "the quality of the soil in southern Chile, on average, is less than mediocre." The publication continued, "The terrain is very uneven, and the climate is humid but characterized by the unfavorable distribution of rainfall, since the rains do not coincide with the period of greatest vegetal development. The dry winds of the south; the low nighttime temperatures, even during warm summer seasons; and the absence of the electrical storms that incorporate nitrogen into the soil aggravate this situation." *Chile Maderero* pointed out that while the southern forests could overcome "these characteristics, which are prejudicial to agriculture. . . . [Agriculture] requires enormous efforts of humans and capital in the form of fertilizers, machinery etc., and as a result of those efforts, only one harvest is obtained. . . . Even forage supplies only one harvest a year." The timberers' association arrived at the increasingly consensual conclusion that "only the forest allows us to truly take advantage of our territory." Yet, *Chile Maderero* contended, "The native forest is not compatible with productive silviculture; in effect, the composition of the Chilean jungle is incomplete from the point of view of forestry because of the absolute predomination of broadleaf species and the absence of the valuable mass producers, which are conifers." The conifers absent in the heterogeneous tangle of native forest supplied more and lighter wood, lower transportation costs, greater consistency in quality, and wood at younger ages. Valuable woods such as araucaria and alerce were heavy and thus expensive to transport, grew slowly, and could be harvested only at an advanced age. The southern soil and climate were thus suited neither for agriculture nor for modern forestry in native woods. Instead, they were ideal for fast-growing Monterey pine.[57]

Declining wheat yields made pine plantations an increasingly attractive investment for landowners. As Rafael Elizalde pointed out, following the initial colonization of Malleco, the province produced yields of 55 quintals of wheat per hectare. By 1932–39, yields had fallen to 7 quintals per hectare, making wheat an increasingly unprofitable crop for most landowners.[58] This was exacerbated by price ceilings on wheat imposed by governments after 1930 to subsidize inexpensive food for burgeoning urban populations. While pine received state incentives and protections such as tax exemptions, wheat prices were regulated and remained stagnant. During the 1930s and 1940s, landowners liked to point to price ceilings on basic food items such as bread to explain the dwindling production of rural estates, ignoring the impact of soil erosion on their declining yields.[59] But, as Arnold Bauer has argued, large estates relied on the intensive exploitation of rural laborers to maintain their revenues. During the 1930s and 1940s, low food prices in cities, a benefit for blue-collar and white-collar workers, were subsidized by the sweat of the campesinos.[60] Indeed, one study from 1939 noted that the burdens faced by landowners—principally, scarcity of capital and high interest rates—were compensated for by the "low salaries that are paid in Chilean agriculture and that are . . . not far from the minimum required for subsistence."[61] In addition, as we have seen, landowners similarly exploited the soil to wring whatever easy profits they could, with little thought to the future.

Cultivating pine plantations made political and ecological, as well as financial, sense to many landowners. As wheat yields declined in the depleted soil of the frontier, pine emerged as a valuable new commercial commodity. In addition, given the new national and international concern with deforestation and erosion, landowners could assume the mantle of modern and environmentally sustainable agricultural practices. As denunciations of the latifundia and pressures for agrarian reform increased during the 1930s and 1940s, the planting of pines allowed landowners to demonstrate that their estates were cultivated according to the modern science of forest management. In addition, forestry estates were productive and contributed to the reconstitution of national forestry riches and the battle against soil erosion. Finally, landowners adopted the mantle of patriotism: they were rescuing the nation's soil from the threat of erosion and participating in the industrialization of Chile by planting pine, since their forests would provide raw material for the pulp, paper, and lumber industries.

By the 1940s, the destruction of the nation's forests had begun to receive widespread attention in Chile. Newspapers and magazines began to devote significant space to articles on the destruction of the forests. As tourism to the south followed the expansion of the railroad, urbanization, and the growth of the middle class, publications such as the state railroad company's *En Viaje* included frequent articles extolling the virtues of Chile's native forests and lamenting deforestation. In 1943, for example, Rafael Elizalde published an article in *En Viaje* titled "La muerte del bosque" (The Death of the Forest).[62] He argued that the destruction of Chile's native forests by fire had led to soil exhaustion, river flooding, and massive erosion throughout the south. The "cancer" of soil erosion had turned Malleco, once the breadbasket of the frontier, into a desert.

Increased travel to the south for tourism brought new attention to the problems of logging, clearing land with fire, and deforestation. The Chilean National Development Corporation made developing foreign, as well as domestic, tourism one of its key programs for southern Chile, creating a tourism department to take advantage of "the country's natural conditions . . . and the possibility of developing foreign tourists' interest in coming to Chile, thus constituting a source of foreign earnings."[63] According to the corporation's official history, CORFO funded "publicity campaigns, credit programs for building hotels, road building in tourist regions, training for hotel personnel and interpreter guides and the conservation . . . of natural parks."[64] In addition, the state railroad company became a major promoter of tourism. *En Viaje* advertised Chile's natural attractions and major tourist spots, with considerable ink spilled on the wonders of southern lakes, forests, and mountains.

In an unacknowledged irony, given the railroad's fundamental role in the destruction of native forests since the late nineteenth century, *En Viaje* became a vocal advocate of forest conservation and published frequent articles about deforestation. Throughout the 1940s, *En Viaje* bombarded its readers with articles carrying titles such as "Chile's Lakes," "The Andes Cordillera: The Enigma of Chile," "Skiing in Chile," and "Chile: Land of Tourism." In 1941, the state railroad company published the "Summering Guide" to aid CORFO in building Chile's tourism industry.[65] Similarly, Chile's new Tourism Department published a guide for foreign tourists titled "Visit Chile" that advertised the south's natural beauty, including forests, lakes, volcanoes, ski resorts, and hot springs. An English version of the tourist guide spoke exuberantly about "the intoxicating beauty of one of the garden spots of the world. . . . the exquisite enchantment of emerald lakes reflecting snow-capped peaks and the purple mists of pine forests." The potential appeal of "virgin pine forests" to foreign tourists compelled a new interest in forest conservation.[66]

For the Chilean state, the development of the tourism industry required the restriction of clearing land with fire to preserve the native forests that were a large part of the country's draw. Government officials built on the preservationist regulations encoded in the Forest Laws of 1925 and 1931 that accompanied the conservationist focus on building a sustainable forest industry. In 1937, for example, the Interior Ministry instructed the intendant of Cautín Province to employ carabineros to enforce prohibitions on clearing land with fire. "You will understand," the ministry wrote, "that the government cannot remain indifferent in the face of the unnecessary destruction of the forests that constitute an important factor in the regulation of the country's water systems and that represent a valuable aesthetic element for tourism, and a significant resource in our national economy."[67] At times, the government's policy of leasing native forest to loggers came into conflict with both its programs to promote tourism and the interests of regions whose economy depended on the nascent tourist industry. In 1949, for example, the municipality of Pucón wrote to the government of President Gabriel González Videla to protest the rampant logging of araucaria pine in the national park near Lake Quillelhue, an important tourist attraction. The telegram underlined the importance of the araucaria forest to Pucón's economic health: "logging constitutes an unqualified attack on forest reserves that remain in the zone. We request that you order the paralysis [of logging and] investigate the proceedings followed by private parties to obtain authorization to exploit araucarias."[68]

The growth of tourism in the southern forests also exposed conflicts between preservationist initiatives and colonization. In 1952, for example, the owner of a "small mountain hostel" near Lake Caburga, in Villarrica, petitioned the state for title to land in the Villarrica Forest Reserve. He was concerned because the land in which he was interested, like much of the land in the reserve, had been targeted for colonization by campesinos. "Like any forest technician who wishes to aid in the conservation of the forests," he wrote, "I only have the goal of defending the forest region, which is clearly a tourist attraction, rather than leave it abandoned to be occupied by elements who would bring about its complete ruin." He concluded that the land "has no value apart from its value for tourism."[69] In this case, the regulation of access to the forests to promote tourism provoked a conflict with campesinos who hoped to settle on public land within national parks and forest reserves.

This new national interest in conservation was reflected in the organization of the Sociedad Amigos del Árbol (Friends of the Tree Society), which held its first conference in Santiago in 1942. While calling for increased preservation of Chile's native forests, the society and its agronomist founders, like Albert and Maldonado before them, also became important promoters of pine and eucalyptus plantations. At the conference, the society focused on reforestation to rescue watersheds, rivers, and the nation's soil. The society urged state and para-state organizations to invest their capital in pine plantations, following the model of the Blue-Collar Workers' Retirement Fund on the Canteras estate: "the social welfare agencies and government insurance companies . . . are called on to adopt preferentially the system of investment and capitalization of their funds in forest plantations." The society also petitioned CORFO to intensify its policy of fomenting tree plantations and called for municipal and communal governments to cultivate plantations to improve the quality of life in towns and cities by creating "hygienic" environments. And they lamented the lack of education in forest science and forestry; the feeble efforts to regulate forest fires, clear-cutting, and logging; and the absence of any forest service and forestry schools.

The society forcefully expressed the sense that the conservation of forests and the establishment of national parks was a patriotic project. At the same time that the Popular Front government was establishing a government commission in "defense of the race," which focused on public health and education, advocates of forests and forestation were emphasizing that trees created environmental improvements that would physically and spiritually strengthen all Chilean citizens, especially the rapidly increasing population of recent migrants to the city from the countryside who lived in squalid slums. José Pinochet Le-Brun, one of the organizers of the Friends of the Tree Society, for example, argued that trees "not only purify our physical environment, exhaling oxygen, purif[ying] our blood and creat[ing] healthy and vigorous organisms, they are also an inexhaustible source of beauty that makes the spirit robust and makes noble in man the most delicate aspirations of the soul." Rather than an obstacle to civilization, forests were a sign of it. Pinochet Le-Brun evoked the "most cultured and progressive countries" with the "most elevated civilization"—the cities of Paris, London, New York, and even Moscow and Buenos Aires—with their parks and forests. Of course, the forests Le-Brun extolled were carefully managed plantations of exotic species, ordered and planned like the European cities and parks he admired. The society's preservationism stood in unresolved tension with its vociferous advocacy of forestation and tree plantations. Indeed, the society's support of tree plantations added yet another voice to the chorus singing the virtues of Monterey pine.[70]

Developmentalism, Foreign Aid, and Forestry

The emergent national concern with conservation and forestation was reinforced by global trends in the dissemination of forestry science. International development programs played a significant role in promoting the forestry industry in Chile. In 1944, CORFO contracted the US Forest Service to perform an extensive study of Chile's forest resources and forestry industry.[71] In 1950, the United Nations Food and Agricultural Organization (FAO) held a meeting of its Forest Products Commission in Santiago, reflecting the widely perceived importance of Chile as a producer of forest products.[72] These international organizations shaped the Chilean state's policy toward the forests by emphasizing the already present tendency to look to pine plantations and export-oriented industries such as cellulose as motors of economic development. International development programs and agencies provided an important stimulus to landowners as they went about the business of transforming rural estates and native forests into monocultural plantations. The US Forest Service report *Forest Resources of Chile as a Basis for Industrial Expansion* reproduced the general tenor of the development policies initiated by the Alessandri and Popular Front governments. It also created the blueprint for future state policies on forestation and the development of forest products industries linked to plantations of exotic species. The report was summarized and disseminated in Chile's main newspapers and in the bulletins of landowners' trade organizations. The report reflected key elements of post-1930 Chilean forestry policy, linking forestation and forest management to industrialization. Forests were to be administered to provide raw materials for emergent new industries. The focus on plantations of exotic species also made sense because forestry science as a discipline concentrated on the management of monocultural plantations and the cultivation of species exotic to Chile, including Monterey pine and eucalyptus. In addition, as we have seen, Chilean buyers had imported North American conifers, especially Douglas fir, since the nineteenth century as an alternative to the wood from native forests because it was less expensive and deemed more reliable in terms of quality. In sum, the report urged Chile to expand its plantations and promote the develop-ment of forest products industries, especially pulp and paper.

The North American foresters noted that the destruction of native for-ests, especially valuable hardwood species, constituted a major problem for the Chilean forestry industry. The central cause of deforestation was the fire used to clear land for agriculture, pasture, and logging. Logging was responsible for the most devastating loss of forests to fire: "a much greater proportion of the total loss is undoubtedly associated with logging oper-ations." Fires used by loggers to clear underbrush spread into the virgin forest "with accumulating momentum and destruction." Almost all forest fires in Chile, the report noted, were "man-made."[73] In addition, logging in native forests in the coastal and Andes cordilleras throughout the south-ern provinces of Arauco, Malleco, Cautín, Valdivia, and Llanquihue had devastated stocks of roble, luma, coigüe, and raulí.[74] The foresters pointed out that one way to limit the rapid destruction of Chile's native hardwoods was sustained-yield logging: "[the] stand structure of many natural forests indicates that selective cutting is the most effective method for the preser-vation of a favorable growing stock and for the recovery of timber values. The possibilities of increasing timber supply through adoption of improved cutting and regeneration methods deserve detailed study."[75]

The foresters argued that the use of forest management techniques to simplify the native forests to produce easily harvestable, valuable species in sustained-yield logging could transform Chile's native forests into a limitless source of wealth.[76] They pointed out, however, that the successful management of Chile's native forests was far from practicable: "it would require many years to bring the wild forest of Chile completely under the forester's control, i.e., to the point where practically all the wood that grows is available for the satisfaction of human needs." The composition of the forests, with their thick undergrowth and diversity of species, as well as the presence of many valueless "overly mature" trees, made logging and management a difficult and expensive business.[77] More important, the lack of forestry knowledge related to Chile's native species made a program of scientific management of native species nearly impossible. As the report noted, there was "an almost total lack of knowledge of the conditions under which Chilean forest trees germinate, survive, and develop best."[78] The re-port argued that, despite the desirability of implementing sustained-yield, or "selective," logging, plantations of exotic species composed an easier al-ternative to the management and regeneration of native forests.

The possibilities for pine plantations seemed boundless. By 1946, Chile already had more than 140,000 hectares of tree plantations throughout the southern region.[79] The report noted that at the current rate of planting and natural reproduction in plantations, by 1966 Chile would have increased the expanse of its plantations, to 450,000 hectares. And, it stated, "A more than ample area is available for the establishment of the [anticipated] plantations." The suitability of plantations for supplying the raw materials for new industries was also important: "the most promising species available in Chile from a pulp and paper standpoint, however, is Insignis Pine."[80] Whereas "at the present time, there is no forest management worthy of the name in the entire natural forest area" of Chile, according to the report, "the plantations of Chile are under reasonably good, in some cases very intensive, management."[81] While nothing was known about how to manage Chile's native species, a great deal of forestry science existed that allowed for the sustainable management of pine plantations. Unlike native forests, pine plantations' homogeneity and age evenness made them easy to manage and harvest. Their speedy growth, in decades rather than centuries, made investment in planting pine trees all the more feasible.

The obstacles to effective management and logging of Chile's native forests and the nature of the forests themselves, combined with the lack of studies of Chile's native species, were compounded by southern Chile's social landscape. As the US Forest Service report pointed out, the logging and milling methods employed in the south were extremely rudimentary and quite wasteful. For the most part, forestry production was still only a secondary activity, subordinated to landowners' primary focus on agriculture and ranching and dominated by merchant houses that monopolized markets and drove down prices. There was little domestic investment in forestry, and the industry was undercapitalized, consisting largely of hundreds of small, portable mills, with which landowners often contracted to log native forests on their estates. This was hardly an organization of production amenable to investment in the techniques of scientific forestry management required to establish a profitable and sustainable forestry industry.

Logging remained an underdeveloped sector of the rural economy. Landowners employed inquilinos and, during off-seasons, contracted temporary laborers in logging and used rudimentary equipment or contracted out the logging of their forests to the owners of the portable mills. Trees were felled using axes, usually during the rainy fall and winter and almost always cut high up the tree. As the US Forest Service noted, cross-cut saws, which were used infrequently in Chile, could save 8–12 percent of the lumber wasted by cutting high with axes.[82] Logs were transported by ox-drawn cart to mills during the dry summer months. Mills also operated no more than one hundred days a year, with crews drawn from the large population of itinerant, seasonal, unskilled agricultural laborers. "As to be expected under these conditions," the report noted, "the quality of lumber is poor, grade recovery low, and stain and other defects common."[83] In addition, the use of circular-head saws produced a tremendous amount of waste—up to 25 percent of each log cut. The vast majority of wood producers did not use driers because, as an article in *Chile Maderero* put it, "driers are expensive apparatuses that require technically trained personnel and are not within the means of most Chilean wood producers."[84] This effectively limited both domestic and external markets for Chile's native woods, reinforcing landowners' tendency to view logging as subsidiary to agriculture and livestock. This situation was compounded by the lack of available infrastructure for transporting logs from remote forests, as well as merchant houses' monopoly of wood markets.

The US Forest Service report concluded that "the entire small mill operation is a model of inefficiency." Yet it had its advantages. It required minimal capital investment and risk, had low labor costs, and could be moved frequently to minimize problems with transporting logs. In short, despite the tremendous waste and destruction of native forests, small portable mills were "an efficient and economical method of logging, particularly in hardwood stands."[85] Minimal labor costs and low prices for the raw materials (the forests themselves) made logging and milling profitable. Thus, the report concluded that, while sustained-yield logging managed by trained foresters in the North American hardwood forests guaranteed the forests' reproduction, in Chile, such a system of forestry production was impracticable under existing conditions.

Only one solution recommended itself, and the US Forest Service proposed it somewhat tentatively: given the difficulty of establishing incentives for landowners to manage native forests, transferring land with forests, especially in the mountains or in environmentally crucial areas (such as near watersheds) to the state would be the only real way to prevent the massive destruction of native forests. The report argued that "more remote forest land, or forest land in mountainous terrain, may be too costly or unproductive to protect and manage to permit private ownership, but may be of great public importance." These forests, it continued, "must be brought into a productive state by skilled and sometimes expensive operations." In addition, public ownership was required to promote tree plantations since, as the report put it, "the forest crop often takes a long time . . . to mature. . . . [I]t is subject to serious injury by fire, disease, or insects, with consequent loss in investment value. . . . [T]he growing of timber as a crop is relatively new and involves the tending and harvesting of plants with relatively low value, in a complex natural environment, by methods that can be properly worked out only after years of study and on a scale that few private owners are equipped to do." For these reasons, "The private owner frequently looks to liquidation of his forest capital at the earliest possible moment. . . . The public has a real stake which it can only protect through public ownership." Of course, this conclusion had already occurred to Chileans who since the first decade of the twentieth century had been proposing the return of land on the frontier held by private owners and concessionaires to the state. However, this policy contradicted the tenor of agrarian policy in Chile. As the US Forest Service report underlined, "Such action will involve a reversal of Chile's traditional policy of transferring public land to private ownership as rapidly as possible through outright grant, homesteading, concession, or long-time lease. As a result of this policy large areas of forest land have passed into private control."[86]

This "traditional" agrarian policy was not, in fact, reversed by the original Popular Front or its successor coalition governments. Indeed, in 1946, private ownership of forest land predominated. In Malleco, for example, 82.1 percent of native forests were in private hands; in Cautín, 64.7 percent; and in Valdivia, 95.5 percent. Even where forests were held publicly, they were leased to private logging concessions. Thus, in Malleco, Cautín, and Valdivia, almost all state forest reserves were leased to private loggers, effectively transferring the forests to private ownership.[87] By the early 1950s, after more than a decade of coalition government led by the Radical Party, the state continued to rent or lease concessions of forest land to private loggers who practiced "forestry mining," in the words of one commentator, "exploiting the soil like one exploits a mine, until the definitive exhaustion of its wealth."[88]

The US Forest Service's report was published in 1946, a crucial year in US-Chile relations. It was the year that the Radical Party's candidate Gabriel González Videla was elected president by a coalition that included the Communist Party and Liberal Party. González Videla's political partners brought conflicting agendas to the new configuration of the original Popular Front coalition. On the one hand, the Communist Party pushed the government to the left and supported the mobilization of workers in mines, urban areas, and the countryside. González Videla met these pressures from below and from the left by immediately overturning a government order that had prohibited the unionization of campesinos. The minister of labor dictated a new decree that would allow peasant unions to organize legally, although it did not establish the legislation necessary for a law passed through Congress. On the other hand, González Videla faced pressure from Chile's landholding class, whose interests were represented in the Liberal Party, to crack down on rural unrest and the waves of labor conflict that had shaken Chile since the 1920s. In response to his Liberal allies, González Videla pushed through new legislation that significantly restricted the rights of rural workers by imposing limits on collective bargaining and strikes in the agricultural sector.

In addition, the new Radical Party-led coalition continued to rely on foreign aid from the US Export Import Bank, as well as foreign investment and CORFO support for private domestic investment to promote economic diversification and industrialization.[89] This dependence on foreign aid and investment meant that Chile's efforts to promote social reform, economic autonomy through diversification, and domestic industrialization were constrained by its increasingly tight economic relations with the United States. In 1947, with the government under intense pressure from the United States to curb labor unrest and the influence of the Communist Party, three Communist Party ministers resigned from González Videla's cabinet. A year later, the party was outlawed. The new finance minister in González Videla's reorganized cabinet was Jorge Alessandri, son of the former president and president of the Compañía Manufacturera de Papeles y Cartones, the country's sole producer of paper and pulp and a major owner of native forests and tree plantations. The inclusion of the head of one of Chile's most important companies, which also had extensive land holdings throughout southern Chile, in the cabinet sent a reassuring message to the United States, Chile's landed oligarchy, and the small group of large industrialists who had emerged during the post-1930 boom in manufacturing.

During the 1950s, CORFO took up the recommendations of the North American mission and the FAO that the state play a major role in developing forestry companies and assist in the management of native forests on public and private property. It formulated a six-year forestry development plan that included replacing the more than seven hundred small, undercapitalized portable sawmills that worked native forests in southern Chile with large, central sawmills equipped with modern technology, in five key forestry regions: the Nahuelbuta cordillera, Panguipulli, Riñihue, Ranco, the area south of Valdivia, and the area around Puerto Montt. The CORFO plan also proposed using mechanical saws, tractors, and trucks, as opposed to axes, carts, and oxen, for logging operations in these native forest regions. And CORFO laid out a blueprint for a cellulose plant in Concepción, where an "abundance" of plantations that could supply raw material already existed. In keeping with the policies of the Radical Party governments, the CORFO plan also called for expanding the production of prefabricated housing to meet the "problem of lack of housing in this country." In addition, reforestation was required in zones such as Collipulli and Traiguén, whose native forests had already vanished, and CORFO aimed to continue its program of loans to private individuals and forestry companies for forestation. Echoing the US Forest Service, CORFO imagined a golden forestry future for Chile: "Chile will be a country with a large volume of forestry production for internal consumption, exportation. . . . For future use, Chile will not only continue an extensive exploitation of hardwoods, but will also count on extensive plantations of conifers, . . . allowing competition in world markets." The agency called for reforesting 200,000 hectares with exotic and native species, "part of which will be done by private parties and on the basis of the incentives provided by the permanent industries, and part of which will be done by the direct action of the state."[90]

Chilean forestry followed the lead of the US Forest Service report by employing a bifurcated approach to forest management. Modern equipment and techniques of management, cultivation, and harvest were introduced in tree plantations, while little investment was made in introducing modern forest management in southern native forests. Like CORFO and North American foresters, *Chile Maderero* promoted pine as a more profitable alternative to selective or managed logging in native forests. Pine enjoyed many advantages for logging companies and landowners. First and foremost, pine was cultivated on land that had been cleared, usually for agriculture, and that lay much closer to railroads and markets in towns and cities than the distant and inaccessible native forests in the mountains. In addition, because Monterey pine was lighter than most native Chilean woods, its transportation costs were about a third of the cost of moving native woods to markets. Topography worked to pine's benefit, as well: "pine plantations are located on undulating terrain near railroad stations, with access to good quality roads. . . . [I]n contrast, the jungles are frequently located on steep and tall mountains 50 kilometers or more from the railroad and in zones that suffer frequent year-round rains." Finally, in a pine plantation the density of the wood per hectare was triple that obtained from native forests. Lower production and transportation costs combined with greater yields made pine a more attractive crop than the difficult to log and mill native forests.[91]

Wood producers looked with envy at their North American counterparts, marking the difference between the temperate rainforests of the US Northwest and the Chilean south: "contrary to what happens in the Northern Hemisphere, the conifers that produce soft woods, as they are commercially designated, are only weakly represented in our jungle. . . . [T]he Chilean jungle is integrated by a mixture of diverse species . . . that differ from the conifers of the Northern Hemisphere in which one or two species predominate."[92] An assembly of wood producers in 1956 passed a resolution recommending "the intensification of Monterey pine plantations that will constitute the basis for the great industry of cellulose, paper" and requested the maintenance of tax exemptions and government support for credits from external lenders, as well as the state bank.[93]

While enthusiastically taking up the US Forest Service's call to develop the forest industry and forestation with pine, Chilean governments and timber companies after 1946 paid little attention to its suggestion that native forests be returned to public ownership. As Brian Loveman has argued, the central condition of the social reforms of the Radical Party-led governments, which benefited urban blue-collar and white-collar workers, was the postponement of agrarian reform and rural unionization. The compliance of landowners and their political representatives with the Radical Party governments' urban reforms in essence was bought at the expense of rural labor. In terms of Chile's forests, this meant that despite concern with deforestation and a new interest in promoting commercial forestry, the state effectively left control of native forests, plantations, and industries in the hands of logging companies and large landowners while supplying a series of incentives, protections, and subsidies through CORFO to promote forestation and industrialization.

During the 1940s and 1950s, sparked by state-directed development programs that were shaped by the technical assistance provided by international development agencies and foreign aid programs, landowners and loggers in Chile's southern provinces continued to reforest their estates with pine and to turn to timbering as an alternative to agriculture. By the 1960s, this partnership of private landowners, the state, and international development organizations had established the largest stretch of tree plantations in Latin America, at more than 250,000 hectares, and laid the foundation for a dynamic commercial forestry industry in southern Chile. Landowners reclaimed eroded land on their fundos by cultivating tree plantations and, in some cases, establishing public forestry companies. In addition, they initiated the clearing of large swaths of native forest to make way for the Monterey pine. Loggers, backed by state development policy, continued ruthlessly to exploit Chile's native forests using nineteenth-century methods of production while implementing modern techniques of forestation and forest management on cleared land planted with North American pine. As the plantations spread, they replaced both agricultural land and native forests. Indeed, state subsidies for tree plantations dictated by the Forest Law of 1931 stimulated forest conversion, or the substitution of native forests with plantations of exotic species. In addition, the scientifically organized plantations replaced local uses of Chile's native forests, transferring the management of the forests from campesinos to the technicians and agronomists employed by the Forest Department and by private forestry companies.

5

Peasants, Forests, and the Politics of Social Reform on the Frontier, 1930s–1950s

In 1963, Juan Sánchez Guerrero published a collection of short stories based on his childhood memories of the Lota coal mines. Sánchez Guerrero grew up in a campesino family who had moved to Lota after losing their land. An early story in the book, one of the narrator's "earliest childhood memories," describes a miner and union activist, Don Remegio, who visits his family's small rancho in the countryside. One day, Don Remegio says that the mine workers are organizing a union so that "one day, I think, it will be possible to recuperate this lost land, sold under coercion, closed in by the pine forest." Don Remegio looks around at the coal company's pine forest and declares,

> It's beautiful, the pine forest, that can't be denied, but there are so many broken dreams and tears buried beneath its roots. . . . I have known this forest . . . since its birth. The plantation began here at the edge of the sea. In those days no one knew any other pine tree other than our araucaria, so generous with its fruit and wood. No one imagined, then, that these foreign plants had so much power and usurped [land] just like man. . . . The pino insigne began here and organized itself in the nurseries, in great number, to invade, accomplishing a true conquest of the region. . . . The plants stayed in line like a true invading army. . . . Its green emerald, permanently advancing, was like an ocean wave. . . . It climbed hills and descended, annihilating everything in its way. . . . In its inexorable and uncontainable path fell the campesino hut, the sacred canelo, the chilco, the maqui, the boldo, the quila and even the imposing roble. It eliminated agriculture; it evicted livestock from the arable land. . . . Today it is a sea of wood, of silence and shade. Nobody remembers anymore . . . exactly where the property is that the pine forest consumed. . . . [T]he small plots were suffocated by the forest and had to be abandoned.

The forest pushed Don Remegio's family off their land and created an alienating landscape. As the company cleared the surrounding land with fire to plant pine, his father and brothers found work in the mines.[1] The narrator's mother describes how their own family also lost their small plot when the fires set to clear land for the pine plantations spread and burned their crops and pastureland, "converting into ashes our only bread."[2]

This remarkable story, rooted in Sánchez Guerrero's memories of land and labor in Lota, reflects the powerful connections between the ecological and social processes that transformed Chile's frontier during the twentieth century. From the first years of the twentieth century, Lota's pine plantations had been heralded as a model for future forest policy. Yet the celebration of the mining company's forestation success story—the recuperation of arid, eroded, valueless land and entrepreneurial brilliance in pioneering Chile's first plantations of exotic species—obscured the history Sánchez Guerrero's story so vividly illustrates. Beginning in the mid-nineteenth century, the coal companies had logged the region's native forests to supply their mines with wood. As pine acquired value, tree plantations began to replace native forests cleared with fire.[3] Sánchez Guerrero evokes the campesinos' sense of an inexorable ecological conquest, the North American pine's replacement of the forest's native species. But he also underlines the social logic of this history. The expansion of the plantations pushed peasants off their land. They lost access to native forests and saw their crops destroyed by uncontainable forest fires, their cattle killed, and their small plots surrounded by pine plantations, which sucked dry watersheds and streams. Their only recourse was to find jobs in the mine. The company acquired both land for its tree plantations and labor for its coal mines. The story captures neatly the dialectical movement of the ecological and social changes that produced the Lota workforce's double alienation: the alienated labor of proletarianizing peasant miners and their estrangement from the transformed landscape covered with pine plantations.

The swelling sea of tree plantations that swallowed up peasants' land was pushed by the ecological necessity of restoring barren soil, as well as economic incentives provided by the state. But the fever to plant pine after 1930 was also part of a broader social process. For many southern landowners, plantations and commercial forestry served as a means to establish political legitimacy in relationship to the state at a time that they faced both the ecological threat of soil depletion and the political threat of rural unionization. When landowners planted trees, they expelled their resident laborers, doing away with the challenges of an increasingly restive rural labor force and recalcitrant campesino population. Landowners who turned to forestation were backed by the state, which viewed the transformation of campesinos into forestry workers as a convenient solution to the clamor for land in the south. For the state, the homogenization of the natural land-scape and the rationalization of forest exploitation were linked to efforts to order rural social relations by transforming often rebellious campesinos into settled citizens and proletarianized workers. This process accelerated during the three coalition governments led by the Radical Party between 1938 and 1952.

The Popular Fronts, Forest Policy, and Social Reform on the Frontier

The "social problem" in the frontier regions devoted to timbering began to receive national attention by the 1930s as the pace of wood production ac-celerated and rural workers turned to unions to represent their demands for improved working conditions and access to land. The plight of the frontier's landless forestry workers was a source of alarm and concern to government officials, especially in the wake of the rural social unrest that shook the re-gion between 1927 and 1934. Unsettled labor and property relations and the continual migratory movement of rural laborers throughout the frontier's valleys and mountains, often across the border to Argentina, made the region, from the perspective of Santiago, an uncivilized place often com-pared to the western frontier of the United States. In 1937, for example, *El Sur* celebrated the high levels of wood production in Cunco, the location of former colonization and logging concessions such as those that belonged to Luis Silva Rivas and Felipe Smith. But the newspaper also recalled Cun-co's "Wild West" history of social turmoil and banditry and denounced the terrible conditions suffered by workers in the forests and sawmills. The paper reprinted an interview with Cunco's priest, who advised *El Sur*'s re-porter that the region's workers were "the most long-suffering people I have known."[4]

Pedro Aguirre Cerda's election at the head of the Popular Front coa-lition of the Radical, Socialist, and Communist parties in 1938 marked a significant shift in Chilean politics. The Popular Front sought to introduce a series of basic social and economic reforms, including the promotion of domestic industrialization and agricultural diversification, combined with labor reforms in urban areas. In addition, it began to build the foundations of a social welfare state with new retirement programs, workers' compen-sation, family allowances, and public housing and health programs to meet the needs of a growing population of rural-urban migrants driven from the countryside and northern mining districts by the world recession of the early 1930s. The Popular Front looked to the southern frontier to begin what would be a decades-long struggle to implement agrarian reform by intensifying the colonization policies introduced under Ibáñez. It intended to leave the well-established large estates of central Chile intact while settling 300,000 campesinos in cooperatives and colonies on vast stretches of public land that had been acquired by estate owners, speculators, and concessionaires and then restored to the state. Colonization in the south would serve as an alternative to agrarian reform.

In an interview in 1939, a Popular Front official discussed the shift in agrarian policy under President Pedro Aguirre Cerda. He noted that "one of the greatest evils that the government is combating is latifundismo; many landowners work only a small part of their haciendas or fundos, leaving the rest abandoned, subtracting from the national wealth and labor the greatest source of production and causing, logically, a scarcity of basic food products." The Popular Front began to intensify the pace of land purchases by the Caja de Colonización Agrícola (Agricultural Colonization Fund), which, the official noted, "acquires fundos or haciendas that fill the required characteristics to be subdivided into lots distributed to colonos." In addition, the Popular Front intended to accelerate colonization of the frontier with "authentic Chilean citizens who with their honest and tenacious labor will constitute the future pedestal of the nation." Finally, confronted with diminishing agricultural yields, the Popular Front viewed reforestation and forestry management as a cornerstone of the nation's industrial and agricultural development. The new government "dictated several measures to prevent the extinction of the trees in six different zones, prohibited burning forests and clearing land with fire." The official linked the "anarchy" of the unregulated destruction of the forests and extraction of wood by the timber industry with large estates whose "acts of vandalism" had "seriously undermined the country's wealth."[5]

This analysis of Chilean agriculture underlined both the continuities and the changes in policies toward the rural sector and forestry under the Radical Party-led coalition governments. First, the condemnation of latifundismo represented the most significant rhetorical assault yet on Chile's landowning oligarchy. Aguirre Cerda had targeted the large estate as one of the obstacles to Chilean progress throughout his presidential campaign. However, he did not endorse agrarian reform. Instead of expropriating large fundos, the Popular Front looked to the frontier to solve Chile's problems of diminishing agricultural production and rural inequality. The state had established a foothold in the region by progressively asserting its ownership of public land since the first decade of the century. It had begun to recuperate land that had been auctioned or leased to concessionaires and even purchased a small number of fundos through the Agricultural Colonization Fund for distribution to colonos. Indeed, by 1938, perhaps as a response to the Southern Property Law and the challenges of rural laborers, a number of southern landowners had willingly offered their estates for sale to the fund.[6] The Popular Front government hoped to use the fund to intensify the expropriation and colonization of fundos on the frontier.

According to Popular Front officials, settling public land and private estates purchased by the Agricultural Colonization Fund would provide small plots for more than sixty thousand mestizo and Mapuche campesino families.[7] Yet this colonization program in the frontier territory did not imply land redistribution. As the president of the Federación Nacional de Colonización Agraria (National Federation of Agrarian Colonization) argued, "The aspirations of citizens who hope to own a piece of land do not go against any private interest, they are not going to damage the rights of others; the colonos . . . are, on the contrary, elements of cooperation and labor; they are elements who will contribute to the modernization of our system of agricultural exploitation. . . . Colonization, we insist, is not the redistribution of land. . . . It is the cultivation of land that is not exploited today; it is to make productive the 27 million hectares that are not worked today." He pledged the support of aspiring colonos for the Popular Front's program of social and economic reforms.[8]

Second, the Popular Front official's discussion of the importance of forestation represented a departure in agrarian policy. The Popular Front linked Chile's declining agricultural productivity to the mismanagement of forest resources by large estates. The estates left land uncultivated and invested little in modernization. In addition, they destroyed and exploited the forests unchecked. This had led to soil depletion, erosion, diminishing crop yields, and climatological changes prejudicial to agriculture. The future of Chilean agriculture depended on reforestation and scientific management of the remaining native forests. Tree plantations would both reclaim devastated soils and provide raw materials for new industries, especially construction, paper, and cellulose.

Finally, the promotion of forestation was linked to other items in the Popular Front's package of social reforms. The Popular Front confronted a large population of rural-urban migrants who had fled stagnating rural sectors, pulled by the growth of industry and jobs in urban centers. Expanding urban working-class slums presented a problem that had festered during the 1930s. For the front, public health and housing programs were essential to dealing with Chile's high levels of infant mortality and poverty. "Hygienic" housing would be a bulwark against disease and urban unrest, including vices such as gambling, alcoholism, and prostitution. Plantations of North American pine would provide inexpensive wood for construction; by the 1940s, wood was in scarce supply as native forest stock dwindled. Building housing for the growing urban working class would also provide jobs in a country where high levels of unemployment were a chronic problem.

Popular Front officials designed agricultural colonies, including forestry colonies, on the frontier partly as a response to southern campesinos' increasingly militant demands for land. In addition, colonization gave the government a way to avoid the demands made by campesinos in central Chile for improved wages and working conditions, as well as enforcement of labor laws in the countryside. A report by the director of the Agricultural Colonization Fund in 1941 argued that forming agricultural and forestry colonies on the frontier would improve the lives of Chile's 300,000 campesino families, most of whom were so poor "that they [could not] satisfy the basic needs of a human being for food, housing and clothing." Especially troubling were the landless workers, the "itinerant, irresponsible workers with their disorganized families, when they are not unraveled," to whom "it does not matter if the harvest is good or bad, since they are paid the same." Transforming landless workers into colonos would make them "direct producers" who would be organized collectively and thus attain "culture and welfare." Collective organization was a moral imperative: "welfare and culture . . . are the fruit of a more just organization of life and work that requires the collective effort of those who feel . . . the organic imperative for social discipline." Settlement in colonies on the frontier would lead to "greater social and economic responsibility."[9]

The Popular Front government defined the problems of the rural poor as much in terms of labor as of unequal landownership. Given Chile's extraordinarily high rates of urban and rural unemployment following the crash of 1929–30 and the decline of the nitrate industry, the Popular Front government viewed the large population of itinerant landless campesinos, workers, and miners in the countryside as part of the general crisis of joblessness. In 1939, for example, the director of the Agricultural Colonization Fund wrote to the intendant of Cautín to ask for a report on "campesino unemployment" in the region. According to information he had received, landowners "were systematically expelling workers from the fundos, reducing sharecropping arrangements, and restricting rentals to smallholder arrendatarios."[10] The center-left government looked to resolve the acute social crisis in the countryside by providing jobs for campesinos rather than land. This program had a moral dimension: employment as full-time proletarians would settle and produce the cultural uplift of a notoriously unstable and exploited southern rural labor force.

The fund proposed purchasing estates and then entering into *mediería* (sharecropping) arrangements with campesinos in which it would provide "hygienic housing," seeds, cash advances, equipment, tools, and clothing, as well as a family garden plot. The medieros would then earn half of their harvests of vegetables, cereals, and grasses and the products of livestock and "industrial crops" (such as forestry). The colonies would be technically managed centers of production and consumption that operated according to the principle of "cooperativism."

The fund rejected the system of distributing individual plots to colonos. The Ministry of Land and Colonization noted in 1939 that over a decade, the Agricultural Colonization Fund had settled only 1,227 colonos and that most had already lost their land because they could not afford to pay the quotas for their plots and had little capital to invest in making their land productive. The solution was for the fund to purchase large estates and establish colonies in which the colonos would work *a medias* (on shares) with the fund on "large centers of organized production" that would provide land and the means to accumulate capital. This system had already been implemented by welfare institutions on twenty fundos, many of them dedicated to logging.[11]

Because a large part of the land in the frontier region was deforested and the soil was eroded, the Agricultural Colonization Fund proposed "a reforestation plan that in a prudent period [would] allow the repopulation of forest land that urgently requires reforestation." However, this key reform was "impossible for individual private landholders." The fund was "the only institution that on an important scale [could] acquire these extensions to form forestry colonies [that could] be distributed to colonos-medieros." The fund intended to purchase cleared and eroded land from landowners and form colonies of landless peasants who would cultivate tree plantations. "In the provinces of Malleco and Cautín alone," the fund's report noted, "it is estimated that the extension of forest land that should be declared in need of urgent reforestation is nearly 300,000 hectares. Of these, 20,000 hectares already have been offered for sale to the fund [by landowners], and we will soon initiate studies to design a project for forming forestry colonies."[12]

The fund's agronomists employed radical rhetoric and illustrations of campesinos with captions such as "The Agricultural Colonization Fund is going to liberate [the campesino] from centuries of exploitation by the landowners. The Popular Front will support him" and "Two laborers enslaved by the oligarchy whom the Agricultural Colonization Fund is going to liberate with its organizational action." For the fund, the emancipation of Chile's campesinos was also a project of racial uplift. Its 1941 program of establishing state-owned colonies of sharecroppers gave a brief analysis of the ethnic origins of Chile's campesinos cloaked in the language of ethnography. Vestiges of indigenous culture were still to be found among mestizo peasants, especially "the absolute lack of respect for others' property" and "the survival of superstitious totemism . . . even in the campesinos' celebrations in the Catholic Church." These vices, as well as the supposed "mental indolence" of Chile's campesinos, however, were not biological; they were the result of the feudal and backward system of landholding that had produced "inertia," "servility," and "lack of responsibility." The need, then, was for a modern and collective organization of work that would rescue campesinos from the squalor of rural life. Replacing large landowners with the state as manager of sharecroppers' colonies would allow the rational organization of production and promote campesinos' cultural uplift.[13]

Foresters and agronomists located the source of both social problems and rampant deforestation in forestry workers' lack of education and training. Forest management and forestation required an educated and stable labor force rather than the transient and seasonal agricultural workers who logged native forests or labored in small, portable mills. As one agronomist put it, the workers in all aspects of the production process in forestry were *"peones ambulantes* [itinerant workers] with a restless character who move constantly like true nomads through the entire territory of Chile looking always for better salaries and adventures."[14] The agronomist argued that training this transient rural labor force was essential to the development of the forestry industry in Chile: "the definitive solution to the problem of the specialized forestry worker is of vital importance for our country because it matters to the future of a source of wealth that could technically surpass copper and nitrates in importance."[15] Settling forestry workers in collectively organized colonies would address the associated social and ecological problems of rural poverty, soil erosion, and deforestation. The article cited the work of the Caja de Seguro Obrero (Blue-Collar Workers' Retirement Fund) on its Canteras hacienda in Bío Bío Province. The fund had founded a "Farm School" focused on the cultivation, care, and harvesting of the hacienda's Monterey pine plantations: "once the forestry workers who come from the farm schools . . . populate and fill the needs of the diverse labors that the wealth of reforestation will inevitably create, that is already a concrete and palpable fact, we will have won for Chile and its children a more elevated standard of living proper to a cultured and civilized country."[16]

Public Forests, State Reserves, and Rural Labor in the South

The election of the Popular Front inspired a new wave of rural organizing on the frontier. In response to the promise of accelerated land colonization, campesinos organized the Agricultural Colonization Union of Southern Chile and stepped up their demands for the expropriation of southern fundos and the settlement of aspiring colonos on public land. In June 1939, committees of campesinos who were "aspiring colonos" from towns throughout the south held a congress of the colonization union in the city of Chillán.[17] The following month, aspiring colonos organized a "March for Land" in

which caravans of carts from southern towns, including "Mapuche delegations," marched to Temuco.[18] Members of the colonization union flooded
the Ministry of Land and Colonization with petitions for the expropriation
of large estates. In October 1939, for example, the Leoncio Chaparro Association of Aspiring Colonos, named strategically for the director of the Agricultural Colonization Fund, petitioned for the expropriation of the Mis-
que and San Ignacio estates and their settlement with landless peasants.[19]
A Land Ministry report noted that "the campesinos march for land, march
around the plazas and avenues of the different pueblos, send telegrams and
letters." The Popular Front, the author of the report asserted, intended to
respond by dismantling the "medieval" system of landholding and intensifying the pace of colonization.[20]

Campesinos' hope that the Popular Front government would initiate
a sweeping colonization program was not, however, fulfilled. Following
Aguirre Cerda's death in office, the Popular Front fractured in 1941, to be
reconstituted as a series of coalition governments led by the Radical Party
that ruled until 1952. After 1941, the promise of the front's radical rhetoric of reform was eclipsed by a developmentalist program that promoted
industrialization and urban social reform but turned a blind eye to the
countryside. Even during his first year in office, Aguirre Cerda had halted
the implementation of the Labor Code of 1931 in the countryside, para-
lyzing unionization campaigns among rural laborers. As Brian Loveman
has argued, the Radical Party-led governments presided over a "compromise state" in which reforms to benefit urban blue-collar and white-collar
workers were negotiated with the traditional right-wing parties of the landholding oligarchy in exchange for limiting reform of rural land and labor
relations.[21] In the south, the coalition governments increasingly favored
private capital over campesinos in disputes over forests and control of forestry production. Forestation and forestry industrialization replaced earlier plans to colonize the frontier with landless campesinos. Rather than
restore campesinos' access to native forests and public land, the Popular
Front governments looked to jobs in forestry, not land redistribution, as
the solution for southern Chile's large population of underemployed and
itinerant landless laborers.

By the 1940s, the coalition governments had begun to view public forest reserves, once submitted to scientific management and reforestation, as
an important source of revenue. To promote forestry, however, they had
to crack down on campesinos who demanded frontier land and regulate
logging by private companies. The Popular Front governments' solution
was to create a system of leases to privately owned timber companies to
exploit public forests while providing a significant portion of their earnings
to the state. Timber production would be carefully overseen by foresters

from the Forest Department, who would guarantee the application of the "scientific" methods of forest management and direct reforestation with pine. Government foresters and agronomists reasoned that campesinos lacked both the capital and the training to exploit the forests rationally. Only private companies, operating with the oversight of the Forest Department, could employ modern methods of forestry production, guaranteeing a sustainable system of logging that would generate revenue for the state. Campesinos would have a steady source of income in forestry work. Instead of implementing plans to expand colonization of the frontier with land-starved campesinos, the Popular Front governments restricted the campesinos' access to land that lay within the growing number of forest reserves or that had the potential to be logged commercially and reforested. Foresters, forest guards, government surveyors, agronomists, and the inspectors of the General Directorate of Land and Colonization began to take on a larger role in determining access to land and forests in southern Chile. Often this implied asserting the interests of the state over the interests of large landowners. However, more often than not, reforestation programs and the exigency of establishing rational exploitation of Chile's native forests came at the expense of the rural poor.

In 1946, for example, the coalition government led by the Radical Party's Gabriel González Videla, which included both the Communist Party (until 1947) and the Liberal Party, confronted serious social and environmental problems in the coastal Sarao cordillera in Valdivia Province. Large landowners had been exploiting forests on public land, leaving behind devastated soil that held little value. The land directorate noted that lack of funding and administrative presence had limited the state's capacity to regulate the exploitation of the "public mountain forests rich in fine woods that have not been declared forest reserves."[22] Numerous campesino families, contracted by timber companies, had begun to exploit the forests without prior authorization. "To avoid the serious social consequences that are produced by mass expulsions of forsaken campesino families," according to the land directorate, the state eventually granted them titles to the land even though the soil was not suitable for agriculture. Once they had obtained titles, however, these putative colonos then sold the forests to large timber concerns. Wood producers thus accumulated large tracts of native forests in the cordillera. After purchasing the titles from the campesinos, the timber companies then sent the campesinos to public forests in other regions, and the process began again with colonos who had received and then sold their titles to forests "two, three, and even four times." As the land directorate reported, "Through this system of *colonización ambulante* [moving colonization], the wood producers are taking the great forest wealth of the unexplored mountains from the national patrimony." Timber companies' "avidity for profit" led them to raze the forests "without concern for reforestation, leaving at the end of their labor eroded, uncultivable soil." In addition, they had created a population of migrant rural laborers who moved from forest stand to forest stand throughout Valdivia's coastal cordillera.[23]

The government confronted a dilemma. Hundreds of campesinos— some of them squatter ocupantes, some transient inquilinos contracted by landowners and loggers, and some colonos granted land by the state in the frontier's typically blurred social and legal landscape—occupied forest land in the Sarao cordillera. Expelling them would provoke "serious social problems."[24] However, the interest of the state and charge of the land directorate was "the conservation of the forests, especially public forests." Ideally, the state would work its forests directly "with the goal of using the totality of its wealth in wood and of lowering the cost of an element indispensable to the solution to the serious problem of working-class housing." To prevent the abuses perpetrated by the timber companies and take advantage of the forest resources, while "taking responsibility for their immediate replacement so that they are not extinguished," the government organized forestry colonies of the peasant ocupantes on two publicly owned fundos in the Sarao cordillera: Chañlil and Raguintulelfu.[25] In addition, it set out to expropriate three fundos—Yerbas Buenas, San Juan, and Esperanza—with the amount of indemnification established by the values declared on their owners' tax returns.[26]

The organization of the forestry colonies was designed to meet a number of potentially competing needs. The González Videla government was less interested in preserving forests in the Sarao cordillera than in the unregulated exploitation of the forest without reforestation. In addition, the government sought to assert its control over the potential revenues generated by forest exploitation. With the organization of the colonies, the state "would continue to be the owner of forest land, which would not be titled privately." In addition, the colonies would allow the state "to conserve the mountain forests and valuable woods through a rational exploitation directed by a technician from the Forest Department." The aim of this scientifically managed timbering was to provide income for the impoverished campesinos in the Sarao cordillera and wood for "inexpensive housing" for urban "empleados and obreros."[27]

González Videla's government also intended the forestry colonies to solve a series of problems caused by competing claims to the Sarao cordillera's forests. The state had been engaged in extensive litigation with Julio del Solar, who had been granted a lease in 1919 to log in the coastal cordillera and had continued his activities for decades, enmeshing the state in lawsuits for years. The colonies would satisfy the demands of all parties, the state, the campesinos, and del Solar. Ocupantes who squatted on land in the coastal cordillera would be settled on the two fundos and given jobs in logging. In return, they would be in charge of reforesting the two fundos with North American pine provided by the Forest Department. Del Solar would be granted a new ten-year lease to exploit the forests under the supervision of technicians from the Forest Department. He would provide the basic capital and sawmills, employ the campesinos in felling trees and in his sawmills, and pay the state a significant tax on the wood he milled. The campesinos would be transformed into forestry workers employed on state-owned colonies by a private contractor.

This solution to the Sarao cordillera's social problem reflected key elements of the Radical Party-led coalition governments' agrarian policy in the southern forests. Fundos on public land without well-established legal titles would be purchased, and their land would be distributed to campesinos. On estates with commercially valuable forests, campesinos would be settled as full-time forestry workers. The state would regulate and oversee forest exploitation to guarantee revenue for the national treasury and income for impoverished campesinos. However, despite this assertion of state control over public forests, the Popular Front governments looked to private capital to make the arrangement work. Landowners would make out well by selling their fundos to the state, and forest exploitation would be contracted to private timber companies. The state would supervise production in the new forestry colonies in the Sarao cordillera, but del Solar would provide the capital and machinery and, in exchange, get a percentage of the earnings generated by the colonies. The idea was to please all parties—the campesinos, landowners, timber companies, and the state—in a mutually beneficial arrangement that would guarantee the rational exploitation of public forests and reforestation with Monterey pine. The deal in the Sarao cordillera had another favorable aspect for González Videla's government: it transformed landless campesinos who exploited public forests into full-time, settled workers employed in industry.

This solution failed to satisfy the estates' ocupantes. They had occupied the land for more than a decade and demanded titles from the state. Their interest in cultivating crops and extracting wood from the forests was frustrated by the concession to del Solar. In a petition for legal titles in 1947, they denounced the contract, which they viewed as a threat to their rights to the forests. Ultimately, they sought to reproduce the basic elements of the campesino economy, including control of forest land for pasturing animals and collecting wood. The prospect of jobs as forestry workers did not satisfy their sense that they had rights derived from their material occupation of the forests and from the forests' status as public property.[28] The logic of their rejection of the arrangement was articulated well when the Communist Party's *El Siglo* advocated on their behalf, arguing that although the soil that once was covered by forests and now was cleared by logging was not suitable for agriculture, once it was planted with pasture, it would become "as agricultural as the rest of the flatlands." The land's "pseudo-owners" had left cleared forests "abandoned and uncultivated," pillaging the nation's alerce forests and leaving only a barren wasteland covered in stumps, while the campesinos had contributed to the public good by making the land productive and planting crops.[29]

The owners of the expropriated fundos, by contrast, were probably quite content with the settlement engineered by the government. The titles of their fundos had dubious legal status whose legitimacy the government had contested as it sought to recuperate public land. In addition, landowners could not exercise effective dominion over either the forests on their fundos or the campesino labor force. When one landowner in the region attempted to log forests on his fundo, for example, he was met by "groups of ocupantes who, with all kinds of threats, prevented him from working, declaring that they occupied public land." Workers in the landowners' logging operations were attacked by forty or fifty people armed with axes and clubs. The intendant of Llanquihue Province reported that, "because of the continuous conflicts that are provoked daily in the Sarao cordillera, it would be necessary to purchase the land." The agreement negotiated by the government placed the financial burden of this "social crisis" on the state's shoulders and gave the region's landowners an easy and profitable way out of a situation that threatened to deteriorate into violence and years of litigation. In addition, the landowners had extracted quick profits by logging the native forests on their property, extracting the valuable alerce and leaving behind stumps, burned-over forest, and invasive second-growth underbrush (see figure 5.1). They were ready to move on to other forests.[30]

Despite this solution, the conflict in the Sarao cordillera was not completely resolved. In 1954, the Ministry of Land and Colonization reported that more than 1,500 campesino families without land titles still survived by logging the Sarao cordillera's alerce forests, constituting a "true volcano." The only way to prevent an "immediate [social] explosion," government officials reasoned, was to expropriate the large San Pedro estate in the cordillera and establish a state-run logging central that would "exploit the alerce rationally and provide permanent jobs for the Saraos [*sic*] cordillera's ocupantes."[31] Rather than distribute the Sarao cordillera's forests to the ocupantes, the logging central would provide stable work, settling a transient and unruly rural labor force. It is worth noting that more than a decade before the agrarian reform, in response to unregulated logging and endemic social strife, land officials had dictated the expropriation (or purchase) of large estates working native forests in Valdivia's coastal cordillera.

5.1 South of Valdivia: Burned over Alerce Forest (Courtesy of Felipe Orrego)

A similar conflict over forests on large fundos erupted in Panguipulli, in the Andes cordillera, in 1947. A committee of seventy-three campesinos (the Committee of Medium and Small Agricultural Producers and Aspiring Colonos of Licán Ray) petitioned the state for the expropriation of the Trafún estate because the land was suitable for agricultural production and the owners, Víctor and Eduardo Kunstmann, had left 33,000 hectares of the fundo "abandoned" (in a private forest reserve). The committee asserted that the Kunstmanns had legal title to only 5,000 hectares; the rest of the fundo consisted of public land appropriate for colonization. The petition was rejected, however, because a surveyor's report found that the land, which lay at the foot of the Quetrupillán volcano, was "not agricultural-quality but, rather, covered with forests, and some parts [were] suitable for raising livestock." He concluded that expropriation was "not advisable."[32] In this case, the state opted to support the development of private commercial forestry production rather than campesinos' demands for land. The Kunstmanns enjoyed legitimacy in the eyes of the state because of their well-managed forestry operations. They were taking full advantage of the forests and building a profitable industry that would help develop a notoriously impoverished region. Campesinos could not be trusted to work the forests rationally and therefore could not be settled as colonos on Trafún.

The state's interest in promoting reforestation with pine plantations also shaped its colonization policies, restricting rural laborers' access to public land during the years of Popular Front government. A case from 1939 illustrates well the ways in which the ongoing replacement of araucaria forests with plantations of Monterey pine in Chile's coastal cordillera undermined campesinos' access to land and petitions to be settled as colonos. That year, Aguirre Cerda's government decided to rent the 20,000 hectare Trongol estate, which belonged to the Junta de Beneficencia de Concepción (Welfare Society of Concepción), to the Oelckers timber company. The estate's previous lessee had sublet the estate to campesino arrendatarios, a typical arrangement in the south, who had then sought the subdivision and colonization of the land they worked. In this case, as in so many others, the estate laborers' petitions were rejected, and Oelckers won the lease. As part of the contract, the company was required to plant 20,000 Monterey pines a year. As *El Sur* noted, this arrangement "impeded the aspirations of the numerous residents of the estate to colonize [the land]."[33] Trongol's laborers refused to recognize the new contract and engaged in a minor rebellion. Oecklers was unable to take possession of the estate because "the inquilinos have risen up in rebellion, refusing to recognize the lessees as such and, even worse, refusing to recognize the legitimate rights of the owner of the land, the honorable Welfare Society." Many of the inquilino families on the Trongol estate had worked the land for as long as sixty years. The Ministry of Land and Colonization reasoned that Trongol was not suitable for agriculture and that the estate's resident laborers depended on extracting wood in the cordillera's araucaria and roble forests to survive. Rather than entrust the forests to the campesinos, the government argued that the rational exploitation of the forests "would require the investment of large amounts of capital, an enterprise that can only be done by a firm that is capable of an integrated and intensive production."[34]

As with other state-owned forests, the government chose to lease logging rights to large companies that could exploit the forests "rationally" and that would commit to reforesting with Monterey pine. The campesinos who had lived and worked on the Trongol estate for years could find full-time work in Oelckers's sawmills. Despite the government's assertion that the campesinos posed a threat to the remaining araucaria forests, their dependence on collecting piñones led them to a conservationist stance in their confrontation with the logging company. The campesinos petitioned the province's intendant to prohibit Oelckers from exploiting Trongol's araucaria forests, since "the forests constitute a resource for them with their fruit." Instead, they asked that the forests be "conserved as forest reserves."[35] Their use of the forests to collect piñones, pasture cattle, and collect firewood, their petitions implied, contrasted with logging operations, which destroyed the forests altogether. Nonetheless, the government rejected the campesinos' petitions, and Oelckers initiated logging in Trongol's araucaria forests. New in the arrangement between the Popular Front government and Oelckers was the logging company's commitment to reforest with Monterey pine. A decade earlier, Oelckers simply would have moved on to another estate, another araucaria forest, another cordillera.

This shift in focus from distributing land for colonization to campesinos to developing commercial forestry determined the state's policy against settling colonos in forest reserves. In 1939, for example, a committee of ocupantes petitioned to be settled on land on the Tres Bocas estate in Valdivia's coastal cordillera. The former owners of Tres Bocas had cleared the estate, using fire in their logging operations. By 1929, it had been completely deforested and abandoned and was then taken over by the Agricultural Colonization Fund to settle colonos. However, surveyors determined that the soil, now covered by blackberry bushes, was ideal for forestation, and Tres Bocas was transferred by the fund to the land directorate to establish tree nurseries as part of the Valdivia Forest Reserve. The campesinos' petition was rejected in favor of a competing petition by the Compañía Siderúrgica e Industrial de Valdivia (Iron and Steel Company of Valdivia), which had won a thirty-year lease for public land on the nearby Cufeo and Corral estates, as well as for a lot within the Valdivia Forest Reserve, to extract wood to be made into charcoal for the ovens of its iron and steel works. The company agreed to take responsibility for replanting Tres Bocas with Monterey pine.[36]

The land directorate argued that colonizing the Valdivia Forest Reserve with smallholders would end in failure because the land was not suitable for agriculture. Instead, Tres Bocas was "admirably suitable for replanting with Monterey pine, an industrial species of great value especially for the province of Valdivia." A cellulose factory was planned for Valdivia, and Monterey pine was "the best primary material" for that industry. The land directorate underlined "the importance of fomenting this class of plantations, which [would] ensure the permanent functioning of the two principal industries referred to."[37] In this instance, the state's interest in promoting tree plantations and commercial forest exploitation on its reserves came into direct conflict with campesinos' demands to be settled on public land.

In another representative case, Francisco Vega Cáceres petitioned Aguirre Cerda's government in 1939 for land in the Nahuelbuta National Park in the coastal cordillera, created that same year to protect the coastal cordillera's remaining stands of araucaria (see figure 5.2). Vega's request was rejected by the Ministry of Land and Colonization, "taking into account the reforestation plan that the government is developing and with the understanding that land declared parks and reserves must be untouchable and, in any case, dedicated to plantations . . . that in the future will be a great social utility."[38] It is notable that logging in the Nahuelbuta cordillera's araucaria forests had already led to state-directed initiatives to reforest

5.2 Nahuelbuta National Park: Cleared Forest and Soil Erosion (Courtesy of Felipe Orrego)

sectors of the national park, a repository for Chile's endangered autochthonous species, with Monterey pine.

The fact that the state was not only restricting their access to forests and public land but also granting concessions to logging companies and large landowners was not lost on most campesinos. This was particularly true in the case of national parks and forest reserves. In 1955, for example, a delegation of ocupantes traveled to Santiago to denounce the eviction of 121 families from public land they had occupied for more than thirty years in the Malleco Forest Reserve. They had organized a Committee of Aspiring Colonos to petition for formal titles to their land, but the regional governor had sent in carabineros to stop them from meeting. As in the case of the Sarao cordillera, many of these smallholders had arrived in the reserve's forests as inquilinos contracted by timbering concessions. The campesinos faced the threat of expulsion from land they had worked for decades because the government intended to lease the land to a new logging company. Deputy Ernesto Araneda Rocha of the Liberal Progressive Party stood up on the floor of Congress to defend the campesinos. "Many of these people have occupied this land for more than thirty years," he said. "Some were there before the dictation of the decree that created the forest reserve. . . . Others came there with the concessionaires who logged the forests. . . . For approximately thirty years, they have been cultivating the land that has been exploited by the concessionaires."[39]

This case revealed a common social and ecological process in southern Chile. Timbering concessions in state-owned forest reserves hired ocupantes as inquilinos or recruited inquilinos from outside. When the forests were cleared, the concessionaires moved on, often leaving their former workers behind to scratch out a living on the now logged and cleared land in the uncertain legal status of ocupantes. These peasants could make a living exploiting the remaining forest for coal and firewood; pasturing cattle in the remaining forest, sub-forest, and underbrush, especially the quila bamboo that invaded after logging and fires; and cultivating any crops they could in soil that was unsuited to agriculture. Despite the campesinos' claims that they had built a thriving agricultural economy that supplied local markets, government officials denied their claims by arguing that "the agricultural exploitation of this soil will provoke erosion, which will damage the forest reserves." In addition, the government planned to reforest already logged land in the reserve with Monterey pine and viewed the campesinos as a threat to future tree plantations. Nonetheless, belying the government's characterization of the campesinos as a danger to the remaining stands of native forest and to its reforestation projects, the campesinos had left intact a number of valuable forest stands on the hillsides that the government now wanted to lease to a private timbering firm.[40] This provides a glimpse of the reality behind the chronic characterization of campesinos as irrational and ecologically destructive. The sources indicate that, in locations such as Trongol, the Malleco Forest Reserve, and Nahuelbuta National Park, campesinos not only sought land to cultivate crops and pasture livestock but also struggled to protect native forests from logging because they relied on forest products for subsistence.

Mapuche Communities, Forests, and Commercial Logging

The Popular Front governments' forestry development model also shaped many Mapuche communities' approach to native forests that lay within their reducciones. A number of communities sought to capitalize on state-sponsored forestry development by contracting with timber companies to log stands of native forest. In general, these communities were driven by extreme poverty to lease logging rights to their forests. Squeezed onto small extensions of often arid and rocky land, Mapuche communities frequently were unable to produce sufficient crops for subsistence. Many male members left communities to find work elsewhere. For many reducciones, small patches of native forest remained one of their few resources. Despite legal restrictions on the alienation of indigenous land, the state made frequent exceptions during the 1940s and 1950s to authorize logging on Mapuche reducciones. At times, Mapuche communities' contracts with logging companies created deep schisms between and within communities. The goal of creating a steady source of jobs and revenue for the communities conflicted with many community members' reliance on harvesting piñones and other forest products. Indeed, many logging contracts reflected Mapuche communities' efforts to preserve the forests by imposing detailed restrictions on logging companies' operations in their forests.

In Lonquimay, the remoteness of the region's araucaria pine forests had provided protection from logging. Until the late 1930s, lack of roads and railroad lines made logging operations difficult. But in the mid-1930s the El Raíces Tunnel, which provided an important artery to Argentina, the primary market for Chilean lumber, was completed, and a number of roads had been built. In 1938, the Argentine Mosso company established Chile's largest plywood factory in the city of Curacautín, exporting almost all of its wood to the other side of the Andes cordillera and engaging in extensive logging operations in the cordillera. Mosso also provided an important market for timber, and a number of other landowners and logging companies began to make their presence felt in the high-altitude araucaria forests. Until the 1940s, many of the araucaria forests and the fundos high up in the cordillera had been left largely unclaimed. Many Mapuche reducciones continued to rely on them for sustenance and maintained a significant degree of autonomy, collecting piñones, pasturing their small herds in the invernadas during the winter and then in veranadas that could be used for pasture only in the summer months, when they were not covered with snow. After 1938, however, local merchants and landowners began to look to Lonquimay's araucaria forests, many of them claimed by the region's Mapuche reducciones, to initiate logging operations.

In 1947, the Bernardo Ñanco community in Lonquimay requested authorization to enter into a contract with the Fressard timber company to log forests of araucaria pine, roble, and coigüe on the reducción. The community had fallen into debt by acquiring food and clothing at Fressard's business in Curacautín, and many of the male members of the community had been forced to leave the region to look for work. The community's lonko, Domingo Camargo, had guaranteed the loans by promising Fressard revenue from wood on the reducción. The community's strategy of contracting the logging to private companies coincided with the Popular Front governments' general approach of leasing logging rights to public forests to private forestry companies and met with quick approval. The members of the community lived in abject poverty, a report by the Juzgado de Indios (Indian Court) of Victoria asserted, relying on only "a few natural products and the little they get from their fields." However, the court also noted, "They are the owners of natural resources that, exploited technically with large capital investments, could improve living conditions, as well as increase production and earnings for private concerns and the public treasury."[41]

As a sign of the state's increasing penetration of the frontier region and regulation of logging, the contract required the logging to be supervised by the office of the Inspectorate of Forests, which would determine which trees were appropriate for felling and thus guarantee "the conservation and regeneration of the forests that are logged." In addition, it stipulated that the company could not cut female trees "to protect the harvest of piñones, a basic source of food for the indígenas." The legal agreement represented the new state forestry policy. While logging on public reserves and Mapuche reducciones by private companies would continue, this was the first time that the state would assert its authority to approve logging contracts and oversee logging operations with the goal of producing a scientifically managed forest exploitation that would limit clear-cutting, preserve native forests, and stimulate reforestation. In this case, while driven into the logging arrangement by poverty, the community was able to impose—at least, on paper—requirements that protected the forest from clear-cutting and that guaranteed the supply of piñones. For state officials, the logging contract had other benefits. The forestry enterprise would provide jobs and wages for the destitute community and thus contribute to its material and cultural uplift: "industrialization and exploitation of the forests will provide jobs for the indígenas, separate them from their vices, [and] incorporate them into civilization, since they live in isolation with their timeless customs." As part of the contract, Fressard agreed to provide a school and teacher to the community and employ community members in its logging operations.[42]

In the end, however, the logging contract brought conflict and strife rather than progress and "civilization" to the community. While Fressard provided jobs in his logging operations, the Bernardo Ñanco community reaped little reward from the timbering. Community members recall that Fressard paid low wages and that many ended up receiving their pay in goods purchased at Fressard's store, installed on the borders of the community.[43] During the 1950s, renegotiated contracts with Fressard required the company to build houses for the community members, respect female and young araucaria trees, and allow supervision by the indigenous affairs directorate. By the late 1950s, however, the community had demanded that timbering be suspended because Fressard had failed to respect these requirements. Venancio Coñuepán, head of the Mapuche organization Corporación Araucana, and Minister of Land and Colonization in the second government of Carlos Ibáñez (1952–1958), went to congress and the press to denounce the fact that the company owed the community money for two and a half years of logging. In addition, he maintained, Fressard had violated both the Forest Law and the logging contract by felling trees that were small, young, or female and by engaging in clear cutting. The conflict remained unresolved; throughout the 1960s the community continued to accuse Fressard of illegally logging the native forests on its land.[44]

In a similar case, in 1955, the Pedro Calfuqueo community, also in Lonquimay, protested against logging in the extensive araucaria, raulí, and roble pellín forests on public land it had occupied for decades. The members of the community, which had grown from 40 people in 1905 to 250, were pressed for land and depended on the forests on the borders of their original land grant to survive. The community signed contracts with logging companies during the 1940s and early 1950s with authorization from the government, but in 1955, it petitioned to end the agreements because the loggers, primarily the owners of the neighboring Galletué estate, were destroying the araucaria forests on which community members relied to collect piñones. Apparently, the companies had also failed to satisfy the terms of the contracts. The community's petition of 1955 described "the fever to possess land with forests or simply to dispossess us of our land, both land granted us by the state in the título de merced and land we have occupied since time immemorial. . . . We must remember that the authorities have recognized our ownership and the Indian Court of Victoria, the Ministry [of Land and Colonization], and the Supreme Court have authorized logging contracts celebrated between the community and private parties, contracts we wish to end because they are unjust." For the community, the contracts supported claims to public forests its members had occupied for generations by implicitly recognizing their property rights.[45]

In this case, as José Bengoa has pointed out, Lonquimay's indigenous communities had been occupying high-altitude veranadas since they were settled in the cordillera after 1883. However, the government settlement commission had tended to include only low-altitude invernadas in the reducciones, leaving the zones of the cordillera at higher altitudes unsurveyed. The settlement commission granted communities titles that failed to reflect the territorial reach of their pastoral activities, including their reliance on both veranadas and invernadas and on the araucaria forests, limiting their reducciones to the low-altitude valleys of the cordillera. High-altitude land that was not included in the reducciones' titles was then auctioned off, but this did not have an immediate impact, because large landowners who acquired fundos in the region used the land to pasture livestock. By the 1940s, however, these fundos had become more valuable because of their araucaria forests. The new owners of the Galletué estate, purchased in the 1940s from the Caja de Crédito Hipotecaria (State Mortgage Bank), formed a timber enterprise that initiated a new era of logging and conflict with Mapuche reducciones in Lonquimay's cordillera that would last until the close of the twentieth century.[46]

This pattern of Mapuche communities selling wood to sawmills or contracting with timber companies to log their stands of native forest was repeated in mountainous regions throughout the south. Near Llaima and Cunco, for example, Mapuche communities also sold logging rights to forests during the 1940s and 1950s, often in exchange for flour, as a recent history of Llaima has shown. In addition, community members often found jobs working for timber companies and large landowners, a key survival strategy. In Llaima, as in Lonquimay, small land allotments and a history of usurpation pushed Mapuche communities to the brink of starvation.[47] Similarly, communities in the foothills of the Andes cordillera near Panguipulli contracted with small sawmills to log their native forests or sold trees to be made into sleepers (railroad ties) to the state railroad. Moises Durán, a former forestry worker, recalled that over the years laboring in Panguipulli's forestry fundos he had accumulated enough money to buy a pair of oxen, which he used in logging. During his chronic bouts of unemployment when estates fired their workers, he often worked *a medias* with Mapuche communities in the Panguipulli and Liquiñe region, helping to log their forests and then transporting the wood with his oxen to a sawmill in Liquiñe. The forests provided necessary income to the communities and work to Durán when local estates dismissed their workers. Durán's history reflected a general pattern of the owners of small portable sawmills or oxen contracting with local Mapuche communities to log the Panguipulli region's native forests.[48]

Mapuche communities' logging activities often led them into conflicts with neighboring estates. For example, during the 1950s, the Vicente Reiñehuel community near Panguipulli clashed with Víctor Kunstmann, owner of the Trafún estate, over logging rights to native forests. In the late 1940s, the community denounced the fact that Kunstmann had built logging roads and felled trees in forests within the borders of the reducción. The conflict had arisen, it appears, because the community had entered into an agreement with Calixto Iglesias in 1955 to log the forests. Under the terms of the contract between Iglesias and the community, which presumably was similar to earlier contracts and was authorized by the government, Iglesias would pay the community for the raulí extracted from its forests, invest in roads and bridges, build a school, provide wood for houses, hire a forester to oversee the management of the forest, and reforest the land with Monterey pine. In this case, Kunstmann lost his legal claim to the forests, and the community won government approval of its contract with Iglesias in an arrangement that contained all of the elements of the reigning forestry philosophy. The logging operations on Mapuche land would be overseen by a forester, and the native forest, once logged, would be replaced by Monterey pine. The community would receive a hefty sum for its wood, houses, a school, and jobs in Iglesias's timbering operation. The community's case against Kunstmann was bolstered by a surveyor's report that found that Kunstmann had fudged the borders of his property to encroach on land contained within the community's original land grant title.[49]

In the Andes cordillera, from the Bío Bío River farther south to Valdivia, many Mapuche communities depended on felling trees and cutting sleepers to sell to the state railroad. In many cases, the communities logged their woods and manufactured the sleepers themselves to exert control over the forests and maintain a certain degree of autonomy.[50] In other instances, as in the case mentioned earlier, communities leased forests on their land to small logging contractors or mobile mills. Mapuche communities also continued to squat on land within national parks and forest reserves to which they claimed rights based on occupation since time immemorial. For example, foresters confronted continual invasions of the Villarrica Forest Reserves. In 1954, members of the Huaiquifil community occupied a stretch of land covered with forests that they regarded as belonging to their reducción, another chapter in a long-standing conflict between the community and the reserve that dated back at least to 1932 (see chapter 3). As occurred in 1932, the reserve's administrator called on the government to intervene because, he contended, the Huaiquifils were "destroying . . . the forest wealth that still exists there and that constitutes a national resource." Two decades later, foresters still were unable to prevent members of the community from occupying forests within the reserve. Despite efforts by the governor of Loncoche to expel the members of the Huaiquifil community, they had continued "with their eagerness for destruction, felling trees and cutting second-growth forest" and "causing serious damage to the public interest, since they are destroying species that are difficult to recover and destroying new species in formation, and above all on land that is steep and of which there is no hope for any agricultural benefit of any kind." The reserve's forester called for carabineros to expel the "destructive and stubborn" alleged squatters from the reserve.[51]

This case reflected well the ways in which conservationist policies had begun to shape land disputes. In part, the Huaiquifils' "squatting" was driven by confusion over borders that was typical in southern Chile. Neither the community members nor the forester knew where the borders between the reducción and the forest reserve lay. Given the many cases of superimposed boundaries, it is probable not only that the Huaiquifils had occupied the forests as part of a transhumant economy for years, but also that their community had been granted land folded into the reserve. Because no clear map or definition of the community's borders existed, the reserve's forester called for a surveyor to be sent to delimit the line separating the community and the reserve "to protect the public forests." The forester denied the Mapuches' rights to land within the reserve because they were a "destructive" scourge of public forests. Their logging and roces meant that, even in a case of indeterminate borders, the imperative of regulating logging in the Villarrica reserve outweighed any claim they might have based on years of occupying the forests or on the vague boundaries set out in their título de merced.

Forestry, Forestation, and Estate Labor

The conflicts over access to public land and forests in the south also played out within the borders of large estates in the region of the coastal cordillera. The march of forestation on large fundos initiated with the Forest Law of 1931 accelerated during the 1940s and 1950s as landowners and the state invested in plantations of Monterey pine. When they turned to pine, landowners who had often left cultivation of their estates' soil to inquilinos, medieros, and arrendatarios sought to redefine land and labor relationships. The *chacras* (garden crops) of estate laborers represented land that could be profitably put to use by planting pine. In addition, forestry estates enjoyed a privileged status in the eyes of the state. They constituted a modern form of agricultural production linked to industrialization and received state subsidies and incentives.

In Concepción Province, the center of the pine boom, conflicts between campesinos and fundos engaged in forestation spread rapidly during the 1940s. In 1947, *El Siglo* reported that "fundos that once planted wheat are planting pine." The paper joined the San Rosendo Association of Small Agricultural Producers in attacking local landowners for mistreating their inquilinos and medieros: the landowners were replacing wheat with pine and had begun to crack down on the customary regalías that had defined inquilinaje contracts for many years. The landowners "refused [the inquilinos] land for pasturing their animals . . . , a right that has existed for many years." *El Siglo* described three fundos with "2,000 hectares of land that are suitable for agriculture and, when planted with pine, will leave unemployed more than sixty inquilinos, who with their families make three hundred people." The inquilinos demanded the expropriation of the fundos, invoking the importance of agricultural production, rather than forestry, to the national economy.[52] Another issue of *El Siglo* declared that "the pine plantations are taking thousands of hectares out of cultivation in Concepción." The paper interviewed a San Rosendo campesino who declared that the expansion of the pine plantations was throwing a large number of campesinos out of work.[53] A lengthy headline in *El Siglo* summed up the conflict between the peasant agricultural economy and commercial forestry:

"There Will Be Evictions on the Fundo San Juan to Plant Pine: [Land on the] Fundo That Belongs to La Papelera [the Compañía Manufacturera de Papeles y Cartones, the country's sole paper producer] in Tomé [Is] Suitable for Garden Crops [*Chacarería*]."[54]

Even publications that represented landowners' interests occasionally expressed doubts about the effects of pine fever. In 1948, an article published in *El Sur* and reprinted in *El Agricultor,* the organ of the southern landowners' association, denounced a growing scarcity of basic goods because landowners in the frontier provinces were planting with pine. The article invoked the dire social consequences of forestation: "I have observed that property that is very suitable for agricultural or livestock production is dedicated only to pine plantations, making them [the plantations] an enormous evil for the obreros and inquilinos who . . . lived from their work with the soil for generations [and] are now rounded up by the new owners [and] forced into exodus to the towns, thus diminishing [agricultural production] with consequences that are easy to imagine, especially with the population increasing significantly." The lack of food supplies and rising cost of living in the region had been caused by the "enormous enthusiasm for planting pine, because it is a safe short-term source of wealth." The article noted that "anyone who has any capital is investing it in buying property with this goal [planting pine] and are organizing public companies to exploit these forests."[55]

In a case that typified the spread of pine in southern Chile, in the late 1940s and early 1950s Juan Bautista González Reyes, the owner of the Pilpilco fundo, a large forestry estate in Lebu, Arauco Province, began to evict most of the older inquilinos from the fundo and to eliminate the benefits he had provided to the remaining inquilinos. He refused them the right to cultivate land, denied them access to roads and egress from the fundo to market their goods, and threatened to expel them altogether. Pilpilco had been reforested with pine over the years, and González Reyes viewed the presence of the inquilinos as an obstacle to expanding his plantations. In 1953, a committee of inquilinos-medieros petitioned the state for Pilpilco's expropriation and the distribution of the land to its workers. Like other estate laborers, they argued that most of Pilpilco's land was public and had been acquired by fraud. In this case, the petitioners themselves were "eyewitnesses," since, they claimed, they had followed González Reyes's orders years earlier and "aided him in taking land from other individuals." They asserted that the "majority of the land occupied by Señor González is not his property." Thus, they had participated in landowners' well-established frontier practice of appropriating land from the state, colonos, and Mapuche reducciones by settling inquilinos on it and then asserting material possession with the inquilinos themselves as witnesses.[56]

In the case of the Comunidad Yrarrázaval-Mac-Clure, the careful reconstruction of land titles can provide a sense of the history of social conflict obscured by the seamless blanket of pine and the exalted rhetoric of entrepreneurial foresight. This task was taken up in 1944 by Deputy Eduardo Rodríguez Mazer of the Socialist Party, who took the floor of the House of Deputies on behalf of three hundred campesino families who were being evicted from land in Pangal del Laja by carabineros on behalf of the Comunidad Yrarrázaval-Mac-Clure, which was fencing its land to plant pine plantations. The incident began when a 1,230-hectare piece of land between the Itata and Laja rivers in Bío Bío Province was auctioned off. Through a series of land transfers following the auction, the Las Mercedes hacienda was formed. In 1930, the hacienda was divided into three parts with new titles. The three parts together consisted of 5,404 hectares, more than tripling the property's original size. Based on these fraudulent titles, Gaspar French Powell received a loan from the State Mortgage Bank that was never paid off. The bank then placed an embargo on the property, but before it could auction the property off again, the region's smallholders, many with titles to their plots, contested the estate's newly invented boundaries, which had expanded to include pieces of their land.

When the State Mortgage Bank registered the property with the local Property Registry, however, it included 3,865 hectares, ignoring the claims of the region's smallholders. The bank then sold the land to the Comunidad Yrarrázaval-Larraín and sought the campesinos' eviction with the backing of Arturo Alessandri's government. The company then began to erect wire fences around the disputed property, even though the "Pangalinos" had gone to court to make a case for their legal rights to the land. While the conflict wound its way through the courts, the Comunidad Yrarrázaval-Larraín sold plots on the estate to the Caja de Empleados Públicos (Public Employees Retirement Fund). In this case, the pine trees were planted on land acquired through a long history of fraud, with a healthy injection of capital from public credit and retirement institutions, and resulted in the theft of land from hundreds of local campesinos. The state's support of the company by using its coercive powers to purge the property of campesinos was indispensable to the future prosperity of its pine plantations.[57]

When this long and well-documented history was presented to Congress, Deputy Carlos Izquierdo Edwards of the Conservative Party argued that the company had purchased land from the State Mortgage Bank that was arid and worthless, built canals to irrigate it, and, "with immense sacrifice, was able to improve the land and transform it so that it can be considered a source of national pride." Deputy Belisario Troncoso of the Liberal Party joined Izquierdo by arguing that the Pangalinos had violated the rights of the Comunidad Yrarrázaval-Larraín by destroying its fences and celebrating the community's "marvelous work of progress . . . for the nation . . . , because in reality it is a marvel to transform an ugly, sandy, bleak plain . . . into a massive green stretch of pine plantations."[58] In this case, the Pangalinos stood little chance. They labored on overworked and eroded soil and had contested land claims tangled in years of litigation while the company boasted of its modern production methods. It was helping recover and restore value to soil whose productivity had dwindled over the years, and it was contributing to national industrial and agricultural progress. Its tree plantations allowed it to adopt the mantle of progressive modernization. Unlike the impoverished Pangalinos who had produced only soil erosion, the Comunidad Yrarrázaval-Larraín had the capital to work an ecological miracle on the land, converting eroded desert into verdant tree plantations.

In addition to opening up land on estates to plant pine and expanding estates onto the land occupied by smallholders, replacing resident laborers and ocupantes with pine trees was a useful strategy for undermining an increasingly organized and militant rural laboring population. In the context of movements to unionize campesinos and promote agrarian reform, evicting workers and cultivating plantations that required relatively minimal labor made a lot of sense. During the late 1930s, with an increasing tempo after the election of the Popular Front in 1938, rural laborers organized unions to demand a host of improvements in working conditions, including access to schools and health clinics, contracts, wage increases, and, above all, improved regalías. Landowners countered this wave of unionization throughout rural Chile with mass dismissals, particularly targeting union leaders and members.[59] In the south, forestation became both a tool for evicting resident laborers and a pretext for eliminating unions from estates.

In the summer of 1946, for example, seventy workers on the Pillimpillim and Chacay fundos, owned by the Hernández timber company in Curanilahue, Cañete, denounced the effects of "forestation policy" in the region. The company had expelled the workers, many of whom had worked on the fundos for more than thirty years, to plant pine. The campesinos asserted that they were not "opposed to forestation," but argued that the land the company was planting with trees could instead produce wheat, beans, lentils, oats, and potatoes. They argued that more than forestation, the company's goal was to expel one hundred inquilinos from their land: "[the company] is surrounding the pueblos with the object of depriving them of land for cultivation and pasture of their animals, forcing them to abandon their land." The Hernández company had restricted the inquilinos' and medieros' regalías to .5 hectares per worker. The campesinos petitioned the government to send in a surveyor to determine which land was suitable for cereal production and which for forestation, "obligating the Hernández company to form plantations on the land where it has exploited forests rather than on the land campesinos had cleared during many years of labor."[60]

For the campesinos, the expansion of tree plantations disturbed the balance of commercial forestry, wheat production, and the peasant chacra economy. In addition, forestation constituted part of a more general assault on campesino land, both the plots granted to inquilinos on the fundos and the parcelas of neighboring smallholders. It is perhaps not a coincidence that, during the previous years, the workers had organized a union and demanded unpaid wages from the company, which responded by denying the workers food from its pulpería. The company had also brought in carabineros, who threatened the union leaders.[61] In response, the Hernández company's workers petitioned (unsuccessfully) for the expropriation of the fundos. The company's interest in expanding its pine plantations was aimed at getting rid of a labor problem, as well as conflicts with neighboring smallholders who competed for access to public land, water, and forests.[62]

In a similar case the following year, the Concepción branch of the CTCh denounced the eviction of workers on the El Rétamo estate. The workers had organized a union and put together a petition for higher wages and increased food rations. In response, according to the CTCh, the estate's owner had "decided to plant 3 million pines on land that is suitable for the cultivation of wheat, a commodity that is needed in this country." The replacement of wheat with pine on El Rétamo, the union pointed out, would provoke the dismissal of the rebellious workers, since "the pines do not require all the people who are now working."[63] Pine plantations offered El Rétamo's owner a way to rid himself of the burden of maintaining inquilinos on his estate and put an end to the labor conflict with his workers.

The Lota coal-mining company also met the challenge of an increasingly restive and organized labor force on its rural properties by replacing workers with pine. This was the case for sixty arrendatarios on its estates in the Nahuelbuta cordillera. The arrendatarios felled trees in the company's forests but were also allowed to cultivate crops on small plots of land. In November 1946, the workers, invoking the Labor Code of 1931, petitioned the company for more land for their crops, social-security contributions, accident insurance, a family allowance, wood for their houses, and a schoolhouse for their children. The workers' fundamental demand was that they be dealt with as full-time obreros, which would give them a series of legal guarantees, rather than arrendatarios, which left them without legal protection and basic benefits. Indeed, unions denounced the fact that throughout the 1940s landowners in the region had redefined inquilinaje contracts as rental contracts and their workers as arrendatarios to evade the obligations of the Labor Code.[64] When the company rejected their petition, the workers initiated a work stoppage. The company responded by dismissing the workers "under the pretext that it was going to cultivate plantations of pine and eucalyptus on its land" (to replace the cordillera's logged-over native forests). The local CTCh *consejo* (council) in Lota noted that this was a first step in evicting all of the resident workers on the fundos and demanded government support for the campesinos. The labor union phrased its attack on the mass firing in terms of the conflict between tree plantations and agriculture.[65] The company, however, justified the dismissals and its attempt to undermine workers' organizations by invoking the ecological imperative of reforesting with pine.

A similar conflict between rural workers' demands for land to cultivate crops and timber companies' logging and forestation operations animated a confrontation between campesino unions on the Pillén Pillén and Chacay fundos and the Bosques y Maderas company in 1947. Fifty members of the Sindicato Apolítico Profesional Agricultural (Apolitical Professional Agricultural Union) who were inquilinos or medieros with sharecropping arrangements on the fundos were evicted by the company, according to a union petition, "as a reprisal against the constitution of our union." The workers requested that the government expropriate the fundos because the timber company exploited the land not directly but, rather, through contract workers. They asserted that they held rights to the land because they enjoyed material possession and had introduced improvements. In contrast to the timber company, they would engage in "intense exploitation" that "for next year and years to come would maintain the security of [their] food supply." Not surprisingly, the campesinos' petitions were rejected, and Bosques y Maderas continued its logging operations.[66]

By the 1950s, conflicts over access to forests within the borders of private estates and public reserves had begun to pose challenges to timber companies. Land invasions, loggers claimed, were a major obstacle to the development and progress of their industry. In 1956, for example, the timber companies' organ *Chile Maderero* published an article denouncing "illegal usurpations and theft of wood." Colonization laws that favored putative colonos who could wield their rights of possession to acquire titles proved an encouragement to ocupante squatters: "impunity makes the audacious valiant, as does the weakness of the authorities, and means that every day these crimes are repeated, especially the usurpation of a little corner of a property or the felling of a few far-off forest stands, which, without immediate sanction, brings on the complete usurpation of the property either by the same individuals or by a small group at the head of a greater legion of others." The Ministry of Land and Colonization, *Chile Maderero* editorialized, "authorizes any audacious individual to usurp land [and] extract wood, and we have even seen cases of homicide in which the criminals have in their power written documents or certificates from this ministry that were little more than authorization to kill whoever bothered them. . . . It does not matter that the occupied property is a national resource for public use, a valuable public forest that requires rational exploitation or—as occurs in most cases—private property with perfect titles. . . . The Ministry of Land [and Colonization] always processes the land petitions . . . and ensures the impunity of the usurper." *Chile Maderero* argued that settling campesinos as colonos in regions with native forests undermined rational forest exploitation: "how can the state direct and establish norms for the wood and forestry industry if its technical agencies give the worst example with public forests handed over to the first audacious person who occupies them without capital or tools, incapable of scientific and commercial exploitation [of the forest]?"[67]

Two years later, *Chile Maderero* pointed to colonization policy as the culprit behind frequent forest fires in southern Chile: "the erroneous colonization policy is responsible. . . . [T]he smallholders, lacking capital, consider the forests an obstacle to using the land they occupy for livestock and agriculture." The article called on Congress to approve a law that would "put an end to the illegal occupation of forests and the invasion of forest reserves, especially those in the public domain, that today are the main originators of forest fires."[68] Timber producers relied on arguments that had been made by state officials and foresters since the beginning of the twentieth century. Only businesses with access to capital should be entrusted with native forests and plantations, and only their capital could bring in scientific methods of forest exploitation. Logging companies' rational exploitation of forests and forestation was ecologically sound, while the activities of campesino smallholders and squatters led to criminal violations of property rights and forest fires that laid waste to a valuable natural resource.

During the 1940s and 1950s, these arguments usually won the day, despite the timber industry's denunciations of colonization laws. The state supported the wood producers' view that forest reserves should be handed over as concessions to large forestry companies that could implement rational and commercially viable forms of forestry, including forestation with pine. Campesino squatters posed a significant threat to this arrangement by invading both private estates and public land on reserves. In addition, the fires they used to clear land for pasture and crops often threatened to spread to the native forests and tree plantations worked by timber companies. As a result, the state and the logging companies renewed their efforts to crack down on ocupantes' illegal exploitation and the use of roces. However, while timber interests prevailed over campesinos in the struggle to gain access to forests on both privately held and public land, the state now imposed stricter requirements on and oversight of private companies' logging operations, an important intrusion into the activities of loggers and landowners who hitherto had enjoyed unrestricted control of frontier forests. In addition, as the complaints of the timber industry's trade association made clear, neither the state, on the one hand, nor logging companies and estate owners, on the other, had consolidated control over frontier territory. A weak state presence continued to allow poor campesinos to invade public and private land. As they had during the early decades of the century, campesinos either invoked colonization laws to claim colono status or, in the case of Mapuche communities, historic use rights, even as they were denounced as a threat to the forests and modern forestry.

6

Agrarian Reform and State-Directed Forestry Development, 1950s and 1960s

In 1966, *El Diario Austral* published a report describing social and environmental conditions in Curacautín, a center of the logging industry in the foothills of the Andes cordillera (see figure 6.1). A large number of colono smallholders, repatriated from Argentina during the early twentieth century, had settled in the region, but soil degradation due to deforestation had forced many to sell their parcels to large landowners. As *El Diario Austral* described, "The colonos began by cutting down the forests. . . . [T]hen came clearing with fire, from which they obtained more or less acceptable harvests, thanks to the potassium that the soil contained. Then the soil was exhausted by the lack of fertilizers and indiscriminate exploitation, and today we find that the city is surrounded by land that is no longer forested and is not the object of adequate agricultural production." As the forests disappeared, colonos could no longer earn money making charcoal and selling wood to local sawmills or the Mosso plywood factory, and their depleted mountain soil produced only meager harvests. Many were forced to sell their small plots to wealthy landowners, who assembled large estates in the region. As the paper noted, "Then the situation changed substantially. The forests were farther away; carts were replaced by trucks; the soil was exhausted by irrational cultivation; and today we find that a large majority of the colonos have had to sell their plots, which have been grouped together in extensive fundos, which in only a few cases are cultivated adequately, but the rest of the smallholders suffer the most absolute misery."[1] In the countryside surrounding Curacautín, deforestation and soil erosion had led to the dispossession of colono smallholders; the formation of large, inefficient haciendas; and the creation of a mushrooming population of landless campesinos and unemployed laborers.

By the 1960s, the need to halt this spiral of ecological degradation and deepening rural poverty became all the more pressing because of the radical decline in agricultural production. Once the country's breadbasket, southern provinces such as Malleco now presented a fairly desolate

6.1 Logging in Araucaria Forest, n.d. (Courtesy of Museo Histórico Nacional)

landscape dominated by large, inefficient estates, with few remaining native forests, eroded soil, and an impoverished rural population that continued to move across the cordillera to Argentina in search of work. In 1965, roughly one-quarter of Chile's national territory and 61 percent of its agricultural land suffered from soil erosion. The provinces of the Ninth (Araucanía) Region, once the most forested in Chile, were now the most eroded.[2] Chile could no longer feed itself, and its dependence on importing basic foodstuffs intensified a long-standing trade imbalance. In response to the stagnation of Chile's agricultural sector and profound inequalities in landholding, the conservative government of Jorge Alessandri (1958–64) initiated a limited land reform program in 1962, transforming the Caja de Colonización Agrícola (Agricultural Colonization Fund) into a new agrarian reform agency, the Corporación de la Reforma Agraria (Agrarian Reform Corporation; CORA). Under Alessandri's agrarian reform, only a small number of state-owned properties were broken up and distributed to inquilinos. However, the Christian Democratic government of Eduardo Frei (1964–70) implemented Chile's first significant policy of expropriating privately held estates and redistributing land to estate laborers with an expanded agrarian reform law in 1967.

The agrarian reforms implemented by Frei represented both a rupture and continuity with previous development programs in southern Chile, building on forestry and land colonization policies established during the

1940s. While the Christian Democratic government set out to transform the rural landscape by building a large population of yeoman smallholders who would engineer Chile's agricultural modernization, its agrarian reform policies left out estates dedicated to forestry production with either native forests or tree plantations. Frei's government proposed accelerating forestry development to create jobs and generate growth that could ameliorate southern Chile's ecological crises and poverty. Rather than redistribute the land on southern forestry estates to their workers, the floating population of squatters and would-be colonos, or to Mapuche communities, Frei's government invested in building the infrastructure of a modern forestry industry, expanding tree plantations, establishing new pulp and paper plants, and imposing state oversight of logging in Chile's remaining native forests. Forestry development constituted a technocratic solution to southern Chile's deepening ecological and social problems, sidestepping the issue of agrarian reform in the forests.

As the example of Curacautín illustrates, the imperative to implement agrarian reform was driven in many zones of the southern frontier by ecological crisis. In southern Chile, ecological degradation provoked by the rampant clearing of forests since the first generation of colonization was a major factor in declining cereal yields. In addition, soil erosion, drought, and flooding had contributed to the inequality in land ownership, driving many smallholders throughout the south to sell their land and labor to large estates. Second, in a number of regions of the south, both campesinos and large estates were responsible for destroying native forests. As numerous reports from the south underlined during the 1950s, members of Mapuche communities and non-Mapuche campesinos relied on producing charcoal and firewood (leña), as well as selling wood from their forests to local sawmills, to supplement the meager production of their small, overworked, and overgrazed plots of land. In part, campesinos' ecologically destructive practices derived from their increasing dependence on the forests. Often relegated to rocky, infertile land in the cordillera or its foothills, campesinos quickly exhausted their soil. Because of their limited extensions of land, pasturing cattle and cultivating crops such as wheat led inexorably to soil erosion and increased pressure on native forests.

Campesinos' role in destroying native forests contributed to the representation of both Mapuche and non-Mapuche campesinos by President Frei's agrarian reform officials as a threat to the forests and incapable of rational forest management. In 1965, for example, a congressional commission charged with developing new forest legislation cited the famous conservationist work of Rafael Elizalde, La sobrevivencia de Chile (Chile's Survival), to place the blame for deforestation and soil erosion squarely on the shoulders of colono and ocupante squatters. These poor campesinos, the commission argued, invaded forest reserves and national parks, burned their forests, produced decent cereal harvests for a few years, and then, once their soil was exhausted, moved on to occupy and burn new stands of public forests. They posed a threat to "neighboring legitimate landowners," conservation of the forests, and the national economy.[3] This conservationist narrative of campesinos' destructive ecological practices bolstered a more general perception, implicit in scientific forestry as it was practiced in Chile, that forestry industrialization required centralized management directed by the state and large-scale production in the hands of private capital. During the 1960s, foresters and agrarian officials contended that redistributing land to campesinos was irreconcilable with both conservation and the development of modern methods of production in Chile's remaining native forests and in regions covered with tree plantations.

Landlessness, Unemployment, and the Ascent of Monterey Pine

In southern Chile, persistent problems with rural unemployment drove the pressure to reform the structure of landholding and organization of agricultural production. As I have shown, southern estates relied less on a permanent labor force of inquilinos than did their counterparts in central Chile. In the forestry sector especially, workers composed a floating migrant population that made its way from estate to estate, forest stand to forest stand, in search of temporary jobs. Periodic waves of migration across the cordillera in search of work in Argentina were a permanent feature of rural social relations. During the 1950s, under the governments of Carlos Ibáñez (1952–58) and Jorge Alessandri, the Popular Front's expansionist development policies rooted in import substitution industrialization and basic social reforms, which included increased public spending and hikes in wages and benefits for urban workers, were reversed. In the face of stagnant economic growth and galloping inflation caused by bottlenecks in import substitution industrialization, including restricted internal markets, inefficient protected industries, and lack of private capital investment, both the Ibáñez and Alessandri governments resorted to a series of austerity measures. In 1955, the North American consulting firm Klein-Saks proposed a series of economic policies to combat inflation, including cutting public spending, limiting wage increases, liberalizing trade, and restricting credit. The policies were largely successful in reducing inflation but had profound social consequences, including significant increases in unemployment.

The impact of the austerity measures introduced by Ibáñez, and largely maintained by Alessandri, on the forestry sector was devastating. Periodic surges in unemployment had been a permanent feature of the industry since its inception, but the economic crisis of the 1950s and Klein-Saks's free-market remedies took a devastating toll on rural workers in the south. First, the new economic policies led to a severe constriction of the construction industry, reducing demand for lumber. Second, the reduction of credit effectively put many small and medium-size sawmills and logging companies out of business. In 1957, *El Siglo* reported that sawmills throughout the south were shutting down in response to the credit crunch and contracting markets for wood in the construction industry. Workers were imploring mill owners to keep the mills in operation and offering to work for food.[4] Several months later, the paper reported that of eight hundred sawmills in the south, only one hundred were working: "the wood industry is being destroyed. . . . In Pucón, of fifty mills not one is working, and in Villarrica, there are only two or three still working. . . . In Malleco, all the mills in the Nahuelbuta cordillera are paralyzed. . . . [I]t has been calculated that forty thousand unemployed Chileans have crossed the frontier to Argentina through Lonquimay, while the workers from Cautín have traveled across the cordillera in Villarrica."[5] Similarly, in 1957, the forestry industry's trade publication, *Chile Maderero*, reported that of six thousand workers on Panguipulli's forestry estates, only 1,200 had kept their jobs.[6]

A chronic labor surplus in the region allowed landowners to impose onerous work conditions on their estates. An article written in the early 1950s described the lot of hundreds of itinerant forestry workers in Panguipulli. The workers labored from 4 AM until 9:30 PM, with a break of only two-and-a-half hours, and were forced to travel from fundo to fundo logging and milling wood. In addition, "The workers come and go with their families by all the roads. There they stay on the side of the roads. They have many needs, and their children's hunger is even greater." Because of the high level of unemployment in the region, when workers complained about conditions, a logging company's manager responded, "He who wants to work works, and he who doesn't can leave."[7] Former forestry workers in Panguipulli recall that the region's landowners paid their workers every two or three months, and workers often were paid not with money but with tokens that they exchanged for food, clothing, and other basic goods.[8] In 1957, labor activists denounced the fact that logging companies frequently fired workers without paying wages or family allowances; many workers received their wages in tokens redeemable at local stores or company stores on estates. Many forestry workers in the Panguipulli region fell heavily in debt to their employers and were bound to the estates in a system of labor relations that resembled debt peonage.[9]

Forestry workers' frequent unemployment stretched into the 1960s. In 1966, for example, *El Diario Austral* reported that the lack of markets for wood had led to the dismissal of one thousand forestry workers in Curacautín and that "mills [were] firing workers throughout the region."[10]

In Lastarria, where campesino smallholders logged native forests to sell sleepers to the state railroad company, markets had contracted, leaving small producers without outlets for their wood. Throughout the region the state railroad was either paying its suppliers late or paying with tokens that the smallholders could exchange for food in local stores, often at inflated prices.[11] *El Diario Austral* observed that forestry workers were departing for industrial centers where labor markets were already saturated.[12] In Curarrehue, a center of logging in the Andes cordillera, the paper reported that "the majority of sawmills . . . have been dedicated to fabricating sleepers for the state railroad, but the recent paralyzation of operations has obligated the producers to enter a period of inactivity, provoking very serious economic difficulties."[13]

For many Mapuche communities, which relied on selling trees from their reducciones to local sawmills or selling sleepers to the state railroad, the contraction of the forestry industry posed a major problem. In 1966, for example, *El Diario Austral* reported that restrictions on credit were pushing sawmill owners around Villarrica out of business. One mill owner said, "The situation of small producers who depend on credit is bad. We haven't been able to work with the banks. If we had enough credit, we could increase production and employ more people. Our big market is the railroad, and it is very weak. The construction industry is almost paralyzed and doesn't buy wood." The importance of the hundreds of small sawmills to many Mapuche communities in the foothills of the Andes cordillera was made clear in an article about mills in Coñaripe: "on both sides of the highway that connects Villarrica and Coñaripe the traveler can see the remains of numerous installations that functioned until recently, milling the robust logs extracted from the forests that arrived at the edges of Lake Calafquén." The Pucura mill, for example, "was installed to work exclusively with indigenous communities which possess forests. . . . [H]undreds of Mapuche smallholders . . . have found in the sale of their wood a great relief for the economic necessities in a zone where life is difficult. . . . The money allows the communities to survive during the winter."[14]

Many of these communities now confronted a serious crisis exacerbated by deforestation. As reserves of native forest dwindled or receded into ever more remote mountain areas, Mapuche communities in certain zones, particularly in the stretch of Andes cordillera from Villarrica to Panguipulli, no longer had wood to sell to the mills. As *El Diario Austral* observed, "Now the mills are working the last reserves of forests [on Mapuche communities] that are almost extinguished." The owner of one mill noted that reforestation was becoming an urgent necessity on indigenous land: "the Mapuches do not have second-growth, regrown forest, they fell [the trees], they cut them, and the remains are converted into firewood."[15]

6.2 *Saw Mill in the Ninth Region (Courtesy of Museo Histórico Nacional)*

Another mill owner in nearby Licán Ray observed that, on land owned by Mapuche communities, "the forest has been logged, burned and the tree trunks removed to convert the remains of the forest into firewood. There is no second-growth forest, which could have been the basis for the recuperation of the land, and the land has begun to erode dangerously, losing its nutrients and conditions for agriculture or livestock."[16] Similarly, another report noted that farther north in Lonquimay and Malalcahuello, native forests logged by campesinos were located farther and farther into the cordillera, and "the second-growth forests often are not adequate for logging."[17] These accounts suggest that the crisis in the logging industry was caused, in part, by the destruction of the native forests and the difficulties of gaining access to remote mountain stands of commercially valuable woods. They also reflect the social reality of campesinos forced to log their small patches of forest and sell the wood to sawmills (see figure 6.2) or to log the less commercially valuable second-growth forests to produce firewood and charcoal to purchase provisions during the winter, when they had little income and no food.

The crisis in the logging industry during the 1950s and early 1960s was provoked by one other factor. The trees on pine plantations cultivated during the 1930s and 1940s were mature and ready to be harvested. They now supplied an inexpensive alternative to wood extracted from Chile's native forests. In fact, it appears that by the 1960s the spread of pine plantations had begun to relieve pressure on native forests. Pine was sold at a price two to three times lower than that of even the cheapest milled native wood and thus drove down prices in an industry in which profit margins were slim and business was almost always precarious.[18] By 1965, some 200,000 hectares of pine plantation already stretched between Maule and Cautín.[19]

Pine's growing dominance in national wood markets had dire consequences for campesinos and the small mobile sawmills that worked native forests. In 1966, *El Diario Austral* observed, "Every day, native woods are being replaced [by pine], and markets are paralyzed. Many mills have closed their doors. . . . [U]nemployment is tremendous and threatens to keep growing."[20] At the Chan-Chan estate near Lake Panguipulli, the owners halted their logging operations in the Valdivian forests, which had become "unprofitable," in 1966. They could no longer compete with "inexpensive species" (North American pine) and shut down their sawmills while expanding their livestock operations. This implied both clearing forest to plant pasture and dismissing workers, since raising cattle required a smaller labor force than logging.[21]

Throughout the south, the decline of forests and forestry exacerbated existing problems with rural unemployment and landlessness. In 1967, *El Diario Austral* reported that the "problem [of unemployment] because of contracting markets for native Chilean woods . . . is getting worse, and many producers have decided to totally paralyze the sawmills. Daily, the employment offices and provincial Labor Inspectorate see new cases of the massive dismissal of workers who for many years found work without any problem in our numerous sawmills. . . . In construction, . . . pine is preferred because it is cheaper."[22] Many of these forestry workers drew on long-standing traditions of migration to cross the cordillera, where they found "better wages and living conditions" in Río Negro and Neuquén.[23]

Pine plantations produced unemployment in another way. During the late 1950s and 1960s, landowners in the south continued to transform their agricultural estates into forestry enterprises. In the process, they expelled inquilinos and encroached on the land of both Mapuche and non-Mapuche smallholders. In 1958, for example, leaders of the national labor federation, the Central Única de Trabajadores de Chile (CUT), denounced the fact that in Arauco and Concepción landowners were "sabotaging" food production by planting only pine: "the large companies and landowners . . . have dedicated all their land to pine plantations. Reforestation is done the same way. This is why they do not engage in agricultural production. They throw campesinos into unemployment, and the cost of basic food becomes more expensive." The CUT leaders also contended that, to plant pine, "these companies and landowners are trying to appropriate all the land and commit all kinds of abuses, especially against the Mapuches, against whom they run fences on their reducciones, make false land claims, and entangle in court cases to legalize the theft of their land." By the late 1950s, the expansion of pine had already provoked conflicts between Mapuche communities and forestry companies in the foothills of the coastal cordillera in Arauco Province. In addition, forestation with pine had led to a significant loss of jobs among estate laborers, many of whom also came from the region's Mapuche communities, since the estates required fewer laborers as they converted from growing cereals to cultivating pine.[24]

For organized labor, planting pine represented not so much a shrewd business choice to move from cereal cultivation to the more sustainable and profitable pine as a tool that allowed landowners to shed their laborers. In the area around Concepción in 1964, estate workers complained that Lota's Forestal Colcura was planting pine on land suitable for cultivating wheat: "the company is not fulfilling its promise to sell 40 kilos of wheat monthly to each of its workers in this city [Curanilahue]. It sells them only 20 kilos . . . claiming scarcity of wheat. But curiously, this company is planting pine on the fundos it owns whose land is good for wheat production. This is what happened on 400 hectares of the Los Ríos fundo, where [the company] planted pine last year. The campesinos on [the] El Descabezado, Quilachanquin, and Los Rios [fundos] have resolved to oppose the company's continuing to plant [pine] on land suitable for wheat."[25] The same process occurred farther south. In 1959, the owners of the Chacamo fundo in Nueva Imperial, Cautín Province, fired the majority of their inquilinos, many of whom were born on the fundo and others who had worked the land there for more than twenty years. *El Siglo* reported that "the administrator has given as a reason for firing the people that the land will be reforested and dedicated to pine plantations, and thus [the fundo] will no longer need inquilinos to cultivate and plant the land."[26]

The State, Economic Development, and the Forestry Industry

Neither forestry companies nor the Chilean state was satisfied with pine's ascendance during the late 1950s and early 1960s. Pine plantations could replace native woods in domestic markets, but this was a far less profitable use for their primary material than that for which planters, who had viewed pine as the basis for a new export industry in pulp and paper products, had hoped. State officials also saw pine both as the basis of new industries that could reduce massive rural unemployment and a valuable new export commodity. In this context, the Corporación de Fomento de la Producción (National Development Corporation; CORFO), with the support of international development organizations, embarked on an ambitious program to develop paper and pulp industries that would provide a market for the thousands of hectares of pine planted over the years and restore demand for wood to the mills that worked native forests, thus driving up prices for both native woods and pine.

An important ecological and economic logic drove the push to develop forestry industries during the 1950s. Chile now had 200,000 hectares of pine plantations that were maturing. The owners of these plantations, many who came from Chile's economic and political elite, needed markets as their trees reached harvestable age. In addition, in a number of cases, urban white-collar workers had sunk their retirement funds into pine plantations with the promise of pensions that would defy inflation, as in the case of the parcelas-bosques projects pioneered by Larraín and Yrarrázaval and the numerous forestry estates held by the Caja de Empleados Públicos y Periodistas (Public Employees and Journalists Retirement Fund). If pine earned only minimal prices on domestic wood markets, these "investors" could lose their pensions, and the government would be confronted with a serious social problem among an important political constituency, the urban middle class. The plantations' maturation made developing cellulose and pulp plants the logical and imperative next step.

Yet this was not a task Chilean landowners and investors were willing to take up on their own. Instead, they looked to the state for assistance in building new industrial enterprises to which they could sell their trees, just as they had looked to the state for tax exemptions and subsidies to forest their land. By the 1950s, planners at CORFO, timber companies, and international development agencies, including the United Nations Food and Agriculture Organization (FAO), the United Nations Economic Commission for Latin America (CEPAL), and the International Monetary Fund (IMF), had begun to advocate the construction of cellulose and pulp industries in Chile that, as one economic analysis put it, would "be the most important source of future earnings from foreign trade."[27] During the early 1950s, CORFO began to follow the recommendations of the US Forest Service, the FAO, CEPAL, and the IMF by making plans to build new cellulose and paper plants directed toward export markets, particularly in Latin America, where, as CEPAL reports underlined, demand was mushrooming without any corresponding increase in regional production.[28]

The forest policy proposed by the US Forest Service's mission in 1946 and then reproduced in CORFO's six-year plan was reinforced by the activities of the FAO, which held its forestry conference in Santiago in 1950. The following year, E. I. Kotok, a former chief of research with the US Forest Service, visited Chile as the head of the FAO's Forestry Mission Technical Assistance group. That year, the Ministry of Land and Colonization allotted 506 hectares within the Llancura Forest Reserve, in Valdivia Province, to the University of Chile for "experimental exploitation and research in

silviculture through the School of Agriculture."[29] The goal was to introduce modern forestry machinery, funded by the FAO, to investigate the
most efficient and productive methods of exploiting Chile's native forests.

In 1952, the FAO established new assistance programs in Chile designed
to promote education in silviculture with the University of Chile. The FAO
provided professors to the university's forestry program, a director for the
research center in the Llancura reserve, and equipment for a demonstra-
tion sawmill and logging project.[30] The agency aimed to train a new generation of Chilean foresters among students from the University of Chile,
Temuco's Technical School, and workers and foremen employed by private
forestry companies.[31] It worked with the University of Chile, the Ministry
of Agriculture, and CORFO to establish experimental forestry stations in
Valdivia Province.[32] The University of Chile, with support from the FAO,
produced the foresters who would provide the technical management of
private forestry enterprises. In 1954, *Chile Maderero* reported optimistically
that "in two or three years the young professionals from the University of
Chile's Forestry School will be able to lend their services to the forestry
and timber companies, which will help solve the problem of the lack of
responsible and superior, technically prepared personnel."[33] This served as
yet another subsidy to the industry supplied by the state and international
development organizations.

The FAO and CEPAL employed their technical assistance programs to
promote pine plantations and the paper and pulp industries as motors of
economic development. They performed studies in the early 1950s to provide a blueprint for forestry development. These studies underlined that
world demand for cellulose would grow rapidly over the following decades
and argued that by 1965, Chile could take advantage of its 200,000 hectares
of Monterey pine plantations to become one of the continent's largest cellulose producers. As an FAO study put it, "This enormous potential, united
with the easy accessibility of the plantations, as well as the privileged conditions in which *pino insignis* reproduces, [makes Chile] probably the most
important country in Latin America in terms of immediate possibilities for
great developments in the paper and cellulose industries."[34]

In 1953, the United Nations Paper and Cellulose Advisory Group sent a
mission to Chile, which, with CORFO, prepared a new report emphasizing
the urgency of developing the paper and pulp industries. The development
of these industries, the report argued, was the "only way to use and even
save the value of the tree plantations in the central-southern zone of the
country." Twenty percent of the privately held pine plantations belonged
to "small investors from the cajas," and there would be a "huge impact in
the country if the plantations [were] lost."[35] During the 1950s, this international support led to the organization of Chile's first major project for

building a cellulose plant, the Empresa Nacional de Celulosa, with the financial backing of several government organizations, including CORFO, which served as the guarantor of loans from the Export-Import Bank and the North American company Parsons and Whitmore. The project was based on studies done by the FAO, CEPAL, and CORFO and involved leading figures among Chile's political and landholding elite, including the vice-president of the Banco de Chile; former President Gabriel González Videla; and Juan Echavarri, the head of the Corporación de Madera and the owner of the Neltume estate and plywood factory near Panguipulli.[36]

Empresa Nacional de Celulosa did not, however, get off the ground. Instead, the Compañía Manufacturera de Papeles y Cartones (CMPC) took on the role of building the nation's first important cellulose plant, employing studies performed by CEPAL, the FAO, and the IMF. The reasons given were that, despite loan guarantees from the Export-Import Bank and CORFO, the project would have difficulty acquiring financing and competing with the CMPC. CORFO's board, which was stacked with Alessandri's supporters, contended that La Papalera would have an easier time obtaining credits from international agencies.[37] Thus, the CMPC, with the security of a loan guarantee from CORFO and loans from the International Bank for Reconstruction and Development, built a cellulose plant in Laja in 1959 and a modern newspaper plant in San Pedro, Concepción, in 1958. It advertised that it was providing markets for the abundant pine plantations around Concepción and heralded its own entrepreneurial drive in constructing the infrastructure of Chile's modern forest industry.[38]

It is unclear why the CORFO project did not materialize even as the agency lent its support to the development of La Papelera's cellulose plant. In 1958, Jorge Alessandri, a former director of the CMPC, was elected president of Chile. Once in office, he implemented a package of economic austerity measures in line with Klein-Saks's recommendation that sharply reduced state spending and sought to limit the role of the state in the development of industries. For Alessandri's right-wing government, an alliance of the Conservative and Liberal parties, CORFO's role was to support private enterprise with credits, as in the Laja project, not to develop industries itself, even in partnership with foreign or domestic private capital. It is also true that halting CORFO's cellulose project reinforced the CMPC's monopoly position in the pulp and paper markets. In fact, this monopoly allowed La Papelera to drive down prices for raw materials, an important subsidy for its newly installed industrial operations.[39]

For a number of years, the CMPC was able to rely on a steady source of inexpensive pine for its paper and pulp factories, since it faced no competition in national markets. This meant, too, that the crisis of the wood industry brought about by low prices for both pine and native woods, which had thrown thousands of forestry workers out of work during the late 1950s, was caused, at least in part, by the CMPC's monopolistic position in the wood markets, a position bolstered by the fact that it enjoyed a close connection to President Alessandri. The CMPC's monopoly of paper and pulp production was combined with access to inexpensive labor due to the massive unemployment in the forestry sector during these years. The security granted by the availability of cheap pine and low-paid workers, loans backed by CORFO, and the support of political allies such as Alessandri made La Papelera's new paper and pulp projects highly profitable and the company the dominant player in the forestry industry.[40]

Christian Democratic Agrarian Reform, Forests, and the Forestry Industry

In the presidential elections of 1964, Eduardo Frei of the Christian Democratic Party won an overwhelming victory over Salvador Allende's Frente de Acción Popular (Popular Action Front; FRAP) coalition of leftist parties, thanks to the support of a coalition of right-wing parties. Frei offered an important, if vaguely defined, social reformist alternative he termed "revolution in liberty" to the conservative policies of his predecessors, Ibáñez and Alessandri, on the right, and the revolutionary program of Allende and FRAP on the left. Most important, Frei's platform called for an expansion of agrarian reform that had been introduced reluctantly and in an extremely limited manner under Alessandri. The Christian Democratic government sought to break up the large estates that dominated the Chilean countryside and redistribute the land to their resident workers. The goal was to build an obstacle to social revolution in the countryside by creating a large population of smallholder farmers. In addition, land redistribution would create an expanded internal market that could stimulate industrialization and economic diversification. Most important, breaking up inefficient estates where production stagnated would give a boost to sorely needed agricultural modernization.

The statistics on landholding in Chile exposed vast inequalities that had impeded agricultural development. Nationwide, 7.5 percent of rural properties (*latifundios*) controlled 84.4 percent of agricultural land by value, while small properties (*minifundios*), composing 84.4 percent of all properties, owned only 11 percent of the value of agricultural land. Four hundred landowners possessed more than 20 percent of land in terms of value, and 1,600 landowners possessed 40 percent.[41] The Agrarian Reform Law called for the expropriation of unproductive and inefficient large estates of more than 80 hectares of basic irrigated land and distribution of the land to the estates' inquilinos. Estate owners were allowed to retain possession of their machinery and livestock, as well as a reserve on expropriated estates of up to 80 hectares of basic irrigated land. In addition, landowners received payment, determined in relation to the productivity of the estate, of a certain percentage of the estate's value, from 1 percent to 10 percent, with the rest paid in quotas by government bonds over twenty-five to thirty years. Estates also could be expropriated if the social conditions of their workers were demonstrated to be subpar. Estates that did not pay regular wages, provide decent housing and access to health care, or pay workers' social-security contributions could be expropriated, as could estates that were inefficient in terms of agricultural production. The campesinos who received estate land would be organized in *asentamientos* (settlements) for three to five years, after which the land would be subdivided, and they could sell their plots, work them individually, organize the land into a cooperative of the *asentados* (members of the asentamiento), or work them in a mixture of private ownership and a cooperative.[42]

The Frei government understood the problems of land redistribution in southern Chile to be linked to the problems caused by the destruction of the native forests. The Christian Democrats saw in forestation and the development of modern forest industries the solution to the south's entwined ecological and social crises. Frei's agrarian reform substituted the development of forestry industries and forestation, as well as state support for the scientifically managed forests on privately held land, for land redistribution. The Christian Democratic government viewed pine plantations as an essential component of both agricultural rejuvenation and industrialization in southern Chile. Agrarian reform thus included inducements for forestation that built on the subsidies encoded in previous forest legislation. Most important, the Agrarian Reform Law excluded from expropriation all fundos forestales—that is, property that was engaged in productive forestry, was reforested or engaged in reforesting, or that was native forest worked in accordance with a "forest exploitation plan" approved by the Ministry of Agriculture. The Agrarian Reform Law provided a significant incentive to southern landowners to both reforest their properties and engage in more sustainable exploitation of native forests, supervised by state foresters.[43]

While Frei promised to redistribute land to 100,000 campesino families, by 1970 only 28,000 had received land. The government expropriated 1,300 estates comprising 3.4 million hectares, which represented only 14.5 percent of Chile's productive agricultural land. The vast majority of the campesinos who benefited from the reform were inquilinos who occupied estates now as asentamientos. *Afuerinos* (nonresident estate laborers) or peones (landless seasonal laborers) received few benefits from agrarian reform and were often contracted by asentamiento or asentados in a new rural hierarchy in which asentados, who were male heads of households, became the patrones of landless campesinos.[44] In the forestry sector, many rural laborers worked seasonally and therefore did not receive land during this period. In addition, because forested and forestry properties were exempted from expropriation, neither inquilinos nor landless laborers who worked in sawmills and forests on large estates received land.

A report on agrarian reform by CORA in 1969 noted that "approximately two-thirds of the useful land in Chile was suitable for forestry" and that "on this enormous surface live and work nearly half of the country's campesino population." The working and living conditions of these campesinos were "worse than those of the campesinos who work in the agricultural sector, since these are regions that are farther from urban centers, and generally they labor in inferior socioeconomic conditions." Yet only agricultural properties had been included in agrarian reform because agrarian reform officials and foresters viewed campesinos as unable to manage forests and forestation: "the discovery of the forest sector has been the great surprise of agrarian reform. It was not included in the initial plans. And many professional foresters maintain the impossibility of agrarian reform on forested land or in forests. It has been necessary to create a new schema." In 1969, CORA began to initiate reforestation projects on asentamientos, with "excellent results." Still, CORA reported, including the forestry sector in agrarian reform remained "a task that remains to be done."[45] By 1970, large estates, many with tens of thousands of hectares of native forests, as well as large estates covered with pine plantations in the region around Concepción, Bío Bío, and Arauco, still had not been touched by the agrarian reform.

The Christian Democratic agrarian reform did, however, extend the state's administrative apparatuses, collected in new institutions such as CORA, the Instituto Forestal (INFOR), the Servicio Agrícola y Ganadero (Agriculture and Livestock Service; SAG), and the Instituto de Desarrollo Agropecuario (Institute for Agricultural and Livestock Development; INDAP) to regulate the exploitation of forests on large estates. In exchange, landowners received technical advice and support from INFOR and INDAP, including the establishment of tree nurseries for forestation on privately held land. The Instituto Forestal began to inventory the forest resources on many large estates in southern Chile, and the estates began, for the first time, to submit forest management plans, designed by forestry technicians, to the Ministry of Agriculture for approval. This was an essential condition for demonstrating that an estate was well managed and that its logging operations were guided by forestry science, an important means of staving off expropriation and acquiring technical aid and subsidies.

In the case of Trafún near Panguipulli, for example, the estate's owners, the Kunstmann brothers, obtained certification from the Ministry of Agriculture for a forest management plan they had developed with the assistance of a professional forester. In addition, the Kunstmanns entered into agreements with INDAP to experiment with reforestation with both native and exotic species, including raulí, North American pine, and eucalyptus.[46] Similar plans were certified by the Ministry of Agriculture and SAG in the case of the 19,153-hectare Galletué estate in Alto Bío Bío. The estate elaborated reforestation and forest management plans for its logging operations in araucaria forests and was thus able to prevent expropriation by the Frei government, despite its long-standing land conflicts with neighboring Mapuche communities and its workers.[47] The Fressard company similarly invoked its experiments with reforesting with Monterey pine and Douglas fir, as well as its work with technicians from the Ministry of Agriculture's tree nursery in the Malalcahuello Forest Reserve, to defend itself against the expropriation of a number of estates in Lonquimay, including the Alaska, Lolén, and Chilpaco estates, where it logged araucaria pine forests.[48] The Ralco logging company, which owned a number of estates in Bío Bío covered with araucaria forests also developed a forest management plan approved by the Ministry of Agriculture during the late 1960s that determined exactly which araucaria stands the company would be allowed to log and which were restricted in the name of conservation. Ralco also elaborated plans to reforest with Monterey pine.[49] The Ralco, Fressard, and Galletué companies thus continued a long tradition, initiated in the Nahuelbuta coastal cordillera by companies such as Oelckers and BIMA, of reforesting areas where they logged stands of araucaria with North American pine. In these cases, agrarian reform policies protected landowners who had decimated Chile's native forests while promoting the substitution of native forests with plantations of exotic species.[50]

In addition to leaving large forestry estates intact, the Agrarian Reform Law included no provision to restitute land illegally usurped from indigenous communities.[51] Because the law focused on redistributing land to inquilinos, it necessarily excluded many Mapuches who resided within reducciones and often worked only seasonally or temporarily on estates to supplement their harvests. In total, the Frei government expropriated only twenty-five properties, with a total surface area of 20,598 hectares, for redistribution to Mapuche communities.[52] As Martín Correa, Raúl Molina, and Nancy Yáñez have demonstrated, these properties were transferred to CORA from CORFO, which owned them; were offered voluntarily by landowners who sought compensation; or were abandoned or very badly exploited. With few exceptions, the Frei government failed to restitute land to Mapuche communities as communities. Rather, Mapuches were settled as individuals on asentamientos on expropriated estates.[53] Agrarian reform thus failed to meet the demands of Mapuche communities that were engaged in long-standing conflicts with forestry estates, as in Lonquimay and Panguipulli. Regions where logging native forests was a dominant economic activity, usually in the Andes and coastal cordilleras, also had large Mapuche populations that had confronted land usurpations for decades. Nonetheless, these regions remained outside the agrarian reform.

The area surrounding the town of Lonquimay reflected the limits of the Frei government's agrarian reform. In 1965, a number of Mapuche-Pehuenche lonkos from the zone traveled to Santiago to meet with authorities from the Forest Department. They decried the fact that logging companies were cutting down araucaria pine forests that they relied on for subsistence. When they returned, the lonkos organized a large meeting with hundreds of members of the region's reducciones. According to an account by Rafael Elizalde, one lonko addressed the meeting: "together we must do everything possible to force these señores who are cutting down the *pehuenes* [araucaria trees] that God gave us for our sustenance, to leave the region, and we must not sell them a single tree, because after logging we will no longer have piñones, and our children will suffer. . . . These forests are what allow us to raise our animals. Without our forests, the soil will dry out." Elizalde reported that, despite reassurances from the authorities in the Forest Department, the conflict continued until 1970 because the intendant of Bío Bío Province, appointed by the Frei government, authorized logging in the araucaria forests: "this time Pedro Maiheun, the cacique of Tracalhue, has denounced the loggers because they are going to destroy the essential sustenance of his community, the flour and piñón of 200 families. There are no other cereals."[54] Authorizing logging in Alto Bío Bío's araucaria forests was consistent with Christian Democratic forestry policy, which imposed new requirements on timber companies and encouraged reforestation but did not ban logging endangered species such as araucaria.

In 1969, the head of the regional office of the Dirección General de Bienes Nacionales (Directorate of Land and National Resources) wrote a lengthy report on conflicts between a number of reducciones and landowners in Bío Bío. He had recently returned from a trip to the region, accompanied by agronomists from SAG, an inspector from the Dirección de Asuntos Indígenas (Directorate of Indigenous Affairs; DISAN), a representative from INDAP, and the sub-delegate from Santa Barbara, to investigate "situations of litigation, arbitrary land occupations, that are produced in this place among the indigenous, Chilean colonos, and logging companies." At stake were a number of hectares of land covered with araucaria and coigüe forests and claimed by both the 22,000-hectare Queco estate and the neighboring Trapa Trapa reducción of Antonio Canío. In 1969, "ex-inquilinos and various Mapuches" occupied part of Queco and demanded the expropriation and redistribution of the land they occupied. Unlike in other cases in which Mapuche and Chilean ocupantes banded together in common opposition to large estates, in this case, the two groups' competing claims led to animosity. The Trapa Trapa community demanded restitution of land usurped by both the estate and the colonos settled in the region. A militant group of community members led by one of its lonkos, José María Tranamil, had invaded part of the Queco fundo, occupied the administration's houses, and set up huts (*ranchas*) alongside the twelve houses of the former inquilinos (who were now ocupantes or aspiring colonos). Members of Trapa Trapa also closed off access to the pasture on one of the fundo's veranadas to the "Chilean colonos (ex-inquilinos)."[55]

As Martín Correa and Raúl Molina have documented, the neighboring 28,583-hectare Ralco estate, which was owned by the Ralco logging and livestock company, was embroiled in a similar conflict with some of the region's reducciones, particularly the Andrés Gallina, Callaqui, and Ralco-Lepoy communities. As with the case of the Trapa Trapa community, the conflict was primarily over veranada pastureland, as well as the landowners' logging operations. The communities sought access to veranadas and forests that putatively lay within the borders of the giant estate. The Gallina community denounced the Ralco company's destruction of stands of coigüe that it claimed lay within the reducción. The Callaqui community had occupied veranadas claimed by the owner of the Ralco, Bío Bío (former Callaqui) and Pitrillín estates, Dionisio González, and the La Leonera forestry company. While La Leonera logged the araucaria forests, it allowed members of the community to pasture their animals on veranadas within the borders of the estate, a common arrangement during the 1940s and 1950s in the region. When González sold the estates in 1962 to Ralco, the company intensified its exploitation of the forests using fire to burn underbrush and clear cutting stands of araucaria that the Callaqui community depended on to collect piñones. In addition, the Ralco company brought in hundreds of workers, provoking increased conflicts with the Callaqui community over access to pasture and forests. The Ralco-Lepoy community (discussed in chapter 3) similarly claimed land covered with araucaria pine forests that it used to collect piñones that was also included within the putative borders of the Ralco estate.[56]

These cases reflected a general trend throughout the Andes cordillera in the Lonquimay region. As forestry development picked up steam and logging operations intensified during the 1950s, loggers abrogated traditional arrangements and temporary agreements between estates and Mapuche communities; communities lost land to which they had enjoyed customary access. In almost all cases, the process was defined by the ambiguities in the titles and property boundaries that separated estates from indigenous communities. Before the 1950s, these ambiguities had allowed estate owners to usurp indigenous land. But because the estates were primarily devoted to raising livestock, not logging, they often permitted Mapuche reducciones to occupy veranadas or to collect piñones, sometimes for a fee.[57] In addition, Mapuche communities, such as the Ralco-Lepoy community, continually contested estates' claims to forests and pasture, occupying and settling on land they considered part of reducciones, theirs by right of use, or simply public. With the growth of a profitable forestry industry in the native forests during the 1950s, however, estates such as Ralco shifted their focus from pasturing livestock to logging stands of araucaria and coigüe, now more valuable and accessible because of the establishment of a basic infrastructure of roads.

In response to the Gallina community's demands, the government inspector obtained permission from Ralco's owners for the communities to pasture their animals in the veranadas, implicitly recognizing the estate's ownership of the land. The inspector also came down on the side of the Ralco company by recommending that the felling of the coigüe forest be allowed but under the supervision of inspectors from the Forest Department, who would designate which trees were appropriate for logging. When it came to the Ralco-Lepoy community, however, the inspector was far less supportive of the company's rights. In his report, he gave a long history of the community and the formation of the estate, whose origins lay in two apparent land sales from Pehuenche lonkos in 1881 toward the end of the final military campaigns in the region. As the report underlined, until the 1950s, despite the sale of land they believed to be their own, the members of the Ralco community continued in their possession of the land: "the indigenous lived peacefully on land of which they were the owners."[58]

This state of affairs lasted until González, who had purchased the estate from the Bunster family, sold the estate to the Ralco company. In the early 1960s, Ralco began serious logging operations, employing modern sawmills and machinery, and brought in hundreds of non-Mapuche workers. Ralco extended its logging operations onto a veranada at the confluence of the Bío Bío and Ralco rivers that the Ralco-Lepoy community had occupied for decades. As a map revealed, the forests and pasture lay both within the borders of the community and within the Ralco estate, a situation of superimposed borders that was not uncommon in the south, where land auctions and sales had often included land already granted to or occupied by indigenous communities. A government report noted that, while the forests lay within the boundaries of the Ralco estate, the members of the community also had an "indisputable right" to them. "We have the duty of remedying past injustices that," the report said, "owing to the hatred and spirit of revenge, [a] product of the War of the Araucanía, [made it possible for] a large number of indigenous families [to be] dispossessed [and allowed] a private party to become the owner of a large extension of land, like those who were granted encomiendas in the time of the colony."[59]

Yet despite this searing historical analysis of the formation of the Ralco company's "encomienda" in Bío Bío, the measures proposed in the report were moderate. Rather than expropriate and restitute land illegally usurped from the Mapuche-Pehuenche communities, the Land Ministry, consistent with the Christian Democratic government's policies, proposed to the director of the Ralco company that it "would establish conditions for the exploitation of the [araucaria pine] wood." It would also establish a capital fund derived from the firm's earnings that would be invested in the reducción. Exploitation of the forest would be overseen, as in the coigüe forest claimed by the Gallina community, by inspectors from the Forest Department, and the logged land would be reforested with Monterey pine. The foresters would determine how much of the firm's earnings would go into the fund to be invested back into the Ralco community.

The company roundly rejected this proposal because it seemed to accord some legal legitimacy to the Ralco community's land claim. But probably more important, although unstated, is that the government's plan called for strict state regulation of Ralco's logging operations. As an alternative, the logging company proposed providing each family in the community with 200 kilograms of wheat, building a new road to connect the community to a road that led from Ralco's mountain valleys across the border into Argentina, and building a school for the community. The Land Ministry representative, in turn, rejected the company's proposal as "paternalistic." The situation remained unresolved and was complicated by the more than three hundred workers who toiled in Ralco's sawmills and forests and who confronted unemployment were logging halted. This would create a social crisis, the report argued, that might be taken advantage of by "political extremists," whom it blamed for counseling and "exciting" the Ralco-Lepoy community to take more intransigent and militant positions.[60]

A year later, the situation in the region had only worsened. A group from the Trapa Trapa community had engaged in a violent confrontation with the former inquilinos/colonos on the Queco estate in which three Mapuches were killed. Unlike other cases of solidarity between Chilean ocupantes or colonos and Mapuches, in this case the competition for access to the mountain pasture had created intense hatred. As the Land Office representative noted, "After visiting the region and communicating with its different inhabitants, I was able to see that differences exist between the Mapuches and the colonos that have characteristics of racial antagonism." The Chilean colonos treated the Mapuches "disrespectfully and, most of the time, with arrogance and superiority." While former inquilinos and forestry workers who were now aspiring colonos on both the Ralco company's estates and the Queco estate in Alto Bío Bío sought work and access to forests they could log, the Mapuche-Pehuenche communities sought to conserve and maintain their access to the araucaria forests, and they opposed the estates' logging operations.[61]

Despite the obvious illegitimacy of the estate's property rights and the colonos' murdering of the three Mapuches, the Land Office representative refused to support the community's land claims. He proposed the "eradication" of the militant Trapa Trapa group, led by Tranamil, from the region because of the need to maintain social harmony. This was also a matter of national security, as the community lay on the border with Argentina and provided excellent access to the Bío Bío region through the Trapa Trapa and Queco valleys, of which "extremist groups" could take advantage. The Land Office representative argued that it was necessary to have the region settled with "a group of colonos who are 100 percent Chilean, a quality that, unfortunately, we have not been able to inculcate in the Mapuche." He also proposed expropriating Queco (22,000 hectares), Trapa-Trapa (12,000 hectares), Ralco (33,932 hectares), and a smaller estate, Los Chenques (3,600 hectares), not to restitute land to the region's Mapuche communities but to settle it with "100 percent Chileans," presumably the forestry workers and neighboring colono or ocupante smallholders. Agrarian officials with Frei's government thus built on colonization policies that dated back decades by implementing agrarian reform designed to populate the region with non-Mapuche smallholders in the name of territorial sovereignty and national security.[62]

By the time the Frei government left office, none of the estates had been expropriated, and the conflicts among the logging companies, their workers, Mapuche communities, and colonos continued to simmer. Despite widespread denunciations of deforestation in the region, the Frei government also continued to allow logging in the araucaria forests, although with increased regulation and supervision by Forest Department officials. The Agrarian Reform Law made it clear that estates with native forests or that engaged in reforestation could not be expropriated. Ralco, like other forestry companies, represented itself as a follower of modern forestry practices, in contrast to the stagnant agriculture sector: "in the forestry sector, the contrary of what occurs in the agricultural sector, intensive and indiscriminate exploitation, is against forestry laws. . . . [A] few years ago, the intendancy of Bío Bío Province prohibited the exploitation of araucaria pine on the aforementioned estates Ralco, Pitrilan, Bío Bío, etc., authorizing logging in only determined parts of the properties." Nonetheless, as a later government report underlined, the company continued to log araucaria stands on the Ralco estate and on land that belonged to the neighboring reducciones.[63]

Forestry Development and Agrarian Reform

In the place of agrarian reform in the southern forests, the Frei government, with support from international development agencies and foreign investors, intensified the push to establish pulp and paper plants that would provide a market for pine. The Christian Democrats' development plan reversed the conservative policies of Ibáñez and Alessandri by reasserting a strong state role in fomenting industrialization and economic modernization. In some ways, this constituted a return to the policies of the Popular Front and the Radical Party-led governments of the 1940s. The state once again assumed a central role in stimulating the diversification of Chilean agriculture and industry, although now with a more established and better-funded CORFO and an array of other agricultural and forestry development agencies, including INFOR, SAG, and INDAP. This developmentalist policy departed from the import substitution industrialization policies of the 1940s, however, by emphasizing a central role for foreign investment and export-oriented growth in new industrial and agricultural sectors. The paper and pulp industries were to be a signature case of Frei's developmentalist policies. The Christian Democratic government returned to CORFO's old plans to build a number of cellulose plants that would absorb the production of existing pine plantations and stimulate reforestation throughout the south. This constituted an effective challenge to the monopoly held by the CMPC over pulp and paper production. But, as government officials made clear, La Papelera would be able to maintain its dominance of national markets. New cellulose plants would produce pulp for export.

The goal of installing cellulose plants in the south was much the same as it had been when the Chilean state began to promote reforestation with pine during the 1930s and 1940s: to produce new industries and exports that could diversify an economy almost entirely dependent on copper revenues. Forestry and land officials frequently described pine and cellulose as Chile's new copper. While copper reserves were finite, they argued, pine plantations were a potentially renewable source of export earnings once processed in a modern industrial complex of paper and pulp plants. For the Christian Democratic Party, as for the left-center coalition governments of the 1940s, the development of the forestry industry in southern Chile would also serve as a solution to pressing social problems and conflicts by providing a steady source of jobs for the growing ranks of the rural underemployed and unemployed who did not receive land under agrarian reform.

Under Frei, CORFO engineered the development of two major cellulose plants in Arauco and Constitución. Celulosa Arauco, in many ways, was a reprisal of the Empresa Nacional de Celulosa project of the 1950s. The company was initiated in 1967, with 75 percent of its capital provided by CORFO and 25 percent from capital from the United States, the United Kingdom, and Australia, most from the North American company Parsons and Whitmore.[64] In 1969, CORFO also began the construction of a major cellulose company in Constitución to take advantage of the abundant pine plantations in the region and provide a dynamic pole of development in a region whose agricultural economy had stagnated. As it did for many CORFO projects, the state provided a large portion of the initial capital but solicited foreign investors—in this case, the French company ENSA. In addition, funds for the project were provided by the Catholic Church's Talca Development Foundation.[65]

The Chilean National Development Corporation also helped to install a paper plant in Nacimiento during the early years of the Frei government. Based on studies done by the FAO, the company Industrias Forestales (whose president was Julio Durán, a senator, the leader of the right wing of the Radical Party, and a presidential candidate in 1964) installed the paper plant in 1965.[66] Industrias Forestales had been organized in 1956 as a private company but was controlled by CORFO in 1964 when the agency purchased a majority of its stock.[67] Like the Christian Democratic Party's other forestry industry projects, the Nacimiento paper mill represented important diversification in an industry that since the beginning of the century had been dominated by the CMPC. Yet as was the case with La Papelera headed by Jorge Alessandri, the Nacimiento project represented the key role played by Chile's political elite in the forestry industry. Even with majority public ownership, Durán continued as the company's president.

The Christian Democratic government combined these forestry industrialization programs with an ambitious program of state-directed reforestation. Levels of reforestation had declined during the late 1950s and early 1960s because of the austerity measures implemented by Ibáñez and Alessandri and because of the constricting effect of the CMPC's monopoly on markets for pine. The Frei government, however, made recovery of Chile's eroded soil an essential component of its agrarian reform and industrialization policies. New plantations would supply the cellulose plants in Arauco and Constitución, as well as reclaim overworked land. In 1965, Frei inaugurated an ambitious reforestation campaign to be led by a new state-run Corporación de Reforestación (Reforestation Corporation) and the Ministry of Agriculture. The goal was to add 450,000 hectares of plantations over four years to the more than 200,000 already cultivated.[68] In

1966, 30 percent of new tree plantations were the work of the state, and Frei's government intended to increase that to more than 40 percent by 1970. The centerpiece of the reforestation campaign was a new Forest Law, passed in 1967, that provided credits and subsidies for landowners who cultivated tree plantations. The law granted a series of tax exemptions, including property, income, and inheritance taxes, for landowners who reforested their soil. In addition, landowners could take advantage of technical assistance from INDAP and credits from CORFO and the Banco del Estado to cover a significant portion of the costs of reforestation. Landowners received fifteen-year credits for the total costs of replanting from CORFO, which were interest-free for six years.[69]

Unlike the Forest Law of 1931 and previous state policies to stimulate forestation, the forest policies of the Frei government also aimed at small and medium-size landholders. Thus, the plan underlined that credits from CORFO "should be oriented not only to the large forestry producers, but also to the small and medium-size forestry properties . . . , since there are thousands of smallholders with land suitable for forestation who cannot reforest because of lack of access to credit."[70] As part of agrarian reform, the plan called for INDAP to organize small and medium-size landholders in cooperatives and committees "to make sustainable management of forests and forestation possible."[71] The institute also provided credits to smallholders for reforestation.[72] Alex Rudloff, a forester who worked with the Reforestation Corporation, renamed the Corporación Nacional Forestal (National Forestry Corporation; CONAF) in 1973, recalls that during the years of agrarian reform, and after as well, CONAF entered into agreements with numerous smallholders throughout the frontier region, from Malleco to Valdivia, to reforest their plots with pine.[73] In addition, CORA and CONAF initiated reforestation on asentamientos and state-owned reserves. In 1965, state-run tree nurseries distributed 135 million plants to landowners, including holders of small and medium-size plots.[74]

By late 1966, the Frei government's efforts to reignite forestation had achieved a measure of success.[75] In Cautín, for example, where there had been relatively few tree plantations, 9,000 hectares were planted, and a number of new tree nurseries had been organized to provide landowners with seedlings.[76] *El Diario Austral* reported that in the Department of Nueva Imperial, landowners were "planting pines in growing numbers." The head of the region's Forests Action Committee stated, "[The reforestation campaign] has gone to the heart of the people. In this campaign there is a lot of enthusiasm, accompanied by a seriousness of purpose, so there is no doubt that this will be the basis for the creation of a paper industry in the department in the not too distant future." The fever to plant pine was especially intense among Cautín's smallholders: "there is a true "pine psychosis" among committees of smallholders who have been the promoters of the reforestation plan."⁷⁷ Similarly, the paper reported that "it is important to note the enormous interest that the small and medium-size agriculturalists of the Department of Villarrica have shown in taking advantage of the opportunity presented to reforest parts of their farms."⁷⁸ The paper observed that in Villarrica, "Small producers participate actively in the reforestation plan: the Association of Small Producers in Villarrica possesses today an important tree nursery in Villarrica on 2 hectares [and has] planted and harvested 1 million *pino insignis* trees, 300,000 eucalyptus, [and] 500,00 cypress, . . . in cooperation with the Ministry of Agriculture's Forest Department in Villarrica."⁷⁹

In Loncoche, a reforestation program elaborated with the US Embassy and the US Agency for International Development established a number of new publicly owned tree nurseries to supply smallholders and Mapuche reducciones.⁸⁰ The US Peace Corps aided in this project, with thirty-eight volunteers working with smallholders to forest their parcels as part of the Frei government's reforestation plan.⁸¹ Some 2,500 smallholders, including Mapuches, in the Loncoche region participated, exhausting the supply of trees in Temuco's state tree nursery, and foresting more than 3,000 hectares by 1967.⁸² Jaime Tohá, the head of the Forest Department, noted that, more generally in the south, "It is surprising to see the importance that is placed by the many different levels of society on the problem of the conservation of natural resources. . . . It is especially interesting to see that the smallholders, above all in the coastal cordillera, have understood through their tragic experience that their soil is disappearing. . . . [T]hey are perfectly aware that reforestation is . . . the only solution that will allow them to conserve their soil. The immense majority of these landowners have decided to reforest."⁸³ The Frei government established *convenios* (agreements) with small and medium-size landholders to reforest their land, with most inputs, including seeds, plants, fertilizer, credits, and technical assistance for managing the plantations, provided by state. As Tohá noted, these convenios were widely accepted.

Under Frei, large forestry companies and landowners continued to enjoy subsidies and protections from the state, but now, for the first time, smallholders and former inquilinos on asentamientos also began to receive state support for forestation and forestry activity. Like the large landowners, they demonstrated their capacity for modern forms of production by actively pursuing forestation with Monterey pine on their small plots, belying the conventional image of campesino smallholders as hostile to forestry and the tree plantation economy. Between 1969 and 1978, when CONAF ended the program of convenios with smallholders, campesinos reforested 200,000 hectares throughout southern Chile.[84]

By 1969, Christian Democratic officials had also begun to approach the idea of incorporating Mapuche communities into programs to reforest regions with severely degraded soil. The intendant of Malleco Province laid out the government's basic philosophy, emphasizing that in his province, Mapuches were "generally more backward than in Cautín." To confront their "backwardness," the government initiated a program to expropriate badly run estates and organize cooperatives of Mapuche communities in Lumaco, including a forestation cooperative. The goal was to "establish factory farms where the Mapuche will be able to work different jobs as a specialized labor force that will train them for a number of productive activities and help them elevate their actual level of life." As during the 1940s, agrarian officials viewed forestry and forestation, along with other agro-industries, as a tool of uplift and development for southern campesinos. Organizing forestry cooperatives would transform "backward" Mapuches into proletarianized and skilled forestry workers, part of a booming pine plantation and industrial forestry economy in Lumaco.[85]

The Frei government, through CORFO, also established a number of public-private companies dedicated to cultivating tree plantations. In 1966, CORFO rescued a nearly bankrupt forestation company, Forestación Nacional (FORESTANAC), by purchasing a large number of its shares. In the eyes of the government, the company could not be allowed to fail because it had been established with the money of thousands of "small investors," much like the parcelas-bosques companies of the 1940s and 1950s. Along with the small investors who owned the majority of the company's shares, the CMPC, Forestal Mininco, and Industrias Forestales (INFORSA) held significant minority shares. The Chilean National Development Corporation took over control and administration of the company and its tree plantations, rescuing thousands of urban white-collar investors, as well as the investments of large forestry companies such as La Papelera.[86] In 1969, CORFO organized the Sociedad Forestal Arauco to purchase fundos with pine plantations and to purchase fundos in Arauco to cultivate pine to supply to Celulosa Arauco. Almost all of the capital was provided by CORFO. In 1967, CORFO acquired a number of forestry estates owned by Forestal Colcura, paying full price for almost 17,500 hectares covered with about 6,400 hectares of pine plantations.[87] In 1969 and 1970, CORFO purchased sixteen already forested fundos in Colcura, Lota, Lebu, Araucuo, and Curanilahue owned by Forestal Colcura and Forestal Cholguan. It also invested heavily in new tree plantations on these fundos.[88] In the case of Forestal Colcura, as with FORESTANAC, it appears that CORFO purchased the forestry estates to rescue the company from bankruptcy. At the time that its estates were purchased, Colucura had failed to pay its workers for a number of months and appeared to be in serious financial difficulty.[89]

In 1967, CORFO organized Forestal Pilpilco in partnership with private investors from the Pilpilco Coal Company, with the goal of reforesting rural properties and managing forested properties, as well as developing wood industries. Unlike Forestal Arauco, which was completely dedicated to planting pine for Celulosa Arauco, Forestal Pilpilco was designed to take advantage of Chile's native forests by providing a market primarily for "small producers," thus "fulfilling one of the objectives that was pursued in the relocation of the plant [to Cabrero]." The Chilean National Development Corporation provided two fundos in Lebu, as well as machinery and equipment, to Pilpilco. The goal was to resuscitate the small producers who worked native forests.[90] In 1969, Forestal Pilpilco and CORFO established Forestal Lebu, with the primary objective of reforesting rural properties in Arauco; CORFO transferred the Colicheu, Victoria, Camarón, Los Altos, and Monte Grande estates to the new company to reforest, as well as to manage the existing plantations. One of the goals of Forestal Lebu, as with Forestal Arauco and Forestal Pilpilco, was to create jobs in a region where rural unemployment reached into the double figures. Because forestry fundos were excluded from agrarian reform, Forestal Lebu purchased estates from landowners and then reforested them. As a company report described these plans in 1970, "Some time ago, our company was informed of the interest of the supreme government in promoting through CORFO a forestry program that would absorb the unemployed labor force." Forestal Lebu's goal for 1970 was to reforest 11,750 hectares in the basin of the Maule River in Constitución to establish a future supply of pine to Celulosa Constitución.[91]

By 1970, the state had reforested 50 percent of all plantations directly, and through indirect methods, such as subsidies, credits, and other forms of support, it was responsible for 50,000 of the 60,000 hectares planted with pine that year.[92] Between 1956 and 1964, about 70,000 hectares were cultivated with tree plantations nationally, an average of slightly under 8,000 hectares a year, a result of the maturing of the plantations begun during the 1930s and 1940s and the CMPC's monopolistic hold on markets for pine. Between 1965 and 1970, 40,000 hectares a year were planted, on average, for a total of 243,255 hectares—a rate five times that of the previous eight years under Ibáñez and Alessandri.[93] Thus, while the Frei government excluded forestry estates from agrarian reform, it invested heavily in building the forestry industry and in reforestation. In addition, its land and forests policies pushed forestry companies and landowners who owned forestry estates to modernize production by adhering to more "rational" forms of logging, overseen by foresters and agronomists from the Ministry of Agriculture, in native forests.

Frei's agrarian reform brought few improvements or changes for workers who toiled in Chile's native forests, particularly in the Andes cordillera. However, for workers in the regions dominated by pine plantations, mostly in the coastal cordillera in Arauco, Concepción, and Malleco, reforms introduced by the Christian Democratic government extended basic labor rights and benefits included in the Labor Code of 1931 to the countryside and forests. In 1967, the Christian Democratic government passed a new Campesino Unionization Law to facilitate the organization and funding of rural unions. The most important aspect of the law was its legalization of unions organized by commune. This allowed agricultural and forestry workers who labored on often dispersed properties to organize larger unions and to coordinate regional labor actions. During the late 1960s, forestry workers, especially in Concepción and Arauco, successfully organized a number of unions and wrested a series of benefits from landowners.

Forestry workers continued to make demands that reflected the persistence of the system of inquilinaje and a campesino economy, even within the boundaries of large estates devoted to logging and forestation with pine. Indeed, the Campesino Unionization Law allowed forestry workers to restore crucial elements of their subsistence economy by negotiating not only for wages and benefits, but also for access to land to pasture cattle and to cultivate the traditional chacra of vegetables, beans, and legumes. While workers on forestry estates did not receive land through agrarian reform, on a number of estates in the coastal cordillera central to Concepción and Arauco's developing complex of pine plantations and cellulose plants, they expanded their access to small plots through guarantees established in collective contracts with newly organized unions, often after militant strikes.

Before 1967, forestry workers' efforts to organize unions and strikes had met with little success. In 1955, for example, workers in Coronel on four fundos owned by the Sociedad Colcura organized a union and presented a petition to the company demanding wage increases, increases in the family allowance paid to them by the company, and the traditional regalías enjoyed by inquilinos, including pasture for their animals and firewood. The strike lasted seven days but ultimately was defeated. The leader of the union was fired, along with seventeen other workers; the striking workers' livestock was confiscated; and scabs were brought in from the town of Curanilahue.[94] Workers on the CMPC's forestry fundos nearby also organized unions during these years and went on strike, including a major strike in 1964 with the support of the coal miners' unions, with similar results. As in the case of Forestal Colucura, the CMPC fired the union leaders and activists who led the strikes and expelled them from its forestry estates.[95]

Under the Frei government, forestry workers' unionization campaigns experienced a significant change of fortune. The Campesino Unionization Law sparked a wave of successful labor organizing in the countryside during the late 1960s and early 1970s. An important example of the trajectory of the labor movement in the forestry sector is the case of the Buena Esperanza hacienda, in the coastal Nahuelbuta cordillera, a center of logging activity in southern araucaria forests. The establishment of rural unions in the region began in 1961 when eighteen inquilino families on a nearby estate, Mundo Nuevo, left after their struggle to win rights to small plots of land on the estate was defeated. A number of inquilinos moved to the Buena Esperanza estate, owned by the Public Employees and Journalists Retirement Fund. By 1964, Buena Esperanza's workers had established a settlement with 112 families and 263 workers and had begun to organize a union. While before 1967 rural unions had been restricted to individual fundos, giving landowners significant leverage over their workers, the new law established rural unions by municipality, allowing for the creation of regional campesino labor organizations that brought together numerous fundos. After 1967, workers on Buena Esperanza traveled around the area organizing workers on neighboring estates into the "Sindicato Nahuelbuta," with about five hundred members. The Sindicato Nahuelbuta then became part of the larger Federación Campesina Caupolicán, based in Cañete, and the left-wing Confederación Nacional Campesina e Indígena "Ránquil" (named for the Ránquil rebellion). By 1970, the Buena Esperanza union had won from the fund the infrastructure of a permanent settlement: a primary school, a medical clinic, transportation, electricity, a union hall, and a community center.[96]

The organization of the unions and the federation of inquilinos on estates dedicated to forestry production in the Nahuelbuta cordillera mirrored the development of the rural labor movement in Chile's southern provinces and forestry zones. Workers organized the first major union federation, the Federación de Trabajadores del Bosque "Liberación" (Federation of Forestry Workers "Liberation") in Concepción in 1967, shortly after Congress passed the Campesino Unionization Law.[97] In 1967, there were only twenty-four agricultural and forestry workers' unions in Chile because of serious obstacles imposed by labor law, including the restriction of unions to individual estates and a requirement that a significant percentage of union members know how to read and write. Campesinos and forestry workers had been organizing unions throughout southern Chile since the 1920s, with rural unionization campaigns especially strong in 1939 and 1946. But decrees first by President Pedro Aguirre Cerda in

1939 and then by President Gabriel González Videla in 1947 had reined in the campesinos' organizing drives.[98] The Campesino Unionization Law of 1967, however, opened the door for a resurgence of rural labor organization, recalling similar moments during the 1920s, 1930s, and 1940s.

By 1970, there were four hundred campesino unions in Chile. The Catholic Church-based "Liberación" federation grouped together five unions of forestry workers with twelve thousand members in the region surrounding Concepción. By 1970, union militants had founded an interregional federation that linked sixteen forestry workers' unions. Two years later, the federation joined with the campesino unions Los Valientes de Concepción and the national Christian Democratic Party confederation El Triunfo Campesino (formed with the support of INDAP officials) to present a common set of demands to 250 agricultural and forestry employers. The move was successful, and workers won a 42 percent wage increase, company-run health clinics, work security systems, education bonuses for their children, and bonuses for births, marriages, and funerals.[99]

Workers on state-owned forestry land also organized during this period. In early 1970, *Ránquil,* the bulletin of the Confederación "Ránquil," noted that one thousand workers in Arauco Province were engaged in strikes against forestry companies, including Forestal Pilpilco, as well as against forestry estates owned by CORFO and para-state organizations such as the Public Employees and Journalists Retirement Fund, including the fund's Buena Esperanza estate in Curanilahue.[100] In April 1970, 2,500 agricultural and forestry workers in Concepción engaged in a massive strike to demand increases in their minimum wages. The workers paralyzed production on eighty-two estates, including the CMPC's Los Pinares fundo.[101] In this case, forestry workers were more successful. Bolstered by the territorial scale of their organizations and sympathetic labor authorities, they achieved both recognition of their unions and wage increases. Organizing by *comuna* (commune), rather than by fundo, prevented employers from engaging in the union-busting tactics of the pre-1967 era and placed enormous pressure on the government to intervene in the strikes. As forestry fundos increasingly were purchased by the state as part of reforestation programs, workers now also confronted not private employers and subcontractors (as on the CMPC estates) but publicly administered enterprises. This also strengthened their hand during the last years of the Christian Democratic government and the socialist government of Salvador Allende.

A contract in 1973 between workers on the Peñuelas estate in Concepción represented by the Federación Liberación and the Reforestation Corporation, which administered the estate, reflected the tremendous gains made by forestry workers during the late 1960s and early 1970s. The workers won a series of rights and benefits that would have been unimaginable a decade before. They included a cuadra of land that they could dedicate to either cultivating vegetables or pasturing livestock, a salary increase, the establishment of a first-aid post, improvements for a schoolhouse they had won in a contract two years before, a tractor to provide transportation for workers, and windows for new houses the workers had won as they began to establish the infrastructure of a permanent settlement. A year earlier, a similar contract between Forestal Lebu and workers on the Peñuelas and Monte Águila fundos demonstrated how far forestry workers had come since 1967. The union won wage increases and a family allowance for (male) heads of households. In addition, workers received plots for their gardens, pasture for three animals, and rights to collect firewood in the forests.[102]

While forestry development before the agrarian reform often led to the expulsion of estate laborers from rural estates to free up land for pine, during the Frei government, many campesinos were able to increase their access to land even in regions where reforestation programs were under way. Despite the accelerated expansion of pine plantations, unionization and labor contracts guaranteed workers on estates planted with pine small plots that were fundamental to the maintenance of a peasant economy. Pine plantations shared space within the borders of forestry estates with land on which campesinos cultivated garden crops and pastured livestock.[103] In addition, by directing credits, subsidies, and technical assistance to small and medium-size producers, the Christian Democratic government made it possible for campesinos to plant pine themselves and to see in pine a viable alternative to the cereal crops and livestock that had depleted the soil on their small plots. Finally, the wages and benefits guaranteed in forestry workers' contracts following passage of the Campesino Unionization Law in 1967 established the basis for the settlement of a notoriously transient landless labor force in stable communities. Health clinics, schools, and decent wages and benefits allowed forestry workers to build rural communities. Forestry development and the rapid expansion of pine plantations during these years did not, then, lead to campesinos' dispossession. Rather, it made it possible for smallholders to plant pine on their eroded land and to mix stands of pine with pasture, garden crops, and commercial crops such as wheat. In addition, for landless laborers, forestry development created full-time jobs on forestry estates with a series of benefits denied to workers who traveled the south laboring for short stints in native forests and facing chronic underemployment.

In the frontier zones that composed the core of the pine plantation-forestry industry complex, mainly in Concepción, Bío Bío, and Arauco, the Chilean state's goal to build a settled population of trained worker citizens and drive regional development by fomenting commercial forestry production appeared to be bearing fruit. Yet as Frei left office in 1970, agrarian reform had yet to reach campesinos in the Andes cordillera in places such as Lonquimay, Alto Bío Bío, and Panguipulli, where forestry estates remained intact and workers labored in conditions that had changed little since the beginning of the century. Bringing the land and labor reforms initiated by the Christian Democrats to these regions would be the task taken up by the socialist Unidad Popular (Popular Unity; UP) government of Salvador Allende between 1970 and 1973. The architects of the UP's forest and agrarian reform policies confronted the still open question of how to meet southern campesinos' increasingly militant demands to expand agrarian reform to include the forestry sector while producing forestry industrialization and introducing sustainable methods of exploitation in Chile's native forests. This was a question Frei's agrarian reform officials had avoided, operating instead under the assumption that state-directed forestry development, managed by trained foresters, would combat poverty by generating jobs in forestry and produce ecologically sustainable growth based on replacing native forests with pine plantations.

1

Agrarian Reform Arrives in the Forests

In November 1970, weeks after President Salvador Allende of the Socialist Party took office, workers on a number of large forestry estates in Panguipulli went on strike. When their petitions for wage increases and improved working conditions went unanswered, they occupied the fundos and demanded their expropriation. The land invasions (*tomas*) in Panguipulli forced the issue of agrarian reform in Chile's native forests on the socialist government. Valdivia's Andes cordillera had been the site of land and labor conflicts dating back decades, and the forestry workers saw in Allende's election an opportunity to find stable work through the expropriation of estates whose dominion over thousands of hectares of forests they viewed as illegitimate. A few months later, Mapuche-Pehuenche communities farther north in the cordillera in Alto Bío Bío also invaded forestry estates to demand the restitution of land usurped since the beginning of the twentieth century and an end to logging in araucaria forests. Like the workers in Panguipulli, the communities had been embroiled in long-standing conflicts with forestry companies and had been left out of the Christian Democratic agrarian reform.

Allende came to the presidency at the head of a coalition government, the Unidad Popular (Popular Unity; UP), which included his own Socialist Party, as well as the Communist and Radical parties and smaller "leftist" Christian parties. The UP proposed to follow a "Chilean road to socialism"—that is, a socialist economy built within the framework of Chile's democratic institutions. The UP program emphasized that Chile's socialist revolution would be constructed gradually and legally, without rupture or revolutionary violence, a stance opposed on the left by the Guevarist-inspired Movimiento de Izquierda Revolucionaria (Revolutionary Left Movement), as well as by radicalized sectors of the Socialist Party, which sought to produce revolution by provoking a crisis in the state.[1] The program called for a series of incremental structural reforms, including the creation of a social property area of major industries that were either deemed monopolies or strategic to the economy. The UP also pledged to intensify the agrarian reform initiated by President Eduardo Frei. To consolidate support from Chile's middle class, which was essential to the UP's political strategy because Allende had been elected by a slim margin and with only a plurality of the vote, the UP's plan for building a socialist economy left small and medium-size businesses and small-scale agricultural properties in private hands. The goal was to create a mixed economy in which the "commanding heights" of the economy—industry, mining, and finance—belonged to the social property area. In the countryside, large and unproductive expropriated estates would be organized in peasant cooperatives and, in some cases, agricultural units known as Agrarian Reform Centers run by local peasant committees. While the UP planned to intensify and expand the agrarian reform introduced under Frei, it did not propose revolutionary land redistribution. Rather, estates would be expropriated and land distributed to campesinos according to the existing laws and through the institutions established during the 1960s, primarily the Corporación de la Reforma Agraria (Agrarian Reform Corporation; CORA) and the Corporación de Fomento de la Producción de Chile (Chilean National Development Corporation; CORFO).

The forestry sector typified the economic problems confronting Chile in 1970. Large estates held most of Chile's stock of remaining native forests and continued to engage in extractive forestry practices without any attention to modern systems of forest management or to the regeneration of the forests. Forest mining threatened Chile's temperate forests with extinction and undermined the sustainable development of southern Chile, the country's poorest region. In addition, the monopoly held by the Compañía Manufacturera de Papeles y Cartones (CMPC) and other forestry conglomerates had produced a bottleneck in forestry development. The Frei government had begun to undermine this monopoly by investing heavily in new forestry companies such as Forestal Arauco and Celulosa Arauco. The Christian Democrats also successfully reactivated the pine plantation economy with significant doses of public investment after more than a decade of stagnation under the conservative governments of President Carlos Ibáñez and President Jorge Alessandri. Nonetheless, in regions dominated by native forests, large estates continued to control forests of valuable native species such as raulí and araucaria pine and to engage in logging with little oversight. Despite the development of new, modern industries and pine plantations in some areas, particularly Concepción, Bío Bío, and Arauco, logging in Chile's native forests continued to be done much as it had at the beginning of the century, with low levels of technology and little application of forestry science. Sustainable forestry continued to be defined in terms of replacing native forests with Monterey pine plantations. The problem confronting Allende's government was how to implement a strategy of forest management that both guaranteed the regeneration of the native forests and provided solutions to the rampant problems of rural poverty and land inequality in southern Chile. At the same time, the UP had to contend with a growing campesino movement for land.

Revolution from Below and Industrial Forestry in Panguipulli

During its first few months in power, the UP government expropriated as many estates as the Frei government had expropriated during its entire tenure. In part, the acceleration of agrarian reform came as a response to what Peter Winn has termed "a revolution from below."[2] Beginning during the last months of the Christian Democratic government, campesinos and Mapuche communities in southern Chile initiated a wave of land invasions, demanding both that usurped land be restored to reducciones and that estate land be expropriated and redistributed to its workers. Following Allende's election, land invasions increased throughout Chile, but especially in the south, which quickly became a cauldron of often violent conflicts between peasants and landowners. Land invasions were so widespread in Cautín in late 1970 and early 1971 that in January 1971, Allende's minister of agriculture, Jacques Chonchol, moved his base of operations to the city of Temuco for two months to oversee the implementation of agrarian reform. Allende and Chonchol made it clear that the UP government would address the pent-up demands for land in southern Chile by incorporating sectors of the rural workforce excluded by Frei into the reform.

As the land invasions in Panguipulli exploded in December 1970, Chonchol addressed a convention of foresters. He emphasized the pressing need to extend agrarian reform to the forestry industry because "the man who lives and works in the forest [is] maybe in worse conditions than the coal miner and, of course, worse than the campesino who works in agriculture, supporting the rigors of winter without permanent housing and far from hospitals and schools." Chonchol argued that "agrarian reform . . . will have an integrated solution only if it includes forests and *terrenos forestales* (land covered with forests or suitable for forestry). . . . The forestry sector is where Chile has demonstrated the characteristics of a merciless capitalist exploitation with no vision of the future of either human or natural resources."[3] The government, Chonchol told the assembled foresters, intended to create 200,000 jobs in its first year in office, and the forestry sector would account for 25,000 of them. The key was accelerating the development of the cellulose industry and increasing the pace of tree plantations to supply raw materials to the existing plants, because Chile lacked "sufficient forests to satisfy this demand." Hugo Bianchi, the head of CORA, echoed Chonchol in a presentation to the foresters: "today, the rural sector presents a scenario of chronic unemployment that industry and the service sector are incapable of absorbing. The forestry sector offers the greatest and probably only source of massive employment for campesinos."[4]

Yet the UP faced certain obstacles to expropriating forestry fundos. Landowners were protected by Frei's Agrarian Reform Law, and the UP was committed, at least officially, to pursuing a legal path to a socialist economy. The government confronted the problem of how to indemnify landowners for their forests. The Agrarian Reform Law held that landowners would be compensated for the value of the wood in their forests, as well as the value of their land. Thus, despite UP officials' commitment to extending agrarian reform to the forests, the government was also cautious when it came to large forestry enterprises because of the costs of paying large indemnifications for expropriated forests that contained valuable species. In the breach between the UP's revolutionary rhetoric and its commitment to a gradual and legal process of change, it was left to campesinos to make the promises of agrarian reform reality in the forests.

In early December 1970, hoping to press the government to expropriate forestry estates, workers invaded a number of large properties in the cordillera near Panguipulli, including Neltume and the neighboring Carranco estate.[5] By February 1971, workers on many other properties in the region, including the Kunstmanns' Trafún estate, had also engaged in strikes and land invasions to pressure the government to expropriate the estates. This was a region where labor struggles had roots going back decades. It was also a place where campesinos held a long-standing understanding that the estates had been built on fraud and that their exploitation of the forests was illegitimate. Workers in Neltume had gone on strike in 1941, 1949, and 1951, suffering mass dismissals and evictions from the logging camps. In 1967, following the new Campesino Unionization Law, they organized a legal union once again. In December 1970, Neltume's new union led a strike for increased wages and improved working and living conditions. When Juan Echavarri refused to bargain with the union and, employing a tried-and-true employer tactic, denied workers access to the company store on the property, the workers occupied the estate and demanded its expropriation.[6] Similarly, on Echavarri's Carranco fundo, workers had also engaged in repeated labor actions.[7] An article on the land invasions in *Punto Final* noted that the strike on Carranco in 1970 was the "fourth time the fundo has been occupied. In 1943 and again in 1953 [the workers] were dislodged with violence by the police. There are still workers from those times participating in the struggle of today. These workers have a tradition of combativeness. Their houses were burned down many times, and they were taken, bound, to the Villarrica jail."[8]

As in Neltume, the occupation of the Trafún estate began when workers

organized a new union following the passage of the Campesino Unionization Law and presented a petition to the Kunstmanns, who refused to
negotiate. A Trafún worker explained the reasons for the strike to *Punto
Final*:

> They paid us every two months. The money went right to the com-
> pany store. Kunstmann has a supermarket, and he is the owner of
> the [grain] mill in Collico [Valdivia]. He brought the food from
> there. He never paid the family allowance in money; he paid it in
> food, and when we complained, he told us that he would take the
> company store away from us. We had no way to go to the village
> to shop, so we had to keep quiet. . . . Most of the time, we came
> out behind with the company store, and we didn't earn a peso of
> wages.

The article noted that the company store records confirmed this and that
"the receipts have on one side a space for thumbprints. That is how a great
majority of the workers note their debts."[9]

For the workers who invaded the forestry estates in Panguipulli, the
urgency of the expropriations was determined by their sense that Allende's election offered a unique opportunity, a rupture with the time when
landowners could count on the support of carabineros to crush strikes and
break unions. An ecological logic also drove their sense that immediate
expropriation of the estates was imperative. By 1970, Panguipulli's native
forests were severely degraded, and workers, who had experienced cycli-
cal bouts of unemployment, faced the closing down of logging operations
throughout the region. They were in a race with landowners' destruction
of the forests. As one campesino involved in a toma of a forestry estate
farther north in the Andes cordillera in Bío Bío Province, put it, "By the
time agrarian reform arrives here, the fundos will be all cleared of forests.
Then we will all be dead from hunger, but the wealthy will have time to
take care of themselves. Telling us that we have to stop the tomas is to give
the wealthy a chance to extract everything and leave us only the shells. . . .
Presenting a [legal] petition for expropriation is a sure way to lose." The UP
government's reluctance to expropriate estates with commercially valuable
forests made it all the more imperative, he contended, for the campesinos
to take matters into their own hands: "the very head of the region for CORA
told us that because it is a fundo forestal, it is not a business for the government. There are 77,000 hectares, and it is necessary to pay a very large indemnification for the coigües and the araucaria. I say this is theft since they
[the landowners] have never planted even a bush. And they bought their
machinery by exploiting the workers and stealing the natural wealth."[10]
Campesinos who invaded forestry estates in the Andes cordillera con-

tended that land reform constituted the only way to ensure that forests were no longer mined and then abandoned, like the workers themselves.

An article in *El Diario Austral* on the toma of Carranco laid out the logic of campesinos' collective action. The article described the deplorable working conditions and ruthless logging operations on the estate. It noted that Carranco employed sixty workers from November to March, then, during the off-season, it reduced its workforce to seventeen, throwing more than forty workers into routine seasonal unemployment. The toma was also motivated by workers' concern with landowners' extermination of the native forests, which signified the loss of their livelihood. The paper commented that "in the illiterate heads of these campesinos grows the same concern of the highly civilized countries to care for nature. They say that the fundo is a state concession to the Echavarri company to exploit tepa, but that they have destroyed everything: pines, robles, araucarias, and raulís. According to one campesino, 'They haven't reforested. They took everything out to Argentina. They destroyed our national patrimony, our land, the trees.' He showed us mountain slopes that look like cemeteries of trunks and blackberry bushes."[11]

The campesinos understood that their survival was inextricably bound to the survival of the forests. In addition, they now denounced large estates such as Carranco for violating the principals of rational forestry management and failing to reforest, making the state's forestry development ideology their own. The workers who invaded Carranco also contended, as campesinos had argued since the 1930s, that Panguipulli's estates and logging operations were state concessions that could be ended. In the case of Carranco, the estate had been leased to a logging company in 1947 and then to Neltume's owners, who logged the tepa forests on the estate to supply Neltume's plywood factory. This defined workers' understanding that the forests were public, national patrimony, not private property.[12] Indeed, campesinos had occupied mountainsides on the border of the estate, claiming it as public land, on at least three occasions in the past.[13] One campesino, who had occupied Carranco's forests before, recalled that when he was expelled from the estate, carabineros had tied him to a tree and burned his hut before taking him prisoner to Villarrica on foot and bound with a rope to one of their horses.[14] As recently as 1966, a group of "aspiring colonos" had demanded an official remeasurement of Neltume and Carranco since, they contended, Carranco was public property. *El Diario Austral* appeared to support the campesinos' claim when it noted that they were asserting rights to "9,000 hectares whose public land has been exploited for a number of years." The paper also described how "The colonos were expelled illegally [from Carranco] through dubious legal maneuvers, burning their crops, their houses and evicting them violently."[15]

The tomas in Panguipulli were as much about work and the preservation of the native forests from the depredations of loggers as about land. Campesinos called for expropriation of estates to guarantee the sustainable exploitation of the forests and thus the security of jobs in a region beset by chronic unemployment and frequent migration to Argentina. Moises Durán worked for many years on Panguipulli's forestry fundos and participated in the toma in Carranco. He recalled that the toma was initiated by a group of sixty unemployed forestry workers who lived in the town of Liquiñe, near Panguipulli, and that Carranco had only seventeen workers laboring in the forests when they occupied the estate. He summarized the workers' motives succinctly: "we did it for work."[16] Similarly, on the Trehuia fundo in Panguipulli, eight inquilinos and twenty-seven part-time workers engaged in a toma to organize a cooperative. The campesinos worked with axes producing sleepers of roble and coigüe for the state railroad in a sharecropping arrangement with the estate's owner. As an article in *El Diario Austral* noted, "only work, they just want work, they don't want to acquire the estate." In Trehuia, the campesinos also demanded "a more rational exploitation of the forests" because logging was done with no view to managing or regenerating the forests. Like many forestry workers, the Trehuia campesinos linked the sustainable exploitation of the forests to their need for stable jobs.[17]

The UP's response to the wave of land invasions in Panguipulli was to search for a legal means of expropriating the forestry estates and fulfilling its promise to include the forestry sector in the agrarian reform. The solution lay in the region's frontier status and its proximity to the border with Argentina. Based on a report from the government's National Security Council, which included members of the armed forces, the UP government dictated the expropriation of more than 270,000 hectares of estate land near Panguipulli and the formation of a state-run forestry complex, the Complejo Maderero Panguipulli, based on the estates' importance to Chile's national security. Officials with the UP argued that regulating forest exploitation and building a modern forestry industry in the southern cordillera was a matter of national sovereignty. The expropriation decree, which actually was issued by the National Security Council, underlined the importance of promoting the economic development of the southern frontier territories to encourage their settlement with "significant numbers of people." The National Security Council described the Andes cordillera in the south as having "a precarious level of consolidation" which necessitated "urgent operations designed to invigorate national integration."[18] Forestry development would incorporate the mountain frontiers into the nation and help people a territory depopulated by the continual exodus of workers to Argentina.

In a speech to workers in Cautín in March 1971, Allende outlined the UP's program for developing a modern forestry industry to integrate the frontier into the national territory. He pointed to industrial forestry as the cornerstone of the UP's agrarian policy in the south:

> In these months . . . we have issued a decree of extraordinary impor-
> tance. It creates a frontier zone to expropriate 270,000 hectares of
> agricultural properties, fundamentally logging estates on the slopes
> of the cordillera and the border with Argentina. We have done this
> to preserve the reserves of native forests that have been exploited re-
> lentlessly due to the drive for private profit. We have done it because
> the Agrarian Reform Law prevented the expropriation of forests,
> and only in creating this zone has it been possible to incorporate
> into the patrimony of the state 270,000 hectares where we will es-
> tablish sawmills, where we will pursue a policy of reforestation, and
> where we will establish a cellulose plant to create jobs in the region
> to make possible the increase in population; to intensify the number
> of men per square kilometer; and to end the misery, hunger, and lack
> of education and culture in which hundreds of Chileans who work
> in sawmills and logging have lived the deepest and most profound
> tragedy, more profound than the tragedy of the campesinos. This de-
> cree, at the same time, has a deep national importance that affirms
> our conception of national sovereignty.[19]

Allende's development plan for the frontier zones included the pre-cordillera foothills and cordillera in Cautín and Valdivia, emphasizing that "in these zones, development must be an instrument of National Security" and "ter-ritorial consolidation." Forestry development would promote "territorial integration."[20] Allende's minister of the interior, José Tohá, underlined the continuities between colonization policy and agrarian reform in the fron-tier region, stating, "We plan to stimulate colonization and economic de-velopment. . . . The plans for reactivating the forestry industry must be put into practice as soon as possible to resolve unemployment and the growing emigration [to Argentina]."[21]

The plan was predicated on both the sustainable exploitation of Valdiv-ia's forests and the development of new industries, including a modern sawmill, advertised as the largest and most modern in South America, a plywood factory, and a new cellulose plant supplied by the forests on the twenty expropriated estates.[22] The Chilean National Development Corpo-ration contracted the Japanese firm Marubeni to produce a blueprint for a cellulose plant in Panguipulli that would draw on Valdivia's remaining native forests. The plan projected a $60 million investment by Marubeni, in association with CORFO, to develop Panguipulli's new forestry infrastructure.[23] The UP also established new state purchasing centers to reopen markets for sleepers produced from native woods throughout the Andes, from Bío Bío Province to Cautín and Valdivia.[24] The goal was to reacticate logging among the many small mills in the Andes' reserves of native forest.[25] Yet, despite this program of exploiting the Valdivian forests to supply small mills, the forestry complex's new modern mill, and a new cellulose plant, the UP's frontier development program was designed to manage the forests, as the province's head of forestry programs put it, in "a rational manner to allow the eternal permanence of the forest and reforestation with native and exotic species."[26] The reigning forestry philosophy of replacing native forests with tree plantations remained the cornerstone of the UP's forest policy, although in Panguipulli, foresters planned to reforest with native species such as raulí, as well as Monterey pine and eucalyptus.

Despite the socialist goals implicit in the organization of the Panguipulli forestry complex, the UP's frontier development plan hewed closely to the developmentalist logic of the 1950s and 1960s, with its emphasis on export-oriented industrialization produced through partnerships of domestic, state, and foreign investors. The US Forest Service report of 1946, a report in 1951 by the United Nations Food and Agriculture Organization (FAO), and a report by CORFO in 1952 laid out much of the basis for the UP's forestry policy in Valdivia. The CORFO report proposed limiting the destruction of the Valdivian forests through the creation of new forest industries that "produce the integral use of the available prime materials and that provide the economic incentives necessary for the application of methods of silviculture and conservation of resources." Valdivia's rapidly disappearing native forests, if managed scientifically to provide the raw materials for an integrated forestry complex, could provide a long-term, sustainable engine of development.[27]

The major obstacle to this future of modern forestry development was the structure of land ownership. In 1951, the FAO had underlined the problems with private landholding. In Valdivia, the report noted, "The exhaustion of the forests is occurring at an alarming rate." Private landowners "extract the best species of greatest value and in the process of exploitation they give little or no consideration to the protection of the rest of the remaining forest. . . . [W]hat is left is so badly treated that there only remain degraded forests of less valuable and undesirable species. . . . [W]e have not found one private landowner who is managing the forests with adequate care and concern for continual forestry production."[28] The following year, CORFO proposed increasing the amount of Valdivian forests in the public domain from 5 percent to 30 percent organized, in reserves of 150,000 hectares and managed by state foresters. Citing the US Forest Service report, CORFO noted that 90 percent of the region's native forests were in private hands. The report had argued that developing a sustainable commercial forestry industry would require placing 50 percent of the country's native forests under state management.[29]

The case of the Pirihueico estate, owned by the Oelckers logging concern, reflected the general state of forest exploitation in Panguipulli. When CORA tribunals reviewed Pirihueico's expropriation, they interviewed numerous witnesses, including foresters, agronomists, and estate workers who testified to the estate's destructive logging practices. As one forester noted,

> According to conversations I had with campesinos from the fundo, there are no longer any woods like raulí, which has been logged. This indicates that they have not implemented adequate management . . . to maintain equilibrium among the species that compose the native forest. I can add that campesinos from the fundo state that it has never made improvements in the forest and, on the contrary, that it has exploited the forest, logging it indiscriminately with the sole objective of extracting the wood with highest value and leaving behind only wood with low value and bad quality . . . without care for the second-growth forests.

The forester concluded that Pirihueico had "engaged in an irrational exploitation of the forests."[30]

The CORA tribunal's reports on Pirihueico were echoed in many respects in a talk in 1965 by Germán Kunstmann about logging operations on his Trafún estate. In the talk, Kunstmann noted that until 1945, the fundo had been employed mostly for raising livestock on land cleared with fire and planted with pasture. Only when roads began to arrive in the region in 1945 did logging in the native forests become profitable. At that point, wood from Trafún could be transported to launches on lakes Pellaifa and Calafquén and then shipped to a railroad station at Lanco, a cumbersome and expensive system of transportation that had constrained the Kunstmanns' forestry activities. As Kunstmann described it, "Because there were various species on the estate and of them raulí has the greatest value, the policy of exploiting the forests was oriented toward that species, for which there was demand in foreign markets, and which compensated for the high production costs and high costs of transportation." While Kunstmann recognized the importance of managing the forests to promote their regeneration, he noted that he had run into a problem shared by landowners throughout the region: when he felled the old-growth forest, the cleared land was overtaken by invasive species, especially coligües, which made it nearly impossible to cultivate new forests or pasture. Landowners like Kunstmann viewed it as more expedient to introduce large herds of cattle that could feed on the brambles and bamboo than to attempt to re-generate the native forest.[31]

Nonetheless, by the mid-1960s, with assistance from the Instituto For-estal (INFOR), Kunstmann had begun to experiment with planting both exotic and native species. The institute sent technicians to Trafún to ex-periment with tree plantations on already cleared land, representing the policy of both the FAO and the Frei government of assisting privately held forestry estates in modernizing their production methods. In keeping with the FAO's and INFOR's interest in reforesting with species for international markets, five of the six trees cultivated on Trafún were exotic species, and only one tree was autochthonous (raulí).[32] Allowing INFOR to cultivate ex-perimental tree plantations on Trafún cost Kunstmann nothing but land that already had been cleared and allowed him to don the mantle of mod-ern forestry practice. Only one other forestry fundo in Panguipulli had ex-perimented with reforestation, and it was not privately owned: during the early 1950s, the Caja de Empleados Públicos y Periodistas (Public Employ-ees and Journalists Retirement Fund) began to reforest its 14,000-hectare Quechumalal estate near Lake Panguipulli with raulí, laurel, and lingue.[33]

Many foresters criticized the extractive logging methods used on Valdivia's estates but found it difficult to find a way to promote sustain-able logging. Most believed that logging and then replacing the forest with pine or other exotic species was a more expedient solution than manag-ing second-growth native forests. The problem, however, was that invasive species that occupied cleared forest also made cultivating pine a difficult business in the Valdivian forests. In an article published in 1968, a forester repeated the well-worn argument for the desirability of "transforming the [Valdivian] rain forest into forests of North American conifers." This ex-change of native for exotic forests confronted the obstacle of bamboo that invaded cleared land. Not only was the bamboo almost impossible to erad-icate, but landowners relied on bamboo thickets to pasture their cattle and lacked the resources "to allow rational exploitation, to build roads, or to reforest to transform the virgin forests."[34]

Landowners and foresters confronted a basic problem in the nature of forestry science, as well. While foresters had been trained in methods of managing tree plantations composed of exotic species, they still had lit-tle understanding of how to manage the Valdivian forest. Forestry science continued to focus on replacing native forests with monocultural planta-tions of exotic species rather than on sustained-yield exploitation of the native forests. Foresters at the Universidad de la Frontera in Temuco noted the difficulties of managing logging in the Valdivian forest to guarantee their reproduction: "the lack of knowledge of the methods of regeneration of the native forest, of species adequate for reforestation, of systems of for-

est management, prevents the creation of additional sources of employment and prevents the improvement or increase of the capital existing in
the renewable natural resources."³⁵ This was a problem for the new state-
run Complejo Maderero Panguipulli. Simply transferring the region's estates to the public sector and organizing an integrated forestry complex
did not solve the problem of how to manage the regeneration of the native
forests.

The UP's first priority was to provide jobs for the region's unemployed
workforce and to develop new industrial facilities, such as a cellulose plant
that would use native woods as its raw material. A government report
noted that, following the expropriation of Panguipulli's forestry estates,
the state-run complex hired one thousand new workers, of whom "a good
percentage [were] indígenas from the neighboring communities, to whom
was also restituted land usurped by the private owners of the estates now
in the complex."³⁶ By 1973, the Complejo Maderero Panguipulli employed
nearly four thousand workers, almost doubling the estates' total labor force
from their pre-1970 levels. The UP had only three years to experiment with
new methods for managing the forests. Neither the project to build a new
plywood plant nor the project to build a cellulose plant with Japanese investment had time to materialize, so it is difficult to assess the impact of
the UP's forestry policies in the Valdivian forest. It would seem, however,
that by increasing production and employment, the Complejo Maderero
Panguipulli probably placed intensified pressure on the region's already
degraded native forests. As the foresters' report noted, "The lack of knowledge we have of the resources of the native forests prevents us from knowing whether there is compatibility between the existence and growth of
this resource [the forests] and the annual volumes of extraction. It is likely
that with the actual modes of extraction we are de-capitalizing the forests
and compromising the future possibilities for developing the region."³⁷

Alex Rudloff, a forester who worked for the Complejo Maderero Panguipulli and now works for the Corporación Nacional Forestal (National
Forestry Corporation; CONAF), recalled that the Complejo Maderero
Panguipulli maintained high levels of production that ultimately were
unsustainable, partly because the UP inherited fundos where landowners
had already logged the forests' most valuable species. However, the UP's
commitment to reigniting the forestry industry to provide a solution to the
region's problems with rural unemployment led to increased logging of the
remaining forests.³⁸ Similarly, by organizing state-run markets for sleepers
produced by smallholders and Mapuche communities, the UP reactivated
logging in native forests where there had been a significant decline in activity since the late 1950s. In addition, expanding forestry development under
both Frei and Allende increased markets for pine plantations. This raised

wood prices more generally, making logging in native forests for regional and national markets profitable once again.

Nonetheless, both the Frei and Allende governments made the first attempts to manage and regenerate native forests. Frei's agrarian reform and forest laws compelled some landowners, including Kunstmann, to work with foresters and organizations such as INFOR to experiment with plantations of native species and to adopt forest management plans. And the foresters who managed the Complejo Maderero Panguipulli's new Forest Management Department after 1970 were concerned with the issue of sustainable logging. When the Complejo Maderero Panguipulli was formed in 1971, Rudloff remembered, "There were estates where the forest was completely degraded, covered with underbrush, quila, coligue [bamboo]."[39] These estates were designated for reforestation with native species. The foresters with the Complejo Maderero Panguipulli, he recalled, focused on cultivating plantations of autochthonous species, especially raulí.[40] This represented a significant shift from the strategy of reforesting with Monterey pine pursued farther north. However, while the raulí plantations may have held the seeds of future sustainable logging of native hardwoods, they signified the reduction of an extraordinarily complex ecosystem to monocultural stands of one species of tree. Ultimately, the forests philosophy that underpinned the UP's development strategy in Panguipulli was rooted in the dominant ideology of rationalization in the name of forestry industrialization and national territorial consolidation.

The Complejo Maderero Panguipulli did, however, bring a series of benefits to the region's poor campesinos. Landless and unemployed workers found stable, high-paying jobs in the forestry complex, as did members of the region's Mapuche communities. Smallholders and Mapuche communities also benefited from expanded markets for sleepers and wood logged in native forests. Nonetheless, the formation of the Complejo Maderero Panguipulli engendered conflicts with Mapuche communities who sought the recuperation of usurped land, and with the residents of asentamientos established under the Frei government who sought to preserve their control of small plots on expropriated estates. In both cases, the UP's forestry program offered secure jobs to Mapuches and asentados in exchange for the inclusion of their pasture and forests in the forestry complex. For some campesinos, this represented a push toward proletarianization, a loss of independence, and a rupture with the peasant economy they struggled to sustain. While the UP's forestry policy in Panguipulli spoke to the interests of underemployed and unemployed landless forestry workers, most of whom did not enjoy the typical regalías of inquilinos, it appeared to threaten those campesinos who retained access to land and who produced their sustenance by cultivating garden crops, pasturing livestock, and selling wood from their small patches of native forest to the railroad or nearby sawmills.

The organization of the Complejo Maderero Panguipulli immediately sparked a conflict with smallholder asentados on the 25,000 hectare Arquilhue estate. The Agrarian Reform Corporation expropriated the estate following a toma during the last month of the Frei government, a sign that by 1970 the Christian Democratic government was beginning to consider including forestry estates in more radicalized agrarian reform. The government then organized an asentamiento among the estate's 380 workers. Months later, the UP officials in charge of the Complejo Maderero Panguipulli decided to incorporate Arquilhue, eliminating its status as an asentamiento and transforming the asentados into full-time workers. As Jaime Tohá, who headed the UP's Forests and Forestry Committee, argued, Arquilhue's forests were indispensable to the Complejo Maderero Panguipulli's operations: "the criteria of our government are very simple: to take control of all of the forest patrimony in this zone of the country [and] to achieve the incorporation of all the fundos that can contribute to the Complejo Maderero, the foundation of a future cellulose industry. . . . Without Arquilhue's forests, there can be no industrial complex." For Tohá, Arquilhue's incorporation into the industrial forestry complex was also vital in terms of "creating a frontier security zone."[41] Similarly, Valdivia's intendant underlined that, "in the case of forestry estates that require large surface areas and long periods of planning to be exploited rationally, state property is the only form that guarantees the adequate use of the existing native forests."[42]

Yet for a number of Arquilhue campesinos, centralized state ownership of the forests represented a threat to their autonomy and ability to cultivate crops, pasture livestock, and collect forest products. As one campesino put it, "Despite the Complejo [Maderero Panguipulli's] expressions of good faith to improve the situation, to rationally exploit the forest . . . we want to work independently, without the tutelage of the complex that intends to take away the forest riches to be processed industrially." To some campesinos, the Complejo Maderero Panguipulli looked like another landowner, a new patrón, whose logging operations threatened their access to native forests, pasture, and land to cultivate. As the campesino put it, "And our zone: what will we be left with? . . . What we will cultivate?"[43] The asentamiento had engaged in mixed agroforestry, logging Arquilhue's remaining forests, pasturing cattle, and cultivating wheat, potatoes, beets, and ramps, but incorporation into the complex implied ending all agricultural work and devoting Arquilhue only to raising oxen to be used in the Complejo Maderero Panguipulli's logging operations while supplying the complex with wood from the forests.[44] In the end, the UP negotiated a compromise.

Arquilhue and roughly three hundred of the estate's workers would join the complex. The remaining seventy-six workers would be settled elsewhere on a new asentamiento.[45]

The long-standing demand for the restoration of usurped land also led to conflicts between the region's Mapuche communities and the Complejo Maderero Panguipulli. In December 1970, for example, sixty members of the Lorenzo Carimán community occupied 150 hectares on the Trafún estate, arguing that the land had been usurped forty years earlier by the Kunstmanns and that "during these years they have taken all our wood without paying us a cent."[46] In response, in 1971, the directors of the Complejo Maderero Panguipulli sought to incorporate a number of neighboring Mapuche communities and their forests into the complex. The head of Neltume's union, backed by the UP's Indigenous Affairs Directorate (DISAN), proposed that the communities join the Neltume cooperative and the Complejo Maderero Panguipulli, which would provide them with land, machinery, and oxen. In turn, the communities would incorporate their land, forests, and livestock into the cooperative. The cooperative would also provide stable work for all adult members of the communities, and the Mapuches would share in the cooperative's profits equally. The Cachim, Linquiñe Alto, Quintumán, Valeriano Cayicul, Fundihuincal, and Neltume communities opposed this plan and demanded that land on the Neltume estate that had been usurped be restored to the original reducciones. Representatives of the Mapuche communities stated:

> The problem originates in the fact that the Neltume estate, now
> expropriated and in the power of the government, has usurped 1,700
> hectares, more or less, from the communities, land that the commu-
> nities want returned in conformance with their títulos de merced.
> Instead of receiving the restitution we expected, they want to take
> the rest of our land and turn us into renters from the state, and we
> only want them to return our land and to keep working as indepen-
> dent workers. . . . Moreover, we consider it very serious that they
> want to prohibit us from having livestock. . . . [I]f they do this, we
> will be deprived of the only thing we have.[47]

In general, Mapuche communities demanded 12,000 hectares of the complex's more than 300,000 hectares. Minister of Agriculture Jacques Chonchol traveled to Panguipulli in 1972 to arbitrate this conflict and recognized the legitimacy of the communities' claims. "Effectively," he said, "it has been demonstrated that the land had been usurped by the old landowners and still has not been restituted."[48]

The tension between local Mapuche communities and the Complejo Maderero Panguipulli was complicated. Many Mapuches found jobs in the complex. In addition, like many smallholders in the region, Mapuche communities historically had also exploited the forests on their reducciones, often selling wood to the mills on the large estates or signing contracts with sawmills to log stands on their property. Mapuches' exploitation of their forests, however, was far less intense than the logging operations of the estates. Rudloff recalls, for example, that an estate in Panguipulli with a modern sawmill could easily eliminate its forest in ten years, while Mapuche communities, which cut an average of two trees a year and sold them to mills, might preserve a forest stand for a century. Mapuche communities exerted more control on logging in their forests and used logging as a supplement, not a main source of income for families, a way to earn money during winter months when food was scarce.[49] If the communities incorporated their land and forests into the complex, they faced not only loss of autonomy and their historical rights to land they had occupied for generations, but also accelerated logging of their remaining stands of native forest.

In part, this tension derived from Mapuche community members' different labor systems. While non-Mapuche workers on the Complejo Maderero Panguipulli's fundos represented a largely proletarianized, if often unemployed, workforce with little access to land for crops, pasture, and forests, Mapuche workers labored mostly on a seasonal basis and maintained a viable, if struggling, campesino economy within their communities. In the complex, many workers labored full time for a salary. Within Mapuche communities, while families might hold land individually, they employed reciprocal labor exchanges known as *la minga*. Families would provide labor for one another on a rotating basis, planting and harvesting. During off-seasons, they would find part-time work on the region's fundos. For members of Panguipulli's Mapuche communities, agrarian reform presented an opportunity to reclaim land usurped by estates such as Neltume, not to obtain secure and high-paying jobs, as it did for many non-Mapuche forestry workers.[50]

Despite the pressure to fold Mapuche land into the Complejo Maderero Panguipulli, during 1971 and 1972 the government restored thousands of hectares of land usurped from local indigenous communities on the complex's fundos that had been taken over by CORFO. A CORFO report from 1979 noted, for example, that the agency had had difficulty exercising dominion over thousands of hectares of the complex's fundos because they were being occupied with the support of DISAN: "in general, problems have arisen because of the legal or illegal occupations of parts of the property by indígenas. . . . Thus, even though CORFO's titles to the Riñinahue estate indicate that the property has 16,000 hectares, in reality it only has 9,215 [hectares] because 6,785 [hectares are] occupied by indígenas based on [a]

decree [issued by the] Ministry of Land and Colonization. . . . In the case
of Maihue, Rupameira, and Trafún Norte, neighboring reducciones have
been occupying adjacent sectors based on an agreement signed between
the indígenas and representatives of the complex in 1972."[51] Thus, the UP
eventually restored thousands of hectares of land to neighboring Mapuche communities either through direct agreements with the Complejo
Maderero Panguipulli, actions by DISAN, or decrees by the Ministry of
Land and Colonization. As a report by the governor of Panguipulli in 1971
pointed out, during the socialist government's first year, "Land usurped by
the private owners of the properties that are now owned by the Complejo
[Maderero Panguipulli] has been restituted" to the neighboring Mapuche
communities.[52]

Restoring land to Panguipulli's indigenous communities was consistent
with UP policy. In early 1971, Chonchol announced the organization of government commissions to re-measure land "that had been usurped legally
or illegally from Mapuches."[53] In part, this was a response to pressure from
below. Beginning in 1969, Mapuche reducciones had initiated a wave of
tomas and *corridas de cerco* (fence runnings) on neighboring estates.[54] In
Cautín and Malleco alone, the UP expropriated 137 estates in favor of Mapuche communities and restored 200,000 hectares of land through CORA
and the Commission on the Restitution of Usurped Land.[55] In the case of
Panguipulli, restoring land to Mapuche communities meant taking land
out of the Complejo Maderero Panguipulli and giving Mapuche communities control over forests. Yet in this case, the UP applied its forestry development plan flexibly and met the communities' demands for the restoration of usurped land.

The UP walked a fine line between potentially conflicting goals, lending
its support to campesinos' demands for land, promoting the development
of the forestry industry to create regional economic growth and to provide
jobs for landless laborers, and managing the forests to prevent their extinction. The UP's forestry strategy in Panguipulli was not preservationist,
but it sought to employ some form of sustainable forestry in the Valdivian
forests for the first time. In addition, by addressing campesinos' demands,
the UP lent its support to a peasant economy that combined exploitation
of native forests with agricultural activities. It also stepped up hiring in
the Complejo Maderero Panguipulli to solve the pressing problem of rural unemployment. For the UP's planners, forestry development based on
logging native forests for export markets and, eventually, a cellulose plant
funded with foreign investment, was compatible with sustainably managing the native forests, as well as with maintaining a robust campesino economy in the region. It is difficult to know how these potentially conflicting
dynamics would have played out over time, but it seems clear that the or-

ganization of the Complejo Maderero Panguipulli did little to conserve the Valdivian forest. On the contrary: had the project of a cellulose plant fueled by Valdivia's native forests gone forward, the impact of forestry development would have been to further degrade the region's forest ecosystems. The UP's goal of managing the forests to produce a long-term, sustainable form of forestry development based on plantations of both exotic and autochthonous species, however, might have worked in Panguipulli. But the UP was in power for only thirty-six months, and this strategy of integrated forestry development was never given a chance.

Agrarian Reform and Mapuche Tomas in the Araucaria Forests

Farther north in the Andes cordillera, in Alto Bío Bío and Lonquimay, the tension among forestry development, Mapuches' campesino economy, and sustainable forest management took on another dynamic. In part, it was driven by Mapuche-Pehuenche communities' historic reliance on araucaria forests for sustenance and the araucaria's national symbolic power. Whereas in Panguipulli the UP planned to log and then manage the regrowth of native forests, farther north it imposed more severe restrictions on logging. Araucaria forests were to be protected rather than managed. In Alto Bío Bío and its neighboring regions, UP officials planned to acquire raw materials for the forestry industry from pine plantations on already cleared land. The major cellulose plants in Arauco and Constitución and the region's sawmills could draw on the hundreds of thousands of hectares planted in the central valley and coastal cordillera. The UP favored expropriating large forestry estates and restoring land to Mapuche-Pehuenche communities. However, because of its concern for preserving the araucaria forests, it did not always redistribute forests to the communities. Instead, in some cases it established national parks and gave Mapuche-Pehuenche communities the right to collect piñones and pasture cattle within the parks' borders.

In 1971, a number of Mapuche-Pehuenche communities invaded estates owned by the Ralco, Fressard, and Galletué forestry companies. Conflicts between the region's reducciones and logging companies had percolated since the 1940s in the mountains where lakes Galletué and Icalma marked the origins of the Bío Bío River. By the 1960s, the tension had come to a head. The communities and logging companies clashed over both veranadas, where Mapuche-Pehuenches had pastured livestock for generations, and araucaria forests, where they historically had collected piñones. Alto Bío Bío's araucaria stands had benefited to a certain degree from their inaccessibility; only in the late 1940s did landowners and forestry companies initiate logging operations in the region. Estates and public land that had been used mostly for their pasture acquired new value for their woods, and the conflicts between logging companies and the region's Mapuche communities intensified. By the 1960s, even *El Diario Austral* had begun to denounce unregulated logging in araucaria forests of the Andes cordillera. In 1969, for example, the paper editorialized that "the indiscriminate exploitation of araucaria pine is creating a serious situation and a visible deterioration in the existence of this species." The paper described the rapid disappearance of araucaria and coigüe forests in Sierra Nevada, Malalcahuello, and Galletué.[56]

Allende's election produced a significant shift in the balance of power in Alto Bío Bío and Lonquimay. In December 1970, 120 members of the Callaqui community, accompanied by a number of non-Mapuche ocupantes, invaded the Ralco estate. *El Diario Austral* reported that "a large majority of the indígenas in the interior of Antuco (Hacienda Ralco) who live there, aided by Chileans, took the hacienda that is also called Ralco. The difficulties arose because the authorities have continued to tolerate logging the araucaria pine that is found in considerable abundance there and because the obreros complain of irregularities in the payment of their wages."[57] As Martín Correa and Raúl Molina have shown, this conflict dated back to the 1940s and 1950s, when regional landowners and the Ralco company began to burn and log forests on which the Callaqui community relied for survival.[58] In response to the toma, the region's intendant traveled to Ralco and met with a group from the Callaqui community, who demanded "the immediate prohibition of logging the araucarias, which produce piñones, the basis of their diets, especially in winter."[59] The Callaqui community was accompanied by eighty of the estate's workers (*obreros*), who also demanded the estate's expropriation.[60] In this case, despite incipient tensions between Ralco's workers and the members of the Callaqui community, the two groups joined together during the land invasion. As *El Sur* explained, "The union with the Callaqui community came about because of the demands of the indígenas for land on the property, of which they declare that they are the owners. [It was] usurped by deceit after it was given over through a rental contract to the owners of Ralco. As a means to recuperate the land, the Mapuches maintain a union with the workers in their petitions for expropriation and redistribution of the land and forest reserves, with the goal of exploiting them communally."[61]

The Ralco-Lepoy community joined the Callaqui community's occupation of the Ralco hacienda, taking control of land it had claimed for generations. In response, Chonchol, accompanied by Jaime Tohá, traveled to Alto Bío Bío to negotiate an end to the conflict.[62] Under pressure from the land invasion, Ralco agreed to indemnify the Ralco-Lepoy and Callaqui communities for wood it had illegally cut on their property. Nonetheless, in

February 1971, the government proceeded with the expropriation of sixty estates in Bío Bío, including many owned by the Ralco company. Because the cost of indemnifying the company for the araucaria forests was high, the government expropriated only a piece of the Ralco estate, which it redistributed to the Mapuche communities and Ralco workers, leaving the company with a valuable reserve of forests.[63]

Finally, in 1972 the UP expropriated Ralco's araucaria forests and established a national park, backed by a ruling from the Supreme Court. The UP drew on the Forest Law of 1931, which gave the president the power to establish nature reserves and national parks to protect endangered species. Establishing the park on land claimed by the Ralco company gave the UP a legal way to halt logging without going through the process of expropriation and indemnification. The national park fell short of the communities' demands for restitution, but it ended logging in the araucaria forests. The UP also granted the Callaqui and Ralco-Lepoy communities use rights to collect piñones and pasture livestock in the park, recognizing the importance of the activity to their subsistence and their historical claims to the forests and veranadas.[64] In this sense, the Ralco National Park differed from the national parks organized during the first decades of the century. Those parks and forest reserves had excluded campesinos, often imposing new restrictions on their access to public forests they claimed based on custom and use or, in the case of indigenous communities, historic occupation. Conservation in this case allowed the UP to avoid the legal restrictions imposed by the Agrarian Reform Law while restoring Mapuche-Pehuenche communities' access to araucaria forests and banning the logging activities of companies like Ralco.

The UP repeated its decision to create a national park out of Ralco's araucaria forests with other expropriated estates in the region. This was the case in the valleys just south of Ralco, in the Andes cordillera, where the Galletué company was also enmeshed in conflicts with Mapuche-Pehuenche communities. A significant chunk of Galletué's Quinquén estate was claimed by a group of families who belonged to the Meliñir kinship line, once part of the Paulino Huaiquillan reducción. As José Bengoa has shown, the community had occupied land in the Quinquén valley at least since the beginning of the twentieth century without a land grant title and continued to occupy it even after the land was auctioned off in 1911.[65] Eventually, following a series of land transfers and sales, including the return of the land to the Caja de Crédito Hipotecaria (State Mortgage Bank) because of unpaid mortgages, the Galletué company purchased the Quinquén estate during the 1940s. In the process, as Bengoa demonstrates, the estate's boundaries expanded illegitimately, incorporating land occupied for decades by the Meliñir community. By the 1960s, as in the case of Ralco, the Galletué company had begun to log araucaria forests intensively on the Quinquén estate, provoking a bitter conflict with the Meliñir community.[66]

Similarly, the community of Pedro Calfuqueo, which, unlike the Meliñirs, did have a land grant title, began to encounter difficulties with Galletué during the 1960s, when the company began logging araucaria forests within the borders of the reducción. In August 1970, following years of complaints by the Calfuqueo community, an indigenous protector ordered Galletué to suspend logging on contested land within the Calfuqueo reducción while he investigated the boundary dispute.[67] The following year, his successor completed the remapping of the boundaries between the Galletué hacienda and the Calfuqueo community and found that, although the community's title stated that it had 1,000 hectares, the map accompanying the title, which outlined the community's boundaries, actually included 5,539 hectares. The re-measurement in 1971 significantly enlarged the reducción and gave it a victory in its contest with Galletué.[68]

In 1971, CORA expropriated the company's Galletué hacienda (19,153 hectares), Quinquén fundo (6,680 hectares), La Fusta fundo (8,870 hectares), and Galletué fundo (4,400 hectares). Galletué argued that its estates should be exempt from expropriation because it had followed the guidelines and practices set out by the Frei government during the late 1960s by developing a forest management plan that was "elaborated by distinguished, specialized professionals and approved by the authorities." According to the company, the estate's plan included a program for reforestation with Monterey pine, "and this was practiced, guided by licensed professionals."[69] The change of government, however, brought new scrutiny of Galletué's logging operations. The company's claims that it had employed sustainable forest management plans were flatly contradicted by an agronomist's report in 1971 justifying Galletué's expropriation: "the exploitation of the forests has been done without a plan for reforestation." In 1973, CONAF also submitted a report in which it described the company's "extermination" of the araucaria forests.[70]

Rather than redistribute the land to the region's laboring population or Mapuche reducciones, however, CORA transferred the estates to CONAF, which created a national park. As in the case of Ralco, this represented an astute legal strategy by UP officials, given that the Supreme Court had acknowledged the government's legal authority to establish national parks. The decision to establish a national park was also dictated by the degraded ecological conditions in the region. As a report by CONAF noted, the araucaria forests on the Galletué company's estates had been logged indiscriminately for twenty-five years, which had led to "an accelerated extermination of the forests with commercial value, leaving behind only some protective forests and hills covered with bushes and undergrowth without commercial value." The mountain valleys had also been significantly degraded by pasturing livestock and the practice of "burning extensive stretches covered with coirón in the false belief that this will increase and rejuvenate the pasture for the following year." Instead, it had produced "major damage to the fragile soil and consequent erosion of the sector."[71] Nonetheless, despite CONAF's concern about preserving the remaining forests, Mapuche-Pehuenche communities won the right to collect piñones and pasture cattle within the new national park. While some land was restituted to communities, including the Calfuqueo community, based on the boundaries included in original titles, other communities, as in the case of the Meliñirs, received usufruct rights to the araucaria forests where logging was now banned.

In Lonquimay, as in Panguipulli, the acceleration of agrarian reform during the UP government was sparked by a series of land invasions. In 1972, for example, the Bernardo Ñanco community invaded 800 hectares held by the Chilpaco estate, as well as by smallholders repatriated from Argentina earlier in the century and settled within the reducción's borders.[72] As we have seen, the Ñanco community's struggles with colonos and landowners in Lonquimay had roots stretching back to the logging contracts of the 1940s with the Fressard company. The conflict between Fressard and the Ñanco community continued throughout the 1960s, with the community persisting in its accusations that Fressard had violated logging contracts and had continued felling stands of araucaria illegally. Fressard, in turn, contended that Ñanco community members were illegally extracting wood and logs from araucaria forests on the border between the reducción and the Chilpaco fundo. In the mid-1960s, Fressard, like the Galletué company, developed forest management plans to "protect the native forest" and experimented with reforesting with Oregon and Monterey pine with the assistance of foresters from the state-run Malalcahuello Forest Reserve, a sign of the company's "constant concern with reforestation."[73]

The Ñanco community was enmeshed in two conflicts: one over pasture with the colono smallholders who had been settled on land within the reducción after it had received its título de merced, and the other over araucaria forests with Fressard's Chilpaco fundo. In both cases, ambiguities in colonization policy had created the conflict. When the community had been settled and given its title in 1907, Francisco Puelma Tupper was in full possession of the hundreds of thousands of hectares that composed the San Ignacio de Pemehue hacienda and included Chilpaco. During the 1930s, the Ñanco community engaged in a legal dispute with Puelma's heirs over land it claimed by right of historical occupation. In the course of the legal proceedings, courts determined that while the community's title granted it 6,500 hectares, it in fact occupied more than 10,000 hectares, which were included in a map of the reducción that accompanied the title. As in the case of the Calfuqueo community, the surveyor's map of the reducción from 1907 conflicted with the description of boundaries in the title, although the boundaries were so vague that they were open to interpretation. This confusion had sparked lengthy litigation with the Puelma heirs and, later, the owners of fundos such as Chilpaco. The other confusing situation was produced during colonization, when the state settled a large number of colonos repatriated from Argentina within the borders of the reducción, sparking generations of conflicts over veranadas.[74]

A report in 1972 by the Commission on the Restitution of Usurped Land described how local landowners in Lonquimay had usurped indigenous land "with the sole goal of exploiting the large quantities of araucaria pine." They had, the commission observed, "eliminated the forests indiscriminately, even deceiving the innocent indigenous people with the direct usurpation of their land, with the purchase and rental of veranadas, and sharecropping arrangements to exploit large tracts of native forest . . . , leaving this land unrecoverable in the short term."[75] Following the land invasion and the subsequent investigation by the Commission on the Restitution of Usurped Land, some of the land occupied by the colonos was returned to the community, and the Chilpaco fundo was expropriated, with a portion of the estate also restored to the community. The Ñanco community thus regained control of stands of araucaria forest it had long claimed from the Fressard company's Chilpaco estate, as well as pasture for its livestock when the borders of the colonos' plots were redrawn.[76]

Driven by land invasions and workers' strikes, the UP expropriated many estates in the Andes cordillera that had implemented forest management plans designed by foresters and that had reforested with North American pine, a direct conflict with Frei's Agrarian Reform Law. For example, Fressard's Lolén (or Porvenir) estate in Lonquimay, which neighbored Chilpaco, was expropriated in 1972 even though Fressard had followed a forest management plan and had reforested cleared native forests with Oregon, Ponderosa, and Monterey pine since 1965. After the eruption of labor conflicts with its workers in 1970, the estate was expropriated, and the owners were given minimal indemnification because CORA studies found that the estate had been badly exploited.[77] In this case, the labor conflict initiated by Fressard's workers, as opposed to a Mapuche toma, gave the UP an opportunity to intervene in the estate's administration and eventually expropriate it.

The UP government provided an important window of opportunity. Colonization laws had produced a tangle of ambiguous and conflicting land claims in the region. Fraudulent estate titles and boundaries; reducciones whose boundaries were described one way, but whose cadastral maps and long histories of occupation defined different borders; and superimposed titles that granted the same land to colonos or estates and indigenous communities made definitive legal solutions to social conflicts impossible. How land disputes were resolved depended on interpretations by state officials and the judiciary. Before 1970, state officials tended to decide disagreements between estate titles and indigenous land claims in favor of the former; during the years of the Frei government, analogous conflicts between colono smallholders and Mapuche communities were decided in favor of the colonos, as in the cases of the Trapa Trapa and Queco communities. This changed with Allende's election. Members of the Commission on the Restitution of Usurped Land and government-appointed surveyors employed by the Ministry of Land and Colonization and DISAN traveled to the mountains, interviewed members of indigenous communities, and reviewed their land claims. In the end, given the myriad ambiguities and contradictions in land titles and cadastral maps, the final arbiter in land disputes was the state, and during the UP period, government officials sided with poor peasants and indigenous communities. But in Alto Bío Bío and Lonquimay, as elsewhere throughout southern Chile, the state did not act on its own. It was forced to intervene by the wave of land invasions initiated by campesinos and Mapuche-Pehuenche communities.

Reforestation in Cautín and Malleco

Panguipulli, Alto Bío Bío, and Lonquimay lie in mountain regions where Chile's last stands of native forest remained. In other parts of southern Chile—in the central valley and coastal piedmont, as well as in the coastal cordillera—most native forest had been already cleared to pave the way for cereal cultivation and livestock or had been felled by loggers. In these areas, the UP intensified the Frei government's policy of investing state resources in reforestation with pine, implementing reforestation programs on expropriated estates and on the parcels that belonged to smallholders and Mapuche communities. During its short period in power, the UP stepped up the pace of reforestation, planting 172,252 hectares with pine, an annual average of 57,417 hectares.[78] Like the Christian Democratic government, the UP made reforestation and the development of the forestry industry a centerpiece of its development strategy for southern Chile. For the UP, as for the Frei government, reforestation constituted a tool for generating employment and providing raw materials for the cellulose plants in Laja, Constitución, and Arauco, as well as a means of restoring eroded soil. In 1971, for example, the UP implemented a 75,000-hectare reforestation program to provide employment for ten thousand workers.[79] Much of this reforestation was focused in the area around Concepción and the basin of the Bío Bío River, where CORFO reforested 20,000 hectares and private landowners, with state subsidies, reforested 10,000 hectares in 1971.[80] In Arauco, CORA, with credits from CORFO, planned to plant 2,200 hectares and to provide credits to peasant cooperatives on expropriated estates to plant 500 hectares more. Forestal Arauco, a subsidiary of CORFO, planned to plant 6,500 hectares in the province. In addition, CORFO and the Servicio Agrícola y Ganadero (Agricultural and Livestock Service; SAG) entered into agreements with smallholders to reforest, providing plants and financing the reforestation of 8,502 hectares on seventy-two asentamientos organized in two cooperatives.[81] Jaime Tohá noted that in 1971, Chile reforested 70,000 hectares, almost all through public investment, since private landowners planted little. That year, the CMPC reforested only 800 hectares on its large estates in southern Chile.[82]

Foresters with the UP government were especially concerned with promoting forestation in Bío Bío, where soil erosion threatened to inundate the Bío Bío River and nearby cellulose plants provided a steady market for pine. In addition, both widespread soil erosion and the recent spread of pine plantations had led to high levels of rural unemployment in Bío Bío. The UP viewed reforestation programs as a key to generating jobs in the province. In 1971, CORA expropriated a number of fundos in Bío Bío, many of them cleared of native forests and covered with weeds and undergrowth, to implement reforestation with Monterey pine.[83] Later that year, CORFO provided credits to INDAP to reforest eighty-eight expropriated properties in Bío Bío that belonged to two peasant cooperatives, Palizal and Los Notros, to supply raw material for cellulose production and recover eroded soil.[84] In 1972, the UP initiated a new program to plant 47,000 hectares with pine in the Bío Bío river basin and slow the advance of dunes in Arauco by planting 9,400 hectares with pine.[85]

State-owned companies such as Forestal Arauco and Forestación Nacional (FORESTANAC) played a major role in the UP's reforestation campaign. In 1971 and 1972, FORESTANAC hired three hundred unemployed workers who belonged to Unemployed Workers Committees in Mulchén, Collipulli, and Chillán to reforest 1,000 hectares to supply the cellulose plants in Laja and Constitución. In addition, FORESTANAC established a new wood factory and a modern sawmill in Mulchén to take advantage of the region's pine plantations and create jobs for the unemployed.[86] Forestal Arauco similarly acquired thirty-seven properties planted with pine to supply Celulosa Arauco, employing 550 workers.[87] In some cases, the National Reforestation Corporation or its successor, CONAF, and CORFO reforested land on estates purchased by the Frei government. This was the case of the Llohué and Casallana estates in the coastal cordillera in Ñuble, where, a CORFO report noted, the soil had been "submitted to intensive use which has caused erosion."[88]

The socialist government's forestation programs had a social logic, as well. The UP employed forestation as a tool for generating jobs in a region with crippling levels of rural unemployment and poverty. As I noted in chapter 6, workers on large properties managed by CORFO and the National Reforestation Corporation (or, later, CONAF), as well as companies such as Forestry Arauco, which had a permanent labor force of 3,000 in 1972, went on strike and negotiated collective contracts that brought unprecedented job security, wage increases, and benefits.[89] Workers on privately held forestry estates planted with pine also went on strike during the three years of Allende's government, flexing new organizing muscle and taking advantage of the favorable political conjuncture. For example, workers went on strike and occupied five estates in late 1971, demanding the estates' owner, the CMPC, put an end to the system of subcontracting it employed in logging and hire them as full-time workers, a status achieved by workers on publicly owned properties managed by Forestal Arauco or the National Reforestation Corporation.[90]

In some cases, forestry workers' strikes pushed the UP to expropriate estates covered with pine plantations. After a work stoppage on the Sociedad Forestadora Nacional's 996-hectare El Guanaco fundo in Ñuble, the UP appointed an interventor (government-appointed manager) to take charge of administering the property. El Guanaco's owners then offered the estate for expropriation. While the estate was covered with a pine plantation, CORA proceeded with the property's expropriation in 1972, paying the owners significant indemnification both for the value of the property and the value of the tree plantations.[91] By 1972, CORA had begun to consider an official policy of expropriating estates with pine plantations, immediately sparking the repudiation of the timber industry's trade association and forestry companies like the CMPC.[92] However, it appears that this policy was never implemented. The high price of compensation for properties covered with tree plantations deterred the UP from intervening directly, as it had in the case of El Guanaco, to expropriate forestry estates held by powerful economic groups such as the CMPC. Despite the demands of striking forestry workers in 1971, for example, no estates owned by the CMPC were expropriated during the agrarian reform.[93] For the most part, the UP expropriated already cleared estates to reforest them with pine, or estates with native forests in the cordillera with limited stretches of tree plantations, as in the case of properties in Lonquimay and Panguipulli.

In Cautín, where Mapuche reducciones were responsible for 86 percent of the province's agricultural production but held only 25 percent of its productive land, the UP also made forestation and commercial logging an important part of agrarian reform.[94] Population growth and the usurpation of

Mapuche land had combined with the small initial allotments to produce overworked soil on Mapuche reducciones. In addition, Mapuches in Cautín, who often relied on jobs on neighboring estates or in cities and towns for subsistence, experienced high levels of unemployment. *El Diario Austral* reported in 1971, for example, that "unemployment is horrible in Cautín because there are 20,000–25,000 unemployed Mapuches living piled together in their communities where there are only a few hectares of land, a few cows, sheep, pigs, and hens, but where there is no work that allows Mapuches to obtain a minimal daily wage to survive."[95] Jacques Chonchol noted that "time has gone by since the reducciones were constituted with 6–7 hectares per person, and today in many cases they have fewer than 2 hectares and in some case less than 1 hectare of exhausted land, of eroded soil [per person]." Chonchol also underlined that the prolonged crisis in logging in native forests, which dated back to the mid-1950s, had limited the jobs available to Mapuche communities in Cautín, "where the sawmills are practically paralyzed."[96] In addition, during the 1960s many landowners had shifted from cultivating wheat to pasturing livestock or to forestry because yields had diminished radically because of soil exhaustion. Estates that had once provided seasonal employment for members of Mapuche communities no longer offered as many jobs because raising cattle required fewer workers.

On land with undulating hills and plains in the coastal region near the Imperial River and around large cities and towns such as Temuco, Mapuche communities engaged in the agricultural activities of typical campesino smallholders. They cultivated wheat on hillsides and garden crops (potatoes, beans, and vegetables) on flatlands and plains. Wheat covered three-quarters of all cultivated land on Cautín's Mapuche reducciones. In addition, Mapuche communities raised livestock, especially in the cordillera and its foothills, where pasture could be found but cultivating crops was difficult. By the 1960s, native forests on reducciones in lowland areas or in the foothills of the mountains had been cleared for pasture or crops, to build fences and houses, or to produce charcoal, although most communities held small patches of native forest on which they relied for fuel, herbs, fowl, and game.[97]

In 1965, the Forests Action Committee of Imperial embarked on a program to reforest 215,000 hectares of the department's land with pine. As the committee reported, almost three-quarters of Imperial's soil was "forestal," even though the forests on most had been cleared or were severely degraded and fragmented. Agricultural land was in a state of "rapid decline in productive capacity . . . because of the erosion of the soil [and] the formation of swampland because of indiscriminate exploitation on minifundios, without tactics or methods, during generations in an increasingly intensive form." This situation had become more acute as small holdings were subdivided over generations. As the committee noted, "Because of the extreme subdivision to which it is submitted, pastureland has lost its productive capacity and is in a condition to only be used in reforestation." Similarly, the "monocultivation of wheat" had led to widespread soil exhaustion "to such a point that the harvests are extremely small, which has created a series of unfortunate consequences for small and medium-size agriculturalists." Part of the explanation for the predominance of the minifundio in the Department of Imperial was that two-thirds of the department's 150,000 inhabitants belonged to Mapuche families, most living in reducciones on small plots and with the majority reduced to "2 hectares of extremely impoverished soil."[98] In 1970, *El Sur* of Concepción explained the wave of Mapuche land occupations as a result of ecological degradation: "there are 3,483 reducciones with 565,931 hectares, of which only 131,410 can be used in agriculture. The rest, 434,521 hectares, are cursed deserts, destroyed by erosion that has left them sterile. This is the tragic truth, then, that today pushes the Mapuche to invade estates."[99]

El Sur employed a convincing ecological argument to explain Mapuches' land invasions, but soil erosion did not occur in a vacuum. On Mapuche reducciones, erosion was the consequence of a long history of land usurpations that led Mapuches to depend increasingly on the intensive cultivation of their shrinking plots. Mapuche reducciones in Cautín had received an average of 5.6 hectares per person during the original settlement process. By the 1960s, this had dropped to 1.9 hectares because of land usurpations and demographic growth.[100] It is difficult to assess the quantity of land lost to illegal usurpation, but one study of Arauco Province conducted by the Inter-American Commission on Agricultural Development in 1963 found that Mapuches owned only 1.2 hectares of land per person because of usurpations. In Cañete, for example, reducciones had lost almost a third of their land to neighboring large estates, some of which held expanses of 14,000–17,000 hectares.[101] The study noted that the pattern of usurpation by large estates in Arauco extended south to Cautín and Valdivia. Land scarcity was exacerbated by demographic growth. While the census of 1907 found that Chile's Mapuche population numbered 107,000, by 1970, that number had grown to more than 500,000.

The UP's program to promote development in Cautín focused on three areas: the intensification of Frei's agrarian reform to redress the fundamental problems of inequality in landholding and to meet the demands of increasingly radicalized Mapuche communities; the promotion of logging in native forests to provide income and jobs for Mapuche communities; and reforestation with Monterey pine. As part of an "emergency plan" for Cautín, the National Reforestation Corporation worked on a program with CORFO to invest in the reforestation of the province. In addition, SAG initiated a program to reforest state-run reserves in Quelle, the Villarrica National Park, and El Peuco, among other locations, with the ultimate goal of employing eight hundred workers, many presumably drawn from Mapuche communities in Cautín.[102] The UP also continued the Frei government's policy of reforesting on asentamientos. During the first months of the UP, government agencies, primarily CORA, CORFO, and the National Reforestation Corporation reforested thirty-one asentamientos in Llaima, Allipen, and Loncoche, regions with large Mapuche populations, with Monterey pine and Douglas fir. This reforestation program employed 328 workers.[103] In addition, campesinos organized a cooperative on expropriated estates with 5,500 hectares and 120 Mapuche families from around Lumaco. The Lautaro cooperative produced wheat, beans, potatoes, chickpeas, and corn, but it also planted 10,000 Monterey pine trees during its first months. The communities in the cooperative entered into an agreement with the government to provide significant chunks of eroded land to be reforested with Monterey pine.[104] According to a study of the agrarian reform by Correa, Molina, and Nancy Yáñez, numerous Mapuche reducciones and asentamientos in both Cautín and Malleco similarly reforested with state aid during the UP government's time in office.[105]

The government also established state markets to acquire wood logged by small producers in the region's native forests.[106] In 1971, the UP established wood markets in Angol and Curacautín, in Malleco Province, and in Pillanlelbún, Carahue, Cherquenco, Cunco, Villarrica, and Ranquilco, in Cautín Province, reactivating logging in the coastal and Andes cordilleras.[107] In addition, Forestal Pilpilco, now a subsidiary of CORFO, opened up wood markets to purchase logs for its mills and wood factories.[108] The UP also installed a modern, state-run sawmill in Temuco, which acquired wood from small producers in Cautín.[109] The goal, as UP officials stated, was to reduce unemployment in Cautín by stepping up logging by small sawmills on campesinos' small parcels and within public reserves, while expanding reforestation projects.[110] In late 1970, CORFO and the National Reforestation Corporation elaborated a plan to employ 4,500 people in logging and sawmills in Cautín.[111]

These policies benefited Mapuche communities, which found markets for wood felled on their reducciones and hundreds of new jobs in logging and mills. Since the late 1950s, a number of Mapuche communities in the region had requested the establishment of sawmills and lumber markets to alleviate their growing poverty. Selling wood to local mills or the railroad was essential to their subsistence as their populations grew and wheat yields plummeted. In Coihueco, Panguipulli, for example, a Mapuche Comité de Araucanos demanded a market where they could sell sleepers directly to the railroad. A newspaper report noted, "Now they have to sell them to unscrupulous intermediaries who exploit them."[112] Similarly, that same year in Puerto Saavedra, the Hueñalihuén community petitioned for the installation of a sawmill because it wanted to exploit the forests "that remain in the zone."[113] Many of Cautín's communities approached the remaining native forests much as their non-Mapuche campesino counterparts did. They survived by clearing land to cultivate grains and vegetables, as well as to create pasture for their livestock. Native forests represented an important supplement to their incomes; the wood could be converted into charcoal and sold in local markets, or forested tracts could be leased to local sawmills.[114] The UP forestry programs aided these communities by providing both markets for their wood and funds for reforestation. In the end, however, the UP's emergency plan for Cautín implied the substitution of the remaining patches of second-growth native forest in the province's agricultural central valley with pine plantations.

The socialist agrarian reform allowed Mapuche communities to see in forestry development a means for guaranteeing the survival of their reducciones. Many Mapuche communities enthusiastically embraced forestation with Monterey pine as an essential tool for recovering their badly eroded soil and as a potential source of future income. On asentamientos and cooperatives that signed reforestation agreements with CONAF, peasants continued to hold land for agriculture and pasture, raising small herds of cattle and cultivating garden crops and wheat. They began to build a mixed agricultural, pastoral, and forestry economy. As with the case of forestry workers on expropriated estates, both Mapuche and non-Mapuche smallholders expanded their access to plots devoted to livestock and garden crops and reforested with Monterey pine on eroded land. State credits, subsidies, and technical support were essential to this process, because few smallholders had the capital or forestry knowledge that would allow them to reforest with pine, manage their plantations, and wait twenty years to harvest their trees.

There were, of course, exceptions. Some campesinos in the 1970s (as they had in the 1950s and 1960s) saw forestation as a force of dispossession. In 1972, for example, members of the Pulitro Mapuche community occupied the Pulememo estate, property of the state's National Reforestation Corporation. The corporation had replanted Pulememo, which lay on the coast near the city of Osorno, with pine. Members of the community, who claimed that the estate's land had belonged to their ancestors, caused considerable damage to the pine plantation, which they viewed as covering land that was rightfully theirs.[115] In this case, the state had made no effort to include the property in the agrarian reform or to incorporate the local indigenous communities into the reforestation program. In a similar case that same year, twenty campesinos who had joined the Lucha por Tierra (Struggle for Land) committee occupied the San José estate near the city of Nueva Imperial in Cautín. Like Pulememo, San José was the property of the National Reforestation Corporation. The campesinos closed off access to the estate, stating that they would maintain the occupation until they received assurances that the land would not be reforested with pine.[116] These cases illustrated the tensions between some campesinos' demands for land and state-directed reforestation programs. The two estates had been expropriated but transferred to the National Reforestation Corporation to be planted with pine rather than to CORA for distribution to their workers or neighboring Mapuche communities.

Nonetheless, throughout southern Chile, the UP sought to balance campesinos' demands for land with forestry development. In areas with few remaining native forests, the socialist government worked to meet campesinos' petitions for land while incorporating them into the rapidly expanding industrial forestry economy. While many campesinos viewed industrial forestry and tree plantations as a threat to their subsistence, both the Allende and Frei governments worked to direct the benefits of forestry development to southern Chile's rural laborers and smallholders. In a number of cases, campesinos responded enthusiastically to state-directed forestry programs because government support made forestation a feasible activity and agrarian reform allowed them to mix forestry with their traditional agricultural activities. Forestry development allowed campesinos, with the support of state agencies such as INDAP, CORA, and SAG, to recover unproductive eroded soil and cultivate a new crop with promising prospects.

The brief moment when industrial forestry development, forestation, and conservation were organized to reduce social inequality in the southern countryside and were balanced with the exigencies of agrarian reform came to an end when the Chilean military toppled Allende's democratically elected socialist government on 11 September 1973. The dictatorship of Augusto Pinochet combined ruthless repression of the left and the labor movement with free-market economic shock therapy designed by the Chicago boys. The neoliberal policies of the economists trained at the University of Chicago accelerated forestry development, particularly the expansion of pine plantations throughout the south, building on the foundation laid by the Frei and Allende governments. But after 1973, industrial forestry no longer worked to the benefit of the rural poor. Instead, the fruits of an industry built with significant state input since the 1930s were directed into the hands of a small group of financial conglomerates, which took over forested land and expelled campesinos from estates throughout southern Chile.

Dictatorship and Free-Market Forestry

On December 31, 1983, Chile's military dictator, General Augusto Pinochet, addressed the nation in his annual New Year's Eve speech. The country had suffered a year of devastating recession, and Pinochet struggled to find some glimmer of hope for an economy battered by the world recession and the radical neoliberal economic restructuring by the regime's economists, the so-called Chicago boys. Pinochet pointed to the forestry industry as the military government's major success story, boasting that the regime's free-market policies had led to a "forestry boom" in exports, as well as to the forestation of 1 million hectares of land.[1] During the late 1970s, the forestry sector grew at double the annual rate of national economic growth, making forestry products a major source of export revenues, third behind mining and manufacturing by the early 1980s.[2]

In 1979, Chile had planted about half of the pine cultivated globally that year, taking advantage of the country's comparative advantage in the tree's speedy rate of growth.[3] That same year, Chile's four major forestry companies established a consortium that, according the national daily *El Mercurio*, constituted the country's second largest exporter after the state-owned copper company Corporación Nacional del Cobre (CODELCO).[4] Between 1973 and 1980, the value of Chile's forestry exports increased by a factor of ten, making the forestry sector a "mini-miracle" and a cause célèbre for the Pinochet dictatorship.[5] "Wood is Chile's new copper," Joaquin Lavín, a minister in Pinochet's cabinet and one of the Chicago boys, exulted in his famous celebration of Chile's "silent" free-market revolution.[6]

The reality behind the so-called forestry boom differed significantly from the rosy picture painted by Pinochet and reflected by the macroeconomic indicators. The rapid expansion of the forestry sector during the late 1970s and the 1980s, frequently cited by regime loyalists as a shining example of the Chilean free-market economic "miracle," had its origins in decades of state planning and investment in both pine tree plantations and lumber, paper, and pulp companies since the 1930s. Indeed, as I have shown, the foundation for the forestation and forestry export booms had been laid during the governments of President Eduard Frei and President Salvador Allende. To be sure, the regime's liberalization of trade stimulated accelerated growth in forestry exports, especially of pulp and logs produced from Monterey pine, but forestry exports had been steadily increasing since the 1960s, and the trees that provided the raw material for the post-1973 export boom had been planted during the 1960s and early 1970s.[7]

Pinochet's boast also concealed the harsh social costs that accompanied the forestry boom of the 1970s. Behind the flattering statistics, the wholesale reorganization of social relations in the countryside following the military coup in 1973 had made the rural zones dedicated to the production of forest products among the poorest in the country, with unemployment rates that exceeded the national average of more than 20 percent (in some zones, it approached 50 percent) and a ballooning population of landless peasants who lived in rural *poblaciones* (shantytowns) along the sides of roads and highways. Workers in the forestry sector suffered the lowest wages and worst working conditions in the country, laboring on a temporary basis for numerous subcontractors. Finally, Pinochet's promise that the forestry boom would provide a model for growth led by the private sector ignored the tremendous ecological costs of this particular version of export-oriented economic development. After 1973, the accelerated destruction of native forests and their replacement by pine and eucalyptus plantations eroded the crucial biodiversity of southern forest ecosystems, while transforming rural labor relations by turning former peasants into a population of landless and itinerant part-time laborers with access to neither land nor the benefits of permanent wage work and unionization.

The effects of forestry development on the environment and on rural labor were not parallel. The erosion of rural ecosystems played a major role in the demise of the peasant economy and the conversion of rural workers, who at one time combined forestry work with subsistence and commercial agricultural production, into temporary wage laborers. The expansion of commercial monocultural forestry production for export was predicated on the availability of both cheap forest resources and an inexpensive labor force. As they were during the state-sponsored forestry development of the 1940s and 1950s, ecological and social processes were connected dialectically: the expulsion of resident estate laborers and smallholders by large commercial forestry companies made agricultural lands and forests available that could be cleared, often with fire, and planted with pine. At the same time, the expansion of the plantations undermined regional ecosystems that had sustained peasant agriculture and forced campesinos to sell their land and join a floating population of migrant temporary workers in shantytowns throughout the countryside. The comparative advantage enjoyed by Chilean forestry exports on the world market after 1973 was not only the speed with which pine grew in Chilean soil. Just as important were inexpensive raw materials, forest plantations subsidized by the state, and cheap labor.[8]

The Free Market and the Forests

The military coup that overthrew the government of Salvador Allende in 1973 led to the complete restructuring of the forestry sector. The military junta viewed the state-run plantations and wood, pulp, and paper industries as emblematic of the policies of the Unidad Popular (Popular Unity; UP) government and sought to promote an entirely different model of forestry development based on the radical free-market policies of the Chicago boys. For the military regime's economic advisers, privatization, the operation of market forces, and the promotion of forest exports would stimulate the development of the forestry industry. In many ways, the regime's embrace of the free-market economics of Milton Friedman (who visited Chile and met with Pinochet in 1976) marked a major departure from the state-centered developmentalist economic strategies of the post-1930 era. However, there were many continuities, as well. In the forestry sector, until the socialist UP government took office, the state had largely handed forests and the forestry industry to the private sector while providing protective policies and subsidies. In addition, since the era of the Popular Front, the state had looked at forests and the forestry industry as a source of export earnings, not just as an element of an inward-looking domestic industrialization program. The military government rolled back the UP's socialist development policies by returning forests, plantations, and forestry plants to private hands, but it kept a key role for the state in subsidizing private forestry companies and promoting commercial forestry, especially tree plantations and export-oriented forestry. In this it maintained certain basic continuities with Chilean forestry policy since the 1930s.

In southern forest regions, as in other rural areas throughout Chile, the regime dismantled the agrarian reform. Forests that had been transferred to the Corporación Nacional Forestal (National Forestry Corporation; CONAF) by the Corporación de Fomento de la Producción de Chile (Chilean National Development Corporation; CORFO), as occurred with other public land suitable for forestry and forestation, were auctioned off, as was land owned by state organizations such as the social-security service. The military regime returned a significant percentage of properties expropriated under the agrarian reform to their owners.[9] The Chilean National Development Corporation began to sell off industrial facilities, plantations, and forests at bargain prices, including the three major companies, Forestal and Celulosa Arauco, Forestal and Celulosa Constitución (CELCO),

Industrias Forestales (INFORSA), and the smaller Forestal Pilpilco and Forestación Nacional (FORESTANAC). These were companies that had been created by CORFO or created with significant input and investment from CORFO. Chile's major forestry companies, Forestal Arauco and Celulosa Arauco, were sold to the former state oil company, Compañía de Petróleos de Chile (COPEC), which had been purchased by the Cruzat-Larraín financial group. Celulosa Arauco, which had begun production in 1972 as a state-financed project, was one of the most profitable and efficient state-owned businesses.[10] In 1979, COPEC had also acquired CELCO, the most modern cellulose plant in Chile, which had been constructed and financed by the state. Industrias Forestales, developed by foreign capital in association with CORFO during the 1960s, was sold to the Compañía Industria (INDUS), controlled by the Vial financial group. Despite the military regime's commitment to the free play of market forces, all of these sales were made at well under the market value of the enterprises' properties and assets.[11]

In 1974, the regime declared Decree Law (DL) 701 to stimulate the privatization and growth of the forestry sector. The new decree law established twenty-year credits for private companies to cover the costs of forestation, including the costs of managing forests (pruning, thinning, and administration). The decree law also dictated a subsidy of 75 percent of the value of forestation; exempted land that was devoted to forestry from the taxes that applied to most agricultural properties; and established cheap lines of credit from the Central Bank, channeled through commercial banks, to foment forestation. A final clause made forested land legally ineligible for expropriation. A second decree law in 1975 allowed the export of forestry products in any stage of elaboration, creating the possibility of exporting logs without any value-added processing.[12]

Large companies, stimulated by the generous financial incentives in DL 701, as well as by the low prices at which public properties were auctioned off, began to accumulate vast tracts of land. Planting pine became profitable in and of itself, and the hectares planted soon outpaced the demands of the companies' industrial plants.[13] By monopolizing land and devoting it to pine plantations, large forestry companies began to squeeze out alternative agricultural activities, as well as the small landholders and resident estate laborers these activities sustained in forestry zones, particularly in the Eighth (Bío Bío) Region. Forestry companies began to expel inquilinos from the private estates they purchased and to buy land from campesinos at low prices, establishing a monopoly over forests. They then earned credits and subsidies from the state by planting pine, which in turn drove up the price of land. The control of land in forestry zones also ensured a cheap source of labor for the forestry companies, since former estate workers and smallholders, deprived of access to agricultural land, turned to wage labor in logging camps and sawmills.

The military regime's coercively imposed free-market model did not simply return the Chilean countryside to the status quo before the agrarian reform. While many estates were restored to their owners—or, at the very minimum, auctioned off and transferred to the private sector—the destruction of the traditional haciendas by agrarian reform allowed for a reordering of landholding. The concentration of forestry activities in only a few hands changed the nature of logging and landholding. Until 1973, midsize and small companies, not to mention the thousands of small sawmills that dotted the southern landscape, were a significant presence in the forestry industry. The three major financial groups not only were owned by families with tremendous political power and close ties to right-wing political parties and the dictatorship they supported; they also had their fingers in an extraordinary array of economic pies and, through their control of banks and financial institutions, enjoyed access to capital that was unimaginable to many pre-1973 logging companies and landowners. In essence, the dictatorship's combination of intense violence and radical free-market policies allowed the big players in the forestry industry to put together vast economic empires of plantations, native forests, sawmills, and pulp factories, setting aside more traditional landowners, who had already been displaced by agrarian reform, as well as small mill owners and loggers. In 1978, for example, the Matte group controlled, sometimes in association with the Cruzat-Larraín group, nineteen major enterprises, including the Compañía Manufacturera de Papeles y Cartones (CMPC); a number of forestry companies, including Forestal Mininco; factories for constructing furniture; and large sawmills, as well as finance companies, banks, and insurance companies.[14]

The reality behind the "forestry miracle" differed strikingly from the military regime's celebratory free-market rhetoric. While Pinochet had indeed presided over the planting of 1 million hectares of pine trees, in reality this had been accomplished through direct state action, state subsidies, and credits and pre-1973 investments in developing the forestry sector. The operation of market forces and private investments played relatively little role in the expansion of plantations and exports. Of the 1 million hectares planted between 1974 and 1980, CONAF was responsible for planting 37.5 percent and for financing with subsidies and credits another 39.4 percent. Private capital financed and planted only 23.1 percent of the pine plantations during this period.[15] In 1977, state investments in forestation totaled $8.5 million, while private sector investments came to only $1 million. In 1980, the state invested $6.5 million, and the private sector invested less than $1 million.[16] In addition, as analysts pointed out at the time, of Pinochet's heralded "1 million hectares of forestation," more than 80 percent had trees that were ten years old or younger and would be ready for production only over the next ten years. The trees logged and exported during the free-market boom of the 1970s had been planted before 1973 as part of the state-directed program of forestry development.

As I have shown, the factories that processed trees into lumber, pulp, and paper had also been founded with significant state input during the period before the coup. By 1989, the year Pinochet finally stepped down, Chile had basically the same number of industrial plants in operation for producing cellulose, paper, fiber board, particle board, plywood, and lumber as it had before 1973, since private investment was directed at expanding existing plants rather than creating new enterprises.[17] Between 1960 and 1974, production in the forestry sector grew at an average annual rate of 9.9 percent. Between 1974 and 1986, however, production increased at an annual rate of 8.4 percent, a rate that was lower than average productivity for the period between 1959 and 1986.[18] The increase in exports of forestry products from $42.5 million in 1971 to $362.2 million in 1983 consisted, to a significant degree, of the export of logs that before 1973 had been used in the production of lumber, limiting the value added through industrial elaboration. Exports of forestry products increased as internal demand, mostly from the construction sector, contracted due to the recessionary crisis that accompanied economic restructuring after 1973.[19]

While small and medium-size landowners and mill owners had played a significant role in the forestry industry in partnership with the state before 1973, by 1984 three major financial groups led by familiar and prominent figures in the forestry industry since the 1930s—Cruzat-Larraín, Matte-Alessandri, and Vial—had come to dominate the industry. Two companies alone, COPEC (Cruzat-Larraín) and the CMPC (Matte-Alessandri), controlled 50 percent of pine plantations and 100 percent of the cellulose industry.[20] The three financial groups taken together owned 75 percent of plantations, 78 percent of industrial production, and 73 percent of exports and received 85 percent of the state subsidies and credits for forest cultivation and management.[21] While they exercised monopoly control over the industry, the three financial groups invested little in new projects or in the expansion of the industrial enterprises begun before 1973. Rather, they profited by purchasing existing companies at bargain prices and by accumulating land to plant pine, for which they received state subsidies and access to cheap credit.

The Transformation of Rural Social Relations and the Environment

The abrupt transfer of forests and forestry companies from the state sector and campesino smallholders to a few major financial groups required the repression of the militant rural labor movement in southern Chile. As with campesino unions, forestry workers' organizations were shut down; their leaders were arrested; and their properties were confiscated. Only three of the seventeen leaders of the Federación Liberación remained unscathed by repression. The president of the union, Mario Ruiz, was detained in a concentration camp for four years, and a number of the union's leaders were simply disappeared; as a union leader looking back on the military coup remembered, "We never knew what happened to them."[22] On the El Morro, El Carmen-Maitenes, and Pemehue forestry estates in Mulchén, eighteen forestry workers employed by CONAF were detained and disappeared in the days following the military coup. In 1979, their bodies were found in a mass grave.[23] Perhaps the most devastating example of the repression of peasant families and communities in the forestry sector was the Complejo Maderero Panguipulli. After the coup, a large number of workers who had joined the parties of the UP, the Movimiento de Izquierda Revolucionaria (Revolutionary Left Movement), and the radical rural labor movement in the region were detained by the military. Forty-four workers from the Neltume, Chihuío, Futrono, Llifén, Arquilhue, Curriñe, Chabranco, and Liquiñe estates within the complex were detained, executed, and buried in clandestine mass graves.[24] Many of the survivors of the repression were forced to abandon the complex and migrated to settlements in Panguipulli and on the outskirts of Valdivia. Because of the complex's reputation as a center of radical political activity, the forestry center's new military managers fired thousands of workers and sent them packing on trucks.[25]

The repression of the rural labor movement and communities of forestry workers through military violence enabled the financial groups that had taken control of the forestry sector to reorder land and labor relations in the rural regions of southern Chile. Thousands of former inquilinos were expelled from estates acquired by large forestry companies, and campesinos were forced to sell their small parcelas through indirect pressure.[26] Rural communities that had been formed on forestry estates or in areas devoted to forestry production were broken up as forestry companies extended their control of forested lands to supply their sawmills, factories, and industrial plants. Small landholders, members of Mapuche communities, and resident estate workers lost their access to land and left farms and estates behind to migrate to cities or to squatter settlements along the sides of roads.

Transferring publicly owned forestry estates to the private sector signified more than an undoing of agrarian reform. For full-time forestry workers like those in Panguipulli and on estates covered with pine plantations, it meant the dismantling of the system of labor relations established after the Campesino Unionization Law of 1967. Workers lost their status as full-time employees and the job security, wage hikes, and benefits guaranteed by collective contracts and a robust union movement. The financial groups that took hold of estates planted with pine were able to revert to and extend the pre-agrarian reform system of subcontracting logging operations. Expanding subcontracting and temporary work in the forestry industry was consistent with the organization of labor relations throughout the forestry sector before 1967. This had been a bone of contention between unions and the CMPC even as late as 1972, as noted in chapter 7. After 1973, male campesinos once again came to constitute a mobile and flexible supply of cheap wage labor to estates, plantations, and sawmills, working for low wages, without benefits, on a temporary basis for subcontractors. They toiled in conditions that differed little from those of fifty years before, while women and children remained behind in rural shantytowns, deprived of access to the agricultural land that had been central to the rural household economy.

The fundos owned by the Complejo Maderero Panguipulli were the most important loci of logging in the Valdivian forests. During the UP government, production and logging intensified, though with a new concern with cultivating raulí plantations and more sustainable forest management. After 1973, the complex experienced a precipitous decline. Following the brutal repression of its laborers and union activists, most of the forestry center's 3,600 workers were fired without any compensation.[27] Nonetheless, the Complejo Maderero Panguipulli's new military managers did not immediately return the fundos to their owners. The military regime maintained an interest in keeping the complex, like the copper conglomerate CODELCO, in the public sector. This meant that the southern zone with the most extensive native forests remained in state hands during the first years after the coup. Nonetheless, many of the estate's former owners, such as the Kunstmanns, received the healthy compensation they had gone to court to demand. For former landowners this was a good deal, since the only valuable resource on the properties—the native forests—had already been degraded by decades of unchecked logging.[28]

Yet because of the regime's lack of interest in promoting a centralized state-run industry along the lines of the UP model, production declined radically. In essence, the Complejo Maderero Panguipulli returned to its pre-1970 forms of exploitation, logging the remaining valuable stands of raulí and, to a lesser extent, coigüe and tepa, and pasturing cattle on cleared land with a significantly reduced workforce that numbered in the hundreds rather than the thousands. Decree Law 701 made pine plantations an excellent investment because of subsidies, tax breaks, and the guaranteed markets provided by cellulose plants. Financial groups favored by the military regime focused their investments there. It would have been a costly and difficult process to clear the complex's native forests and replace them with pine. In addition, the project to build a cellulose plant in Panguipulli to provide a market for the complex's woods, which had been in the planning stages when the coup hit in 1973, never materialized since the military regime made no effort to pursue it. While the Bío Bío Region had hundreds of thousands of hectares of pine plantations already on their way to maturity and preexisting industries that could be bought cheaply, Valdivia's Andes cordillera had no nearby industry to supply and deficient ecological conditions. It also required massive clearing of native forests to plant pine, as well as investments in a new cellulose plant.[29] These were the kinds of investments that the large financial groups, despite their touted entrepreneurial élan, were loath to pursue.

In 1979, in keeping with the regime's free-market philosophy, CORFO decided to transfer the Complejo Maderero Panguipulli's fundos to the private sector. Privatization, it believed, would promote the development of the cellulose industry in Panguipulli (the same project planned by the UP and aborted after 1973). But two years later, CORFO had succeeded in selling only four of the fundos and there was no project for a new cellulose plant. In the case of one fundo, Trafún Norte, part of the estate once owned by the Kunstmanns, the Aledo Transnational Trading Company purchased the property with the goal of logging the forests with a modern sawmill and then establishing herds of livestock and a meat-packing plant (which never materialized). In essence, the company proposed to pursue the traditional path of landowners in the region: logging the forests and then introducing cattle to graze in the bamboo and underbrush that invaded cleared land.[30]

The military regime itself was conflicted about selling off the Complejo Maderero Panguipulli. In 1978, for example, the Dirección Nacional de Fronteras y Límites (National Directorate of Frontiers and Borders) vetoed a proposed sale of a number of the complex's fundos, including part of Trafún. In 1980, CORFO's directorate, which included Sebastian Piñera and Eugenio Matte, proposed selling the estates "to fulfill the policy of the supreme government." Yet the Directorate of Frontiers and Borders refused to approve the sale.[31] Similarly, as late as 1986, the Ministry of Defense imposed a prohibition on the sale of the Pirihueico and Pilmaiquén fundos. In both cases, the logic of national security overrode the privatizing drive of CORFO's directors. In addition, before 1978, the complex's military administrators hoped to make the forestry center a pole of development, along the same lines as the UP. Some were particularly interested in the project of developing a cellulose plant in Panguipulli.[32]

COFOMAP took over the management of the estates' degraded forests, cultivating small stretches of raulí and North American pine plantations. Interestingly, while these plantations, administered by CONAF, constituted an important continuity with the first raulí plantations cultivated during the Frei and Allende governments, on the estates that were sold or returned to private owners, no such investment in reforestation took place. A study by foresters at the Universidad Austral found that during the 1980s, "Privatization signified the elimination of the last existing stands of native raulí and the destruction of extensive plantations of raulí through the introduction of livestock without any control. . . . We know of no privately owned estate that conserves the remaining native forests, where silviculture management is practiced to perpetuate the forestry types and species." The private sector showed little interest in investing in and managing plantations of raulí or second-growth raulí forests. As the study noted, "The tendency of private landowners is to widen the extensions of land with livestock or, in the best of cases, to establish plantations of fast-growing exotic species." The study concluded that "given the low long-term profitability of investments in [reforesting] native forests with species that have a longer rotation than exotic species, no private party wants to take on the business of transforming the already exploited forests into productively managed forests. Only state companies like CONAF and COFOMAP [Complejo Forestal y Maderero Panguipulli] can and should take on this challenge."[33]

In part, this was true because both the Chilean state and international development agencies focused on the production of pine to supply the cellulose and paper industries after the 1940s, a development philosophy maintained by the military regime. From an environmental perspective, the forestry strategy developed by the Chilean state during the 1940s was to reproduce the ecological conditions of North America and to apply the technology and systems of knowledge developed there to produce a profitable forestry industry.[34] As one CONAF forester noted, "On the pine plantations and the sawmills, we have the most modern technology and systems of production, just like in the United States, Europe, Japan, Canada. We have studies and detailed knowledge about everything that has to do with pine and eucalyptus. But we know very little about managing native forests." He pointed out that the state and CONAF had invested little in studying the sustainable management and exploitation of Chile's native forests.[35] During the 1970s and 1980s, it was still cheaper to exchange natural environments, the forest, and natural resources of Chile for imported forests, technology, and forms of exploitation than to develop a forestry industry based on the sustainable exploitation of Chile's native forests. As a report by the Instituto Forestal (INFOR) observed in 1991, "The country has achieved significant development of the forestry sector based on its plantations, which have impelled a dynamic and internationally competitive industry. However, in terms of its native forest resources, it has not known how to take advantage of the potential that these have as a source primary materials and industrial growth."[36]

The idea that the Complejo Maderero Panguipulli could be a profitable state-run company along the lines of the state copper company, CODELCO, eventually ran up against the privatizing logic of the regime. In 1987, having weathered the severe recession of 1983–85, the military regime initiated a new wave of privatizations, which included returning Neltume and Carranco to their previous owners and selling off whichever estates remained. Production on Neltume and Carranco had been paralyzed since 1981 due to a plethora of lawsuits against CORFO by the fundos' former owners. The lawsuits were resolved when CORFO finally returned Neltume and Carranco. Their owners then sold the fundos to a French forestry company.[37]

In general, the Complejo Maderero Panguipulli's fundos had only degraded and second-growth stands of Valdivian forests. Most had been intensively exploited by private landowners before 1971 and by the complex after 1971. While on some of the estates CONAF and the complex continued previous policies of reforesting with raulí and pine following the military coup, on other estates, logging operations were subcontracted to private companies that returned to pre-1971 methods of extracting valuable species with no forest management program or reforestation plan. In the case of the 28,000-hectare Pilmaiquén fundo, for example, a report from 1987 observed that "all of the areas with good forests have been intensively exploited, which is why the quantity of available commercially valuable wood is unknown." Pilmaiquén's forests had been logged by "CORFO with subcontractors."[38] Another report on the Complejo Maderero Panguipulli's Huilo-Huilo fundo indicated that CORFO employed subcontractors who extracted raulí from the forest unchecked by any regulation.[39]

In Alto Bío Bío and Lonquimay, the military coup restored the pre-1970 balance of power between Mapuche communities and forestry companies. Throughout Alto Bío Bío and Lonquimay, estates covered with araucaria forests that had been the object of long-standing conflicts with the region's Mapuche-Pehuenche communities, including the Trapa Trapa, Queco, and Pitrillín fundos, were returned to their former owners. The Galletué fundo, expropriated in 1971, was handed back to the Lamoliatte family in 1974, along with the Quinquén fundo, putting an end to the project of creating a nature preserve in the Quinquén valley.[40] The Meliñir kinship line

was left without a land title and without access to the veranadas and araucaria forests its families had relied on for survival. In effect, they returned
once more to their former status as squatters within the borders of a privately owned estate. During the 1970s, the Galletué company began legal
proceedings to have the Meliñir community evicted from land its families
had occupied for generations. In addition, Galletué reinitiated logging the
araucaria forests until 1976, when the araucaria was once again decreed a
protected species by the Pinochet regime.[41]

In October 1973, only a month after the military coup, a group from the
Huaiquillán community petitioned the director of the Institute for Agricultural and Livestock Development (INDAP) to undo a contract signed
by a "minority group" within the reducción with the Sociedad Agrícola Sierra Nevada, based in Lonquimay, to exploit araucaria pine "at a price well
below the market price."[42] Sierra Nevada had taken advantage of the post-
coup political climate to renew its logging operations in the community's
araucaria forests. The institute's new directors, who were sympathetic to
private-sector interests, agreed to a proposal to sell rights to log the community's forests, settling on a public competition for the leases.[43] They did,
however, stipulate that timber companies could log only those trees that
had diameters larger than thirty inches, and they had to be supervised by
government technicians. In addition, the milling of the wood would take
place on the reducción, and community members would be given jobs
logging and working in the mill. The logging contractor would also allow
the community use of the sawmill. Members of the community would be
appointed to oversee the felling and milling of the trees to enforce the contract's stipulations.

Sierra Nevada won the rights to the logging lease, even though it had
long maintained a conflict with the community over the araucaria forests,
paying the community for its forests with basic provisions that would allow
its members to stave off starvation. In May 1975, the community petitioned
INDAP to be allowed to receive a shipment of flour from Sierra Nevada
as an advance on the payments it was owed for the logging rights.[44] The
following March, a similar payment was arranged between the two parties; the community would receive flour, wheat, and pasta in exchange for
its forests.[45] It is worth noting that this is precisely the kind of arrangement that logging companies had proposed during the 1960s and that
had been rejected by the Frei government's representatives in the Forest
Department.

A similar fate appears to have befallen the Bernardo Ñanco community.
Although under the UP land surveyors and land officials had determined
that 8,000 hectares of the community's land had been illegally incorporated
into Fressard's Chilpaco estate and ordered the usurped land restored to

the community, after 1973 Fressard regained the 8,000 hectares and maintained its control of this land until 1984, when the community finally won restitution following a long process of subdivision. Community petitions from the 1970s reflect a new desperation. In 1975, for example, the Ñanco community requested permission to sell araucaria trees that had been felled by both the Fressard company and the Sierra Nevada company on its land to the large Alaska sawmill owned by Fressard.[46] Many of the community's members had also fallen into debt to the owner of a sawmill in the region and had begun to pay him in wood from araucaria stands that lay within the reducción. In 1979, the community, "seeing the economic needs of the people of the reducción," sold to the mill a number of araucaria logs that had been felled illegally on its land by local logging companies since 1976.[47] The revocations of the land expropriations by the Corporación de la Reforma Agraria (Agrarian Reform Corporation; CORA) thus opened the door to renewed logging in araucaria forests in Lonquimay and reignited long-standing conflicts between desperate Mapuche communities and loggers eager to gain access to valuable wood.[48]

In the central valley and foothills of the coastal cordillera after 1973, government incentives stimulated large companies and financial groups to expand their landholdings to plant pine by purchasing large estates and small farms. With rural labor and the political left decimated by repression, the process of expanding tree plantations and logging operations onto land once occupied by campesinos encountered little opposition. Indeed, the repression of the rural labor movement and the generalized terror implemented by the Pinochet dictatorship served as an additional subsidy to landowners and logging companies that sought cheap labor and land. By the early 1980s, estates throughout southern Chile planted pine trees where they had once cultivated wheat and corn or pastured animals, pushing out resident agricultural laborers in a wave of expulsions that far surpassed the evictions of estate laborers during the 1940s and 1950s. In addition, few small landowners could compete with large forestry companies as the push to plant pine plantations consumed rural areas in the provinces of the Eighth Region, Arauco, Bío Bío, Concepción, and Ñuble. Campesinos could not afford to plant pine and then wait twenty years for their land to produce, since many depended on their parcelas for subsistence. Nor did smallholders have access to the credits and state subsidies enjoyed by the large companies. As a forester with CONAF pointed out in noting the demise of smallholdings in the forestry sector, "The state doesn't provide incentives or subsidies for the small producers. Only the large companies receive credits and incentives."[49]

Planting pine required long periods for the trees to mature, large extensions of land, and a great deal of capital.[50] In addition, large companies' monopoly of sawmills and factories allowed them to purchase wood from smallholders at low prices. As another forester pointed out, "The large monopolistic companies such as Celulosa Arauco and Celulosa CELCO [drove campesinos out of forestry zones by] fixing low prices [for the wood of] of small producers."[51] Similarly, a leader of the forestry workers' federation, the Confederación de Trabajadores Forestales (CTF), noted that "there were many small landholders who, sadly, could not compete, who lacked the resources, and it was easier for them to sell off their land.... [T]hey were smallholders who had small stands of forest that they exploited, but they did not have the resources to make an adequate investment, to buy plants, to contract a technician.... They lacked the training and the capital ..., and thus the smallholders began to disappear.... Many stayed on working for the same businesses that had purchased their land."[52] For union leaders, the lack of state support for small producers and the monopoly of land, mills, and industries exercised by large forestry companies led to deepening rural poverty and pushed peasants to migrate to the outskirts of rural towns and cities, "Where they live in shantytowns in belts of poverty."[53]

Many members of the asentamientos created during the agrarian reform, suddenly deprived of credit and access to basic infrastructure such as seeds and fertilizer, were forced to sell their land to forestry companies. Around Concepción, for example, as one agronomist recalled, investors frequently bought former asentados' lands and planted pine, earning benefits and subsidies under DL 701, then sold the plantations to the large forestry companies.[54] On asentamientos that had entered into agreements with CONAF to plant pine on their land, smallholders frequently were forced by economic necessity to sell their plots. Often the plots were purchased by investors who then mortgaged them to acquire inexpensive credit for other business endeavors. When their businesses failed during the economic crises of the late 1970s and early 1980s, their holdings went back to the banks and were sold off at low prices to big holding companies such as Matte-Alessandri (CMPC). According to Alex Rudloff of CONAF, many campesinos who received land under Frei's and Allende's agrarian reforms had entered into agreements with CONAF to plant pine. By the late 1970s, most of them had sold their plots and pine, allowing for the concentration of the plantations in the hands of the major forestry companies.[55]

In one representative case in 1975, an asentamiento on the former El Volcán fundo in Villarrica petitioned to have its land subdivided so the twenty-two families who had been granted land during the agrarian reform could sell their individual plots. According to the smallholders, they had entered into agreements with CORA and CONAF to reforest the land. Further, CONAF had hired members of the asentamiento to work in the reforestation project, giving them an indispensable source of income. In 1975, however, their economic situation appeared desperate; they no longer had wheat and flour and had no way to buy "their most urgent necessity, food." Wood markets once subsidized by the UP, to which the asentados had sold lumber extracted from their remaining stands of native forests, had shut down. In addition, the military regime withdrew the credits, technical support, and other inputs, including saplings, that the Frei and Allende governments had directed to smallholders. Like many asentamientos, the campesinos were deeply indebted to the state bank and CORA, had little income, and faced starvation. Under these circumstances, subdividing and selling off land planted with pine by CONAF seemed to be the only option.[56] However, the asentados' plans to divide El Volcán met with little success. The following year, CORA revoked the expropriation and restored the fundo to its former owner.[57] In many cases like this one, CONAF was able to get back the money it had put into reforestation, but as Rudloff remembers, in many other cases, CONAF, having invested in reforestation, simply did not receive any money or sold its rights to plantations at below-market prices when the parcels on former asentamientos were sold to forestry companies or returned to their former owners.[58]

The new military officials who staffed CORA's offices and tribunals employed old arguments to justify the reversal of expropriations and the expulsion of campesinos from forestry estates. In Mulchén, for example, CORA revoked the expropriation of the 5,000-hectare Villacura fundo in the Andes cordillera. Before the agrarian reform, Villacura's owners had logged the estate's abundant native forests. An investigation by CORA had found that the landowners had applied no system for managing the forests to promote their regrowth. The owners had never even hired foresters or agronomists to assist them in elaborating a forest management plan. Nonetheless, following the military coup, the CORA tribunal revoked Villacura's expropriation and dissolved its asentamiento, arguing that only the former landowners enjoyed the resources to implement rational exploitation of the forests, guaranteeing their regeneration. A post-1973 CORA report contended that the landowners "understand the magnitude of the problems that a forestry enterprise confronts in this zone and also understand the need of maintaining forestry reserves for the new generations."[59]

The new report on the estate emphasized that a peasant cooperative or asentamiento would not have the same understanding of the need to conserve forest reserves. If the land were granted to individual asentados, the campesinos would not have enough capital or machinery to exploit the forests productively. Moreover, the CORA report contended, "It must . . . be recognized that the campesinos do not understand as they should the idea of the conservation of natural resources."[60] It is notable that conservation and rational management of native forests continued to play a role in evaluating property rights, even as the military dictatorship rolled back agrarian reform, withdrew support for asentamientos, and restored fundos to their owners. Demonstrating an "understanding" of the need to regenerate the forest or reforest with pine now constituted a well-established criterion for establishing property rights in southern Chile. While the UP government had used this principle to define large estates as suffering from inadequate management during agrarian reform, under the military dictatorship campesino smallholders were once again characterized as a threat to the forests.

Campesinos' dispossession in forestry zones was accompanied by the expulsion of resident laborers from estates dedicated to forestry production. As landowners sold their estates to the big forestry companies or turned to planting pine themselves to take advantage of the incentives in DL 701, they began to evict permanent workers and their families. Whereas many rural estates had once engaged in agricultural production that had sustained a significant population of inquilinos, they now substituted both agricultural crops and laborers with pine trees. On the Totoral estate in Coelemu, for example, a disagreement over work hours between twenty workers and the estate's manager led to the workers' immediate dismissal. Shortly afterward, the estate was sold to a large forestry company that evicted the workers and their families, most of whom had lived on the estate for twenty to thirty years and who now found themselves living in a rural settlement on the side of a highway.[61] This situation was repeated throughout the region. As *El Pino Insigne,* the publication of the Vicaría Pastoral Obrera (Workers' Pastoral Vicariate) in Concepción, noted, "In Ranquelmo, the Magosa [estate] joined the show and fired forty workers, not coincidentally the oldest. Why? And the Magaluf [estate], with more than one hundred workers, is shutting down this month. The Puchacay estate . . . shut down here, shut down there—dismissals, unemployment, debts, . . . [N]o union organization."[62]

In Buena Esperanza, where Filadelfo Guzmán had helped to organize the Sindicato Nahuelbuta and then the Federación Campesina Caupolicán during the 1960s, the estate was sold in 1979 to Forestal Arauco, which then expelled the workers and their families. As Guzmán recalled in an interview with *El Pino Insigne,* "Then everything changed. The Forestal threw us all out and kept the [pine] forest that we had planted."[63] One hundred and three workers and their families, who had lived on the estate for fifteen to forty years were accused of being "*revoltosos*" (rebellious) for resisting the expulsion and were forced by carabineros to leave the estate and watch as trucks brought in temporary laborers hired by private contacting firms to work the forests.[64] The eviction brought an end to both the community and the union that had been formed on Buena Esperanza during the 1960s. For smallholders in the region, Forestal Arauco's expansion also spelled disaster. Campesinos in the neighboring villages who had not already sold their plots remained surrounded by the company's pine plantations. The only roads in the region were built for the company's trucks, and campesinos were barred from access because their carts damaged the roads. In addition, the campesinos were forbidden to enter the forests to collect firewood from rotting branches or to collect mushrooms, seeds, and pinecones. Forestal Arauco installed private guards to expel those who entered. The campesinos wound up selling their land to the company.[65]

The expulsion of the rural labor force from forestry areas was intensified by changes in regional ecosystems and strategies of land use. Traditional patterns of rural residence posed a threat to the extension of pine plantations. Large companies sought to displace smallholders and resident laborers from the margins of their expanding estates by establishing security zones protected by private guards on the outskirts of plantations. Forestry companies feared forest fires as a danger to their plantations and viewed campesinos, who relied on fire for fuel, heat, and clearing land, as a risk to their investments. Like Forestal Arauco, the companies began to prohibit campesinos from collecting forest products in their woods and from using paths and roads through their property. Large pine plantations began to encircle campesinos' small plots and cut off their access to water, forests, and transit, making peasant production for subsistence increasingly untenable. As a union leader from the Federación Forestal Liberación noted, "In Concepción Province, there were a lot of smallholders with 5 [hectares] even 20 hectares of land. The holding companies that arrived [during the 1970s], [such as] Arauco [and] Mininco . . . surrounded and isolated the smallholders. . . . They had them trapped and did not let them leave their land. They could not raise animals, since animals damage young pines. The campesinos were pressured off their lands and forced to leave for the city."[66] This assessment echoed a study conducted by the forester José Ignacio Leyton for the United Nations Economic Commission for Latin America (CEPAL) in 1986. He found that "the sale of small farms increases in the zones where the large [forestry] companies operate. First, when they buy large amounts of land the companies close off the roads, putting obstacles in front of campesinos, forcing them to sell."[67]

The testimony of a campesino printed in 1983 in *El Pino Insigne* described the disruption of peasant agriculture as pine plantations replaced native forests in the coastal cordillera and its foothills. During the winter, campesinos would go into the mountains with their carts to bring back firewood for fuel and heat. In addition, they would collect piñones to be used as a substitute for flour. Campesinos also took their livestock up into mountain forests for the winter, where the animals would be protected by trees and feed on tender bamboo shoots until they could return in the spring to pasture on grass in the valleys. Mountain streams provided water for campesinos' small plots and, at times, trout for fishing. Campesinos also hunted small game and birds in the forests:

> Nobody claimed the forests in the cordillera; everyone depended on them and used their products, but they didn't belong to anyone. . . . [B]ut if someone sets fire to the woods or if an ambitious forestry company destroys them, everyone ends up prejudiced because the rivers become muddy like chocolate and those who live below are left with no water; the birds take off and fly farther south, and no one can collect cones anymore or take animals up to pasture in the winter. . . . In our region, we see that the plantations are fenced off; the roads are barred; the water of the Andalién, Bío-Bío, and Trongol rivers is contaminated because the landowners exploit the forests that grow on the banks of the watersheds and they don't care who uses the water below.[68]

Forestry companies replaced the native forests that had been the staple of the peasant economy with pine plantations, eroding the regional ecological biodiversity. In 1983, *El Mercurio* described the scene as forestry companies burned native forests in the foothills of the Andes cordillera to plant pine trees:

> The day was strange. Quilleco's sky [near Los Angeles toward the cordillera] was leaden gray and half pink, even though there were no clouds. . . . But as you looked toward the mountains, you saw what was happening: huge columns of smoke were rising from the hills. Fat raulís with their yellow autumn leaves burned along with gigantic coigües more than thirty meters tall; secondary forests of robles, avellanos and lingues were rapidly consumed by the flames. A small forest deer (*pudú*) jumped among the branches covered in flames, looking for a stream to throw itself into. *Torcazas* [Chilean dove] and *choroyes* [Chilean parakeet] flew away toward somewhere where the pine has not yet arrived. The fire engulfed the estates of Las Lumas, Olvillo, Manzanares, and Los Cuartos, whose land covered 3,000 hectares. Nothing was to be left alive so that next winter, Forestal Mininco could plant its pine trees. . . . They had to be planted to earn the forestation subsidy. . . . It was not a bad business: you could get about 25 million pesos [$9,600] from the state for each hectare planted.[69]

As *El Pino Insigne* noted, "These events are repeated every year in our region. Who hasn't seen the pink clouds rising above the Nahuelbuta cordillera in burning season . . . or in Cayucupil, or Trongol, or Coihuesco, or Antuco. . . . Every year, the companies burn between 7,000 and 10,000 hectares of Chilean mountain forests in this region."[70] In effect, the state was paying private forestry companies to burn Chile's unique temperate forests and replace them with monocultural pine plantations.[71]

While forestry industry advocates contended that the spread of pine plantations served to protect and reduce the pressure on native forests, the military dictatorship's policies stimulated the substitution of pine for native forest by making pine cultivation an inexpensive and highly profitable business. A report in the FAO's publication *Unasylva* noted in 1993 that "there is evidence that at least 100,000 hectares planted involved forest conversion. The loophole seems to be found in a clause in the decree that gives the forestry authority the possibility of approving conversion in areas where the native forest is 'not suitable for commercial harvesting.'"[72] Indeed, DL 701 gave forestry companies both the legal excuse and the financial stimuli to convert native forests, especially second-growth forests on which campesinos depended for forest products, firewood, and pasture, to pine plantations. Between 1978 and 1987 in the Bío Bío Region, 31 percent of the native forests in the coastal cordillera were burned and replaced with plantations.[73]

Pine plantations' corrosive effects on the regional rural ecology undermined the viability of campesino agricultural production. The extensive plantations caused the desiccation of topsoil and diminished the amount of water in the valleys at the feet of planted hillsides, leading to the deterioration of the conditions of agricultural production on which peasants depended for subsistence. As the wife of a forestry worker outside Concepción said in 1981 in a study by the Grupo de Investigaciones Agrarias, "There is great poverty in the countryside. . . . [T]here is no longer any water. . . . You have to go out in the world to look for water, far from the estuary. The pines suck up all the water that ran by here. . . . [B]efore the plantations were here, I remember that I had more than enough water for my garden."[74] To combat the spread of insects, fungi, and animals, forestry companies sprayed pesticides and fungicides in areas inhabited by campesinos, often contaminating streams and groundwater. The chemicals had a disastrous effect on the health of local peasant populations and on their livestock and crops.[75] The fumigation of the pine plantations also contributed to the decimation of the population of wild game that had provided an important supplement to peasants' diets, as well as to the deaths of farm animals. Defoliants and herbicides aimed at destroying the native species that could compete with the pines also damaged crops on land near plantations and contaminated groundwater. Leyton's report for CEPAL found that one of the principal factors in the exodus of campesinos from the countryside was "the difficulty that the small landowners confront in the poison used by the large forestry companies' plantations, which is highly damaging to livestock."[76]

Campesinos who had depended on forests for water, small game, and firewood now faced a threat to their subsistence as pesticides, herbicides, and defoliants made life and labor on the margins of the plantations impossible.[77] As a union leader and forestry worker observed,

> In this zone [Concepción], which was a zone of campesinos, the smallholder produced everything he needed: potatoes, wheat, cereals. . . . [H]e stayed in the forest. But everything changed. There is no longer any agricultural production in the countryside. There is no water for agriculture because the pines require a lot of irrigation. This region produced a lot of wine; there were numerous vineyards on the land of smallholders. Almost everyone had grapevines, but now everything is over because of the lack of water, the fumigations. The forest blocks out the sun that the grapes need to grow.[78]

Similarly, a leader of the CTF described the transformations in agricultural production wrought by the expansion of pine plantations:

> Where there were plantations there were no longer the things [campesinos] lived on—for example, forest animals. They could no longer hunt rabbits for food. They no longer had natural fruit, like seeds, that they could consume. They lost the biodiversity that existed before because there was only pine and eucalyptus, and as we all know, these are species that do not allow other kinds of life within the forest. The native forest allowed rabbits and other species to live there, while the other forests destroyed this biodiversity. As a consequence, campesino families were surrounded and lost a great deal of space.[79]

Faced with the erosion of the regional ecosystem, young campesino men in the early 1980s looked for jobs in government public works programs designed to alleviate unemployment, abandoning their fields. One report from 1983 noted that the Bío Bío Region was "heavily forested" and that in areas traditionally dominated by a balance between pine plantations and smallholders who owned parcelas of 4–8 hectares, most campesinos had left and sold their land.[80] The head of the Departamento Campesino in the office of the Workers Pastoral Vicariate observed, "Today you see something that never before occurred in the countryside. The campesinos are coming to the cities . . . and they no longer cultivate the land or work in the forests."[81]

The restructuring of the forestry industry exacerbated campesinos' displacement. Reorienting the industry away from the internal market for elaborated wood products and lumber to export markets for logs and pulp led to the firing of thousands of sawmill and factory workers. While exports of pulp and logs grew, employment in wood industries declined. Lino Lira, the president of the Federation Liberación, noted in 1983 that since the transition to exporting logs, the permanent labor force in industrial activities related to forestry had dropped by 70 percent and had been replaced with subcontracted temporary workers.[82] Another union leader linked the growth in exports of unprocessed logs to forestry companies' focus on short-term profits and the low costs of planting and harvesting pine:

> A number of factories that produced goods for the domestic market closed down because they had to look to other markets. For us, it is a contradiction that businesses earned so much money exporting prime materials, because in exporting logs they shipped out the primary material that could have been used in our factories. . . . They only exported logs, and why? Because they had the guarantee that they had huge amounts of forest for plantations. They planted exotic species that grew rapidly, that in Chile gave a very good result, in twenty-five years you could harvest what was planted and in other countries it could take forty years. . . . They also need fewer workers for exporting logs, just for cutting down the trees and transporting them. They work the forests rapidly and earn profits fast too. Exploiting forests with other goals, with value added, requires more investment, more machinery, more workers . . . and the results take longer. It was easier and quicker [to grow pine and export logs].[83]

In 1974, Chile exported no unprocessed forestry products. Between 1976 and 1988, log exports grew from 17,000 cubic meters to 2,801,300 cubic meters.[84]

The large forestry companies also displaced small sawmills, substituting inexpensively grown pine products for the native woods worked by smaller mills and factories. Lira observed in 1983 that "before, in Concepción, there were about two thousand [small mobile] sawmills. Today there are no more than fifteen [large mills], which has aggravated the situation of workers who work with axes and chain saws."[85] Similarly, Leyton found that "the emigration from the rural sector in areas with a large proportion of plantations has been produced fundamentally by the closing down of small sawmills that were absorbed by a few highly mechanized and capitalized mills."[86]

Restructuring and the Labor Force

Unemployment caused by the expulsion of campesinos from forested land and the decline in the industrial processing of wood products was exacerbated by the spread of subcontracting in the forestry industry. During the 1970s, forestry companies, unhindered by labor legislation or the presence of unions, began to dismiss their permanent labor forces and replace them with workers employed by subcontracting firms. This process accelerated after a new labor code was imposed by the military dictatorship in 1979 that made it possible for employers to fire permanent workers and to hire temporary workers without paying them benefits or negotiating collective contracts.[87] In 1982, 75 percent of forestry workers had labored for fewer than six months for one employer, and 50 percent had lasted two months at one job. Twenty percent had changed jobs every month or two weeks.[88] In 1972, Forestal Arauco had employed three thousand permanent workers. By 1978, it had cut this workforce by one-third, and in 1982, it fired the last seventy workers, transferring all of its operations to subcontractors.[89] In 1984, the largest forestry companies—Forestal Arauco, Forestal Mininco, CRECEX (INFORSA), and CELCO—had almost no permanent workers.

After 1973, labor conditions in the forestry sector returned to their pre-1967 state, defined by transience, insecurity, and exploitation. *Contratistas callampas* (fly-by-night, or "mushroom," subcontractors) came to dominate the industry. The contratistas competed for jobs from large forestry companies by lowering their costs at the expense of workers' salaries and work conditions.[90] Most of these small firms were undercapitalized and possessed no trucks, machinery, or tools. They hired workers temporarily, according to jobs contracted with the large forestry companies, and covered their preliminary expenses with advances. Until the mid-1980s, these companies frequently took on jobs, hired workers with the promise of a steady wage, operated until they received advances from the forestry companies or until they were paid for a portion of what they had produced, and then closed down, abandoning their workers without pay when they realized they could not cover their debts or costs.

Many workers received no pay or only part of their wages for their weeks of labor in logging camps. In 1983, a forestry worker described to *El Pino Insigne* how about forty workers were hired by a contratista to cut wood on the El Roble estate: "we worked three months for the contratista, who paid us only in vouchers. After two months, we began to complain a little, because we had not seen any cash. He promised us 'the earth and sky,' but when pay day came, there was always another excuse. We were almost finished when one day we found out that the contratista had taken off. . . . Three months almost given away for nothing to a contratista, with my wife and kids at home complaining [of hunger]."[91] The subcontractors survived by winning advances and recruiting a low-wage temporary labor force from the large population of landless young men in the countryside. Far from urban centers, with little oversight from the labor inspectorate, and facing no challenge from a moribund rural labor movement, contratistas reproduced pre-agrarian reform work conditions in their camps to reduce their expenses and compete for contracts.

Contratistas offered no benefits and paid wages by the quantity of wood logged. Working by piece rate (*metro-ruma*), workers frequently made less than the legal minimum wage. Contratistas frequently undercounted the amount of wood cut by workers to reduce their pay and failed to pay social-security contributions. Workers were forced to provide their own tools, boots, gloves, and helmets and lived in *rucas* (rudimentary huts), often sleeping on straw. Workers were responsible for feeding themselves with food they brought from outside or purchased at artificially inflated prices at company stores. Often they were paid in tokens that could be redeemed only at the pulperías, and employers deducted the workers' debts from their wages.[92] Because contratistas looked for unemployed young men with little experience in logging and provided no security equipment, the forestry industry had the highest number of accidents in the country, surpassing even the construction industry.[93] In 1983, Lira pointed out that accident rates were high in the forestry sector because "the owners do not provide even minimal systems of security on the job, since they do not provide helmets, boots, gloves, or adequate clothing."[94]

The restructuring of the forestry industry led to the erosion of peasant household economies and changing patterns of residence and labor. Former peasants and forestry workers migrated to squatter settlements throughout the countryside, usually on the sides of roads and highways, or to squatter settlements in towns and small cities. There they constituted an inexpensive temporary labor force for estates devoted to producing fruit and forest products for export markets. In 1981, a report by the Grupo de Investigaciones Agrarias (the nongovernmental Agricultural Research Group) noted that "along the roads running to the Pan-American Highway, where there once were only fields and copses, one can now see groups of small, multicolored houses, the same as in the shantytowns (*poblaciones marginales*) of the big cities."[95] A union leader noted that the dispersal of the rural population was due both to the peregrinations of men throughout the forestry economy and the lack of income and resources that could support families and allow the maintenance of permanent communities: "the campesinos can't live in permanent communities anymore. To form permanent communities, they need income to support them. If there's no work, there's no community. They have been transformed into permanent migrants who travel to Concepción, Temuco, Valdivia, looking for work, knocking on doors."[96]

The Rural Labor Movement under Military Rule

Military repression, the restructuring of the forestry industry, and the re-ordering of rural labor and land relations provoked an abrupt decline in union activity. As forestry unions struggled to survive during the 1970s, union leaders faced dismissal, arrest, and internal exile. Many unionized companies replaced their workforces with nonunionized temporary workers or simply closed down and moved to other regions, where they opened up under different names. The campesino unions that had been organized at the level of *comunas* and on rural estates completely disappeared after 1973, destroyed by repression and by the breakup of rural communities. By the early 1980s, however, following the reemergence of the national labor movement and a reinvigorated grassroots struggle for democracy in urban poblaciones, forestry workers began to rebuild their unions. In August 1984, forestry workers employed by three subcontracting firms in Cerro Alto and Los Alamos, outside Concepción, formed the first interindustry union, the Sindicato Interempresa "El Araucano," with a seat located in the coal miners' union hall. After two months, the union had 150 members. In October of the same year, a group of ten forestry, wood, and paper unions with more than 1,500 members founded a regional federation in Concepción.[97]

These new unions reflected efforts by labor activists to organize and negotiate contracts by region, rather than company, and rebuild the kinds of regional and industry-wide organizations that had characterized the labor movement before 1973. Organizing workers employed by contratistas in regional unions gave temporary workers more clout and raised the possibility of negotiating collective contracts with groups of employers. In addition, by linking forestry workers' unions to the unions of industrial workers in cellulose, wood, and paper plants, the new regional federation made it possible for unions to negotiate with employers, such as Forestal Arauco, whose activities spanned the gamut of these activities, even if their operations were divided among a variety of subsidiaries and subcontractors. Regional labor organizations created the possibility for workers to overcome the structural and geographical obstacles presented by the dispersal of the workforce and the fragmentation of the industry.

The Catholic Church played an important role in the reanimation of the labor movement among rural laborers and forestry workers in southern Chile. In Concepción, the church's Workers' Pastoral Vicariate sponsored classes for workers on how to organize unions and offered training and education for union leaders. Workers held meetings in the vicariate's offices as they attempted to build new union networks. In addition, the vicariate published a newsletter, *El Pino Insigne*, that documented work conditions and labor conflicts in the forestry sector. The newsletter interviewed workers and union leaders and provided updates on organizing campaigns, strikes, and dismissals. *El Pino Insigne* also provided accounts of the history of the forestry sector and rural labor and land relations in the region. The vicariate organized social assistance programs in the new rural poblaciones, creating small projects to earn residents income and providing classes on how to make do with the resources at their disposal. During strikes, workers held assemblies and cultural programs and organized soup kitchens in the vicariate. In addition, the vicariate provided legal aid to workers. In 1983, for example, it helped bring lawsuits on behalf of 180 workers against twelve subcontracting companies for failing to pay workers' salaries.[98] Caupolicán Pávez, a co-founder of the CTF, observed, "I don't believe that there is a union or a federation . . . that didn't receive the support of the vicariate [in Concepción]. We held union meetings at the vicariate, and even today the old union leaders keep meeting there."[99]

Despite the rebuilding of forestry workers' unions after 1983, repression and the structural conditions of the industry imposed limits on organizing. The military regime responded to the reemergence of rural unions in Concepción by jailing union leaders and sending them into internal exile. On 4 February 1985, Lino Lara, president of the Federación Campesina Liberación, and Carlos López López, president of Sindicato Agrícola Manuel Rodríguez and vice-president of the Federación Liberación, along with four other regional union leaders, were arrested and sent to a military prison in the Atacama Desert for four months, where they were joined by more than twenty other union leaders from Santiago and Valparaíso. They were then "relegated" (banished) for three more months to distant and isolated villages in the north.[100]

Workers' efforts to rebuild the labor movement in the forestry sector were also hampered by the dictatorship's Labor Code of 1979 and the restructuring of the industry after 1973. The new labor laws neither recognized the legal status of industry-wide or regional unions nor required employers to bargain with temporary workers. Subcontractors were under no legal obligation to negotiate with their workers and could use the temporary nature of forestry work to fire labor activists and impede the unionization of their labor force. Until the early 1990s, forestry workers' new unions were unable to organize strikes or win significant concessions from employers. Almost three-quarters of all forestry workers labored for small subcontractors on a temporary basis and thus had no legal right to bargain collectively. In addition, the transience of this labor force made it extremely difficult to organize. Workers moved from work site to work site, employer to employer, region to region almost monthly. In logging camps, workers remained isolated and dependent, cut off from communication with the outside world. The expulsion of rural workers from farms, estates, and parcelas led to the dispersal of forestry families throughout rural areas and the breakup of local communities that had provided the foundation of the militant rural labor movement of the 1960s.

In 1989, as it contemplated labor strategies for the transition to democracy, the recently organized CTF noted the importance of organizing new unions and federations of "modern" workers in the agro-industrial sector. The CTF observed that campesinos and campesino unions had been replaced by a largely nonunionized, wage-earning workforce of former campesinos. Because of repression, the expulsion of inquilinos and campesinos from the countryside, and the regime's Labor Code, the campesino unions that had been the backbone of the rural labor movement before 1973, such as the Confederación Nacional Campesina e Indígena "Ránquil," had disappeared.[101] In 1989, only eleven thousand workers, about 10 percent of the forestry labor force, were organized in three unions: the Federación Liberación, the Federación Nacional de la Madera, and the Federación Forestal Maule. These three unions formed the CTF in 1988.

Beyond employers' ability to intimidate workers, dismiss union activists, and move to other regions, as well as the legal constraints imposed by labor laws, union organizers confronted the effects of military repression and economic insecurity on workers. Caupolicán Pávez noted, for example, that it was extremely difficult to reorganize unions in the early 1980s. He had to travel throughout the countryside, going into the forests to talk to workers. To win their trust, he often was accompanied by old-time union leaders who were known in the region. He observed that forestry workers had their own culture that marked them as different even from other workers, and they were suspicious of outsiders. This lack of trust was exacerbated by repression. "They are still afraid," he said. "They saw union leaders killed after 1973 and think, 'Maybe there will be another coup, and I'll be killed.'" In addition, Pávez pointed out that "often, for fear of losing their jobs, they prefer not to participate in the union and accept the bosses' conditions."[102]

Despite the organization of some regional union federations and the CTF during the early 1980s, forestry workers approached the transition to democracy in a weakened position. The CTF's "Proposal for Democracy" of 1989 reflected workers' sense that only profound changes in labor legislation and state regulation of the forestry sector would improve working conditions and make building a strong labor movement possible. The proposal also indicated the unions' efforts to reverse the social and ecological changes that had provoked the *"descampesinización"* (proletarianization) of the countryside. In many ways, the CTF's proposal called for a return to pre-1973 conditions, when small landholders and mill owners played a significant role in the forestry industry and the state promoted the development of industries and regulated both reforestation and the management of native forests.[103]

Beyond basic demands for the elimination of subcontractors and temporary work, the eight-hour workday, partial employer financing of unions, and national and sectoral contracts negotiated by the CTF and the regional federations, the confederation critiqued the Pinochet regime's neoliberal model of development and proposed state intervention to support small and medium-size landowners and enterprises that produced wood products for the domestic market. Rather than exporting forestry products with little added value, the CTF advocated technology transfer and technical assistance for small producers to "incorporate the maximum value added to forestry products." The CTF called on the state to promote internal markets for industrialized wood commodities produced by small and medium-size businesses. The "Proposal for Democracy" asked for a revision of the subsidy structures in DL 701 to "direct its benefits to small and medium-size proprietors, subsidizing and assigning state financing according to scales of production. . . . A strong credit incentive is necessary for small and medium-size companies, the creators of large numbers of jobs . . . [along with] technical support, technology, and marketing to supply the domestic, not the external, market."

The CTF in effect proposed that the state intervene to reproduce the developmentalist programs of the pre-1973 period and reestablish the social arrangements that had characterized the forestry industry. Thus, it urged the state to establish incentives for campesino smallholders with forests "with the objective of . . . integrating agricultural cultivation, forestry, and the pasture of animals on their farms" to rebuild the vanishing peasant economy. In addition, the CTF proposed subsidies and credits for the owners of small sawmills, as well as training for workers and owners to make small wood producers' use of forests more efficient and ecologically viable. It argued that credit and technical assistance to campesinos with native forests and the promotion of marketing cooperatives would "ensure the management of forest resources." In addition, the confederation called for a "revision and redefinition of the transfer of lands from indigenous communities to private companies and an inventory of forests on indigenous lands." This demand reflected the close ties between forestry workers' unions and Mapuche organizations. Many forestry workers came from Mapuche communities, and as the communities began to organize opposition to large forestry companies' domination of the countryside and usurpation of Mapuche land, they found support in the CTF.[104]

The CTF's proposal hinged on a reevaluation of the social ecology of forestry development. The confederation explicitly linked the post-1973 reorganization of the forestry industry and the restructuring of agrarian labor and land relations to destructive environmental changes. It noted the ecological precariousness of a development model rooted in monocultural export agriculture and the effects of ecological transformations on campesinos and workers. Diversification within the forestry sector would involve not just the establishment of a variety of industrial forestry activities oriented toward the internal market. It would also depend on ecological diversification. By promoting sustainable-yield forestry and a multiuse model of forest exploitation, the CTF's plan would prevent the wholesale destruction of these diverse ecosystems and their replacement by pine and eucalyptus plantations. Thus, the CTF proposed that the native forests of the pre-cordillera be incorporated into regional economies "through the production of wood, firewood, honey, [and] hazelnuts, prohibiting their complete transformation into wood chips, which will ensure, among other things, that they will not be replaced with pines." The confederation noted that its proposals for reforms of the forestry sector were directed at promoting "the preservation of species and natural environments for the future generations." To this end, the CTF advocated a return to the pre-1973 system of state regulation of forestry exploitation and curbs on the substitution of native forests by pine plantations: "we propose that our native forests be submitted to the tutelage of the state to ensure their preservation and to prevent the substitution of the species that we observe today."[105]

The CTF's proposal represented a combination of the concerns of environmentalists and the labor movement. While it acknowledged that pine plantations were an established economic and ecological fact, the confederation linked the spread of the plantations to the social and economic changes that had redefined the terms of labor and land relations and allowed the domination of the industry by a few large companies oriented toward export markets. The CTF recognized the expansion of plantations as a menace to native forests and to the jobs and working conditions of forestry workers. The union's vision was not conservationist; rather, it provided a program for combining commercial and subsistence agricultural production, forests and agricultural crops, pine plantations and native forests, to satisfy the demands of workers and campesinos who depended both on the products of native forests and on the commercial exploitation of native forests by small sawmills. Rather than sustainable forestry defined by converting native forests to tree plantations, the forestry workers called for an entirely different kind of sustained-yield and multiuse forestry rooted in state-directed management of Chile's remaining native forests. The CTF viewed this model of forestry development as laying the foundation for a more socially sustainable use of the forests and allowing local community participation in forestry activities, from logging and milling wood to collecting forest products.

The resurgent, if feeble, rural labor movement of the 1980s was able to make direct links between the extraordinarily exploitative social conditions in the forestry industry and the environmental degradation produced by forestry development. Rural labor leaders began to employ environmentalist arguments and language to buttress their critiques of the forestry monopolies and export-led forestry development. This was not the first time that social and environmental injustice was linked by peasant movements in the south. Since the 1920s in southern Chile campesinos had tied the fraudulent formation of enormous landed estates, such as those owned by the Puelmas, the Bunsters, Silva Rivas, and the Budi concession, to the rampant destruction of native forests, which they defined as public wealth that belonged to the people and the state. They had articulated their rights to land by counterpoising their own forms of production and relationship to the forests with the "vandalism" committed by loggers and landowners. By the 1980s, this general sense of outrage at the destruction of native forests and their replacement by pine found more formal expression in the language of the contemporary environmentalist movement. Rural labor leaders began to speak about the value of "biodiversity" and to link strategies of forest management that would ensure the regeneration of native forests to a system of labor relations and production that would ensure social justice.

Forestry workers' unions now explicitly tied their understanding of a fairer social order to a vision of ecological balance and sustainability. They wove together their sense of social and environmental rights. They also drew on the institutional and ideological resources of the national and international environmentalist movement to articulate their repudiation of the Pinochet regime's neoliberal policies. As the labor movement rebuilt itself in the 1980s, it offered an important critique of the links among the spread of plantations, the destruction of native forests, and social changes in the countryside since 1973. The unions' growing understanding of the close ties between the natural worlds of forests and the social worlds in which labor and land, like trees, were transformed into commodities bridged the concerns of labor and environmental activists and provided an important alternative to the view that the market was the best regulator of both labor relations and the environment.

9

Democracy, Environmentalism, and the Mapuche
Challenge to Forestry Development

In 2002, I traveled around Lake Panguipulli following the road into
the Andes cordillera to Neltume. On the trip up the dirt roads into the forests of raulí, coigüe, and tepa, I was able to appreciate the marks of history
on the mountain landscape. Huge swaths of cleared forest scarred mountainsides. In 1997, the French company logging Neltume had been fined for
destroying native forests, and, despite commitments to the Corporación
Nacional Forestal (National Forestry Corporation; CONAF), continued
to ignore prescribed forest management plans.[1] Neltume typified logging
in the Tenth (Lakes) Region, where only one fifth of logging companies
actually followed the forest management plans they submitted to CONAF.[2]
At the entrance to Neltume stands a memorial to the forestry workers as-
sassinated by the military dictatorship: a statue of a bare-chested worker
with a plaque with the names of the Complejo Maderero Panguipulli
workers who perished after 1973 (see figure 9.1). Many of the estates that
belonged to the complex had been either returned to their former owners or auctioned off during the 1980s. For a time during the late 1970s, the
Complejo Maderero Panguipulli was run by General Augusto Pinochet's
son-in-law Julio Ponce; in 1983, he was forced to resign from his position
at the head of CORFO and the Complejo Maderero Panguipulli because of
judicial investigations into his illicit enrichment during the dictatorship's
privatizations.[3]

Neltume had an air of desolation, isolated in the mountains and guarded
by a carabineros station at the entrance to town. Most of the residents
worked for the French logging company Neltume-Carranco, but none
were actual employees. Instead, they toiled for small contratistas who competed for logging jobs. Like their counterparts in the mountains outside
Concepción during the 1980s and 1990s, Neltume's subcontractors were often former employees of the company or fellow workers. They bid on jobs,
and workers paid the price in low wages and no overtime pay, insurance
against accidents, social security, health care, or job security. Workers sup-

9.1 *Human Rights Memorial, Neltume (Courtesy of Museo de la Memoria y los Derechos Humanos)*

plied their own gear and tools. Often contratistas paid them with chits that could be used to purchase goods at one of the few stores in town; the stores' owners frequently operated as contratistas. Subcontractors hired fewer than twenty-five workers to prevent unionization, and labor inspectors never made it to this isolated mountain region near the frontier with Argentina. Workers who sought legal remedies or complained about abuses to the company were commonly fired. A young man I interviewed, the son of a forestry worker and brother of one of the workers killed by the armed forces after 1973, had worked for more than twenty separate contratistas in the previous decade. In each case, he and his fellow workers had logged the mountains for anywhere from one to six months until the paychecks, always late in coming, began to disappear. After working for a month or so without pay, they would quit and look for work with another contratista who could pay, giving up their lost wages. In Neltume it was hard to see how more than a decade of democratically elected governments had made much of a difference in the lives of forestry workers.[4]

A few years later, I returned to Neltume to find a rapidly changing social landscape. The Huilo-Huilo fundo, whose Valdivian forests had been logged for decades, first by its owners and then by the state-run Complejo Maderero Panguipulli under the Pinochet dictatorship, had new proprietors. They had built two luxury hotels with avant-garde architecture (the exterior of one hotel is a giant waterfall; the other boasts extraordinary views of the imposing Mocho-Choshuenco volcanos), to promote ecotourism in the remaining Valdivian forests. The hotels offer zip-line trips through the forest canopy, snowshoeing on the slopes of the glacier-covered volcano, and hikes along paths that wind through the remaining stands of temperate forest, although logging has eliminated almost all raulí from the region's forests. Huilo-Huilo, once the administrative center of the socialist Complejo Maderero Panguipulli, now serves a wealthy population of Santiago's upper crust and foreign travelers who arrive in sports utility vehicles on dusty roads, driving right through Neltume, where they pass the monument to the complex's fallen workers. It is hard not to see the tragedy in the location of the monument, a testament to the Panguipulli workers' revolutionary dreams of sustainable socialist forestry, at the gateway to hotels that cater to wealthy tourists. Nor is it difficult to escape the irony of an ecotourism complex built not only over the bodies of massacred workers but in forests that have been degraded by logging for decades.

The Forestry Industry and the Concertación Governments

During the transition democratic governments of the center-left Concertación de Partidos por la Democracia (Coalition of Parties for Democracy, 1990–2010), the forestry sector continued to expand at rates that exceeded levels of national economic growth. Between 1989 and 1997, exports of forest products nearly tripled in value.[5] As in most areas of the economy, the Concertación maintained the military regime's policies of market-driven development based on the export of primary commodities to external markets. A significant increase in foreign investment beginning in the late 1980s, before the transition to democracy, and increasing during the 1990s marked the most important change in the forestry industry. Following the economic crisis of 1982–85 and the bankruptcy of a number of large financial groups, foreign investment flooded into the forestry sector, often through debt-for-equity swaps. In 1985, foreign investment in the forestry industry amounted to $350 million. By 1989 that figure had reached just under $800 million and continued to grow during the 1990s. Transnational paper companies from Canada, Japan, New Zealand, Switzerland, and the United States, often in joint ventures with Chilean capital, began to purchase tree plantations and native forests and install wood-chip and cellulose plants.[6]

During the recession of the early 1980s, for example, the military regime was forced to take over a heavily indebted and bankrupt oil company, Compañía de Petróleos de Chile (COPEC), owned by the giant Cruzat-Larraín financial conglomerate. Despite the fervor of its free-market ideology, the Pinochet dictatorship became the owner of the largest private enterprise in Chile: COPEC's financial empire included forestry companies such as Celulosa Arauco and Celulosa Constitución, as well as major financial institutions such as the Banco de Santiago. In 1986, the military regime once again turned to privatization, and a new financial conglomerate, the Angellini group, acquired COPEC and its many subsidiaries at half their market value four days before the company was to be publicly auctioned. Angellini then canceled COPEC's considerable debt in a debt-for-equity swap with the New Zealand forestry giant Carter Holt Harvey, which received ownership of 30 percent of the company. By the 1990s, two of Chile's largest cellulose plants, built with significant state inputs during the 1960s, were held by COPEC and the Angellini conglomerate in partnership with Carter Holt Harvey and the US-based International Paper Company, which purchased 20 percent of COPEC's shares.[7]

Foreign investors also took on an important role in the other major player in the forestry industry, the Matte-Alessandri financial group's Compañía Manufacturera de Papeles y Cartones (CMPC) during the late 1980s and 1990s. The Simpson Paper company of the United States and Anglo-Dutch Shell purchased significant chunks of stock in the Laja cellulose company.[8] Shell had purchased a number of forestry companies in Chile already, including Forestal Colcura, organized decades earlier as a branch of Lota's coal-mining enterprise outside Concepción. During the 1990s, Forestal Santa Fe, a subsidiary of Shell, also invested, with Scott Paper and Citibank, in building a plant that employed eucalyptus plantations as its source of primary material to produce short-fiber cellulose. The Santa Fe project went into operation in 1997, complementing the output from the company's older cellulose plant in Nacimiento.[9] Finally, the CMPC built a new cellulose plant, Celulosa del Pacífico, in Mininco, in Chile's Ninth (Araucanía) region, in a joint investment with Simpson Paper.[10] These investments were driven partly by Chile's competitive location in the global forest products markets. During the 1990s, according to the World Bank, Chile enjoyed production costs that were 30–50 percent lower than those in the United States and Scandinavian countries, a testament to the country's inexpensive labor and loose environmental regulations.[11]

During the late 1980s, Japanese paper companies also began to invest in eucalyptus plantations and wood-chip operations, mostly in Chile's Lakes Region. These companies acquired and planted tens of thousands of hectares of eucalyptus to provide raw material for wood-chip plants.[12] This new wave of foreign investment reflected a boom in the production and export of wood chips. While Chile exported no wood chips in 1986, in 1995 it exported 4,076,500 tons.[13] In 1992, Japan purchased 80 percent of the world production of wood chips (used for making short-fiber cellulose in the production of high-quality computer and fax paper), and Chile was Japan's third largest supplier.[14]

The explosion of the wood chip industry led to the intensified exploitation of native forests in Chile's Lakes Region by Japanese, North American, and Chilean companies. At the same time, these companies also cultivated eucalyptus. Tree plantations began to replace native forests, reproducing the Eighth (Bío Bío) Region's experience with pine. Between 1989 and 1998, Chilean and transnational companies, often in partnership, planted 15,000 hectares of pine and eucalyptus annually in the Lakes Region for cellulose and chip production. These companies also logged the region's native forests and purchased logs from small producers who extracted them legally and illegally from the forests. In 1994, 65.2 percent of wood chips came from native forests; 33.4 percent, from eucalyptus plantations; and 1.5 percent, from Monterey pine, a major shift from the early 1980s, when the chips came almost entirely from pine.[15]

The wood chip boom reinitiated logging in the southern temperate forests and marked an important geographic movement in the industrial tree plantation economy, as eucalyptus plantations spread throughout the coastal regions south of Valdivia. This new motor of forestry development threatened native forests. As a report in *Unasylva* noted in 1993, "The replacement of native forests is becoming a contentious issue since the Chilean forest industry is rapidly increasing its exports of short-fiber wood chips, now using native forests and eucalyptus plantations as a resource base."[16] A study conducted by the environmental accounting group of Chile's Banco Central (Central Bank) in 1995 estimated that between 1985 and 1994, anywhere from 275,000 to 620,000 hectares of native forest had been destroyed, a significant amount for wood chips or conversion to tree plantations. As the study pointed out, Decree Law 701 allowed landowners to clear native forests that contained few commercial species (referred to as "degraded" forest), as well as second-growth forests, and plant exotic species with government subsidies.[17] Forestry scientists such as Antonio Lara of the Universidad Austral, who with the economist Marcel Claude wrote the Central Bank's report, argued that half of Chile's annual loss of native forest was due to substitution with pine and eucalyptus plantations.[18]

The push to produce wood chips provoked a significant backlash that reflected the concerns of Chile's emergent environmentalist movement and the growing importance of ecotourism to the national economy. In 1990, marking a significant shift in tone from the days of the dictatorship, *El Diario Austral* led with a story titled "Alarm in Llanquihue: Depredation of the Forests." In a forum on the increased logging in the forests of the Lakes Region, a forestry scientist concluded that "there is irrefutable proof of an irrational exploitation of the forests without any forest management plan."[19] *El Diario Austral*'s reporting on deforestation reflected the growing influence of environmentalist groups such as the Committee in Defense of Flora and Fauna (CODEFF) and the more militant Defensores del Bosque Chilenos (Defenders of the Chilean Forest) in placing environmental issues at the center of debates over development during the transition to democracy. In 1992, the paper reported that "hills of chips from the native forests will continue to invade the southern landscape in the years to come. . . . [I]n mid-1993, together with the opening of a port in Corral, a new plant will produce chips and extract its prime material from the Valdivian evergreen forest." The article quoted a forestry company executive who argued that manufacturing chips "is the only way to make the native forest profitable and to avoid its deterioration" by removing overly mature trees. The regional secretary of the Ministry of Agriculture similarly celebrated new investments in chip plants, stating, "Producing chips, if done well, permits the management of the forests, because this is the product that you obtain from thinning the native forest to improve it."[20] The contention that the chip industry allowed sustainable exploitation of southern forests met with skepticism from environmentalist critics. *El Diario Austral* included an interview with the forestry scientist Claudio Donoso Zegers of the Universidad Austral, who noted that "few of the companies manage the native forest appropriately. What they do is extract wood for fuel, sawmills, and chips. . . . When they enter a forest, they . . . extract the best-quality trees and leave the worst, altering ecosystems and even clear-cutting."[21]

In the context of debates over forest conversion and the impact of the wood chip industry, the government of President Patricio Aylwin introduced new environmental regulations in 1992 through an Environmental Framework Law. Chief among them were requirements that any major development project have an environmental impact study approved by a new government agency, the Comisión Nacional del Medio Ambiente (National Environmental Commission; CONAMA) or its regional affiliates. That same year, the Concertación began work on a new native forest law. The bill introduced regulations on forest conversion, banning substitution in parks, forest reserves, watersheds, and habitat for endangered flora and fauna. In addition, it restricted landowners' rights to substitute native forests. The law extended the incentives to plant pine codified in DL 701 to the native forests by offering subsidies equal to 75 percent of the costs of managing sustained-yield harvests of native forests or, in degraded forests, 75 percent of forest recovery. Finally, the law redirected some of the financial incentives for forestry management hitherto monopolized by the large forestry companies to small and medium-size producers.[22]

Because of the significant opposition mounted by the financial groups dominant in the forestry industry, the law stalled in Congress, where conservative congressmen, including nine "designated senators" appointed by the Pinochet regime, were able to block legislation put forward by the Concertación government and its majority in the House of Deputies. Indeed, authoritarian "enclaves" in the Chilean political system left in place by the dictatorship proved a major obstacle to the democratic government's efforts to implement new environmental regulations. These obstacles included the nine designated senators (four of whom were former heads of the different branches of the armed forces) and a skewed "binomial" electoral system that guaranteed the right-wing opposition at least half of the elected votes in the Senate.[23] In addition to these institutional impediments, sectors of the Concertación coalition and the government joined the right in its reluctance to hamper market-driven growth with environmental regulations, particularly in an industry in which foreign investment played an increasingly important role and in which Chile's global comparative advantage lay in its inexpensive natural resources. Congress approved the Native Forests Law only in 2010, eighteen years after the Aylwin administration had first introduced it and five years after constitutional reforms finally removed unelected senators from Congress.[24]

Although the Native Forests Law floundered in Congress during the 1990s, the Concertación succeeded in pushing through reforms to DL 701 to redirect subsidies to small and medium-size land owners and producers. As Eduardo Silva points out, state support for smallholders did not conflict with the interests of the large forestry companies, since eventually smallholders who planted trees would have to sell their products to the companies. State subsidies for smallholders to forest with pine and eucalyptus provided inexpensive raw materials to conglomerates such as the CMPC and COPEC for their pulp plants, in essence making investment in new plantations unnecessary.[25] While under the dictatorship CONAF and the Instituto Forestal had served to support the large forestry conglomerates by providing technical assistance, research, and subsidies, under the Concertación the two agencies began to elaborate programs to work with smallholders on foresting with pine and eucalyptus and on designing forest management plans for their small patches of native forest. As during the late 1960s and early 1970s, CONAF and the Institute for Agricultural and Livestock Development (INDAP) entered into agreements with smallholders to forest their land. Since state funds for these projects were limited by the Concertación's fiscal conservatism, financing for the programs came partly from international development programs.[26] In Panguipulli, for example, CONAF developed an extension program to work with smallholders who had requested permission to log the remaining forests on their plots. As Alex Rudloff noted, CONAF extension agents trained smallholders in forestry so they could "manage their forests using technical criteria that allows them to guarantee a degree of future sustainability." Funding for this program came in part from the German government. Further, INDAP and CONAF maintained a nursery in Valdivia to supply smallholders with raulí seeds and saplings.[27]

Thus, while maintaining the Pinochet dictatorship's model of market-driven forestry development, the Concertación governments introduced some changes in the organization of the forestry economy by making the subsidies provided by DL 701 accessible to campesino smallholders. In addition, the Concertación's willingness, in certain cases, to impose environmental restrictions on foreign and domestic logging operations through the environmental framework law and CONAMA represented a shift from the years of radical deregulation under Pinochet. However, the Concertación governments were willing to go only so far in saddling the forestry industry with environmental regulations. After the environmental accounting group of the Central Bank released its report describing logging companies' role in forest conversion during the late 1980s and early 1990s, leading figures in the government of President Eduardo Frei publicly attacked the report and questioned its findings. Government pressure led the Central Bank to distance itself from the report and to issue a new study in which Claude's estimates of past and future loss of native forests due to substitution were significantly scaled down.

Workers and the Forestry Industry during the 1990s

The contradictions that bedeviled the Concertación's efforts to produce high growth rates while introducing new regulations on environmental degradation were reproduced in the arena of social policy. Despite the rebuilding of their unions during the 1980s, the transition to democracy failed to meet forestry workers' expectations. Forestry workers encountered a major impediment to organizing unions in the Pinochet dictatorship's Labor Code of 1979. The laws had been designed by one of the regime's most famous Chicago boys, José Piñera, to "flexibilize" labor markets by restricting unionization and collective bargaining rights. They gave employers almost carte blanche to dismiss workers with no or little notice, often without severance pay. Temporary workers, who made up an increasingly significant sector of the labor force, had no legal rights to form unions or bargain collectively. In addition, workers' right to strike was severely limited. Employers were permitted to fire striking workers after fifty-nine days and to hire replacements during strikes. The Labor Code also proscribed industry-wide contracts, redefining legal contracts in terms of individual plants or enterprises. This weakened agricultural and forestry workers' unions because their industries were organized around multiple small subcontractors, and workers labored only short stints for their employers. The code forbade closed shops and allowed many unions within one enterprise.[28]

The Aylwin government introduced reforms to the code in 1991–92, but they did little to limit employers' power to fire workers at will or to improve workers' rights to bargain collectively. Agricultural and forestry workers were still basically excluded from collective bargaining rights, since few had permanent contracts. In addition, the trend toward replacing full-time workers with part-time temporary workers continued during the 1990s, weakening unions further and significantly reducing wage earners' incomes.[29] The forestry industry provided one of the most acute examples of workers' plight during the transition to democracy. For some workers, conditions in logging camps improved due to pressure from the unions and increased state regulation, but in many camps run by subcontractors, as was the case in Neltume, labor conditions remained the same as during the years of dictatorship.[30] The union leader Caupolicán Pávez noted, "when we go into the Nahuelbuta cordillera we see really miserable conditions where forestry workers still live in huts and depend on a pulpería, that in the majority of cases, belongs to the contratista. The contratista contracts people for one month, two months, three months sometimes; and no one regulates him. . . . there are contratistas who have beaten workers for the sole fact of demanding their wages, demanding their rights." Union leaders observed that often in the worst cases, contratistas preferred to go to court, where cases could take years, and then pay small fines to changing conditions in the camps.[31] As late as 1996, 82% of forestry workers lived beneath the poverty line. At the end of the decade, 75% of all forestry workers still found employment with subcontractors.[32] The case of Neltume's permanently temporary laborers was the rule, not the exception, in the forestry industry.

In its annual report for 1998, the CTF argued that organizing in the forestry sector had been hampered by the lack of support from the Concertación governments combined with the dictatorship's labor code: "In this terrible code it is established that collective bargaining is a privilege for those who have permanent contracts . . . this produces the marginalization of thousands of workers from the right to organize." The CTF report noted that, "In our best moment [1989–1990] we brought together 67 unions and nearly 10,000 workers, today we have no more than 40 and we barely represent 6,000 workers." Union membership in the forestry sector had actually declined during the 1990s.[33] In 2000, one union leader wondered, "How are we to organize workers who have two-month contracts? . . . [W]hen workers try to negotiate with employers extralegally, they are simply dismissed. . . . If we try to organize a work site, after two months [the employers] close down the work site and fire the guys, and then the next month they call the same workers and sign them to new contracts, without the union and at lower wages."[34] He observed that "on a large estate of 30,000 hectares, there could be thirty different contratistas so the workers are divided by activity and company."[35]

Mechanization also undermined the labor movement in the forestry sector. During the 1980s and 1990s, forestry companies introduced new equipment, from chain saws to machinery for planting and pruning saplings, that raised productivity and reduced the labor force. Whereas a logging enterprise working a pine plantation might have employed forty to sixty workers during the 1960s and 1970s, by the 1990s it only employed two to fifteen. As a union leader noted, one worker equipped with modern machinery could do the work formerly done by ten.[36] Although industrial forestry advocates, from the early decades of the twentieth century through the socialist Popular Unity and the Pinochet dictatorship, had viewed tree plantations and logging as a remedy for the poverty produced by rural landlessness, by the 1990s mechanization had severely reduced the number of jobs in forestry, weakening unions' capacity to organize and win benefits from the myriad subcontractors who dominated the southern countryside.

As much as the dictatorship's restrictive labor laws and mechanization, the geographical reorganization of forestry production undermined workers' organizations. Until 1973, forestry workers' unions had been largely built around the demands of rural laborers who prioritized increasing their access to land and the natural resources that would allow them to reproduce a peasant economy. Union petitions made central claims to small plots to grow crops, pasture, and forests. During the Frei and Allende governments campesinos had expanded their access to land within the borders of forestry estates. By the 1990s, however, the changes introduced by the dictatorship had effectively severed most forestry workers from the peasant economy; a large percentage of workers now lived in towns and migrated throughout the south in search of work. They represented a more fully proletarianized, if not fully employed, labor force.

A report by the CTF issued in 1998 noted that during the 1980s and 1990s, the nature of the forestry labor force had been redrawn by proletarianization: "the appropriation of land by transnationals has meant the conversion of thousands of former campesinos into modern forestry and wood workers."[37] However, the confederation observed, rather than creating the basis of new working-class identities and organizations, proletarianization had undermined the rural labor movement. By the 1990s, the remaking of the southern countryside around industrial forestry had severed most forestry workers' ties to the countryside, producing increased rural unemployment and migration to the city and undercutting the central force that had animated their struggles before 1973: the demand for land, pasture, and forests. An entire generation separated workers from the land conflicts and labor actions before the military coup. In addition, the destruction of rural communities after 1973, as in Neltume and the fundos that belonged to the Complejo Maderero Panguipulli, fragmented the historical memory that might have served as a vehicle for political identity and collective action.

The Mapuche Movement

Whereas geography and generation limited forestry workers' and campesino smallholders' efforts to build a challenge to the dominion of the large forestry companies, during the 1990s Mapuche communities throughout southern Chile mounted an increasingly militant challenge to the pine plantation economy. A labor organizer contrasted the decline of unions in the forestry industry with the robust Mapuche movement to recuperate land from forestry companies. He noted that the Mapuche movement of the 1990s was rooted in a strong sense of ethnic identity that relied on a powerful historical memory located physically in the spaces of indigenous communities.[38] Despite the changes introduced during the dictatorship, most Mapuche communities had been able to sustain their ties to the world of peasant production, even if agriculture, raising livestock, and collecting forest products were increasingly constrained by the expansion of tree plantations. Mapuche communities were able to phrase their understanding of ethnicity in terms of historical claims to land that implied a set of ecological relationships rooted in their uses of forest resources. In addition, they were able to draw on the language of modern environmentalism to build alliances with environmentalist groups to challenge forestry conglomerates' hold over the southern countryside.

The first Mapuche protest of the forestry status quo during the transition to democracy erupted in Lonquimay over the Quinquén estate, much of whose land was claimed by the Mapuche-Pehuenche Quinquén community of the Meliñir kinship line.[39] As José Bengoa describes in his book on Quinquén, during the late 1970s the Galletué company had brought a lawsuit against the Chilean government demanding indemnification for the value of the araucaria forests it was prevented from felling by a 1976 prohibition that returned the tree's status to the endangered species category. The legal conflict lasted for a decade until, in 1987, the Pinochet regime ended restrictions on logging araucaria forests in a deal with the logging company. Galletué could reinitiate its timbering operations as long as it desisted from its legal claims against the government. The new decree permitted felling araucaria forests on private property, unleashing a new wave of logging around Lonquimay as Pinochet stepped down and the Concertación took power.[40]

In 1990, the Quinquén community, which continued to occupy land within the estate's putative borders, went to court, supported by CODEFF, to stop Galletué's logging operations. This legal challenge to Galletué's property rights, which invoked the protected status of the araucaria, lost in court.[41] Galletué also succeeded in its legal efforts to remove the Meliñir or Quinquén community from land it occupied in Quinquén. In 1990, the Chilean Supreme Court supported Galletué's petition to have the Meliñirs evicted. The case immediately had an impact at the national level, reflecting the political opening initiated by the election of Aylwin's Concertación government in 1990. Roberto Muñoz Barra, the Concertación deputy for Lonquimay, denounced the decision and endorsed the community's land claims, arguing that the Meliñirs had occupied land in the Quinquén valley for more than a century. In addition, he invoked Galletué's destruction of the araucaria forests to argue, in effect, that the company had relinquished its property rights, an argument that had been used during agrarian reform: "in this zone, the exploitative hand has entered in an irrational way: they have cleared the cordillera of forest and today, once again, they are felling the ancient araucarias."[42]

Members of the Quinquén community were able to appeal to a wide base of support by casting themselves as the defenders of the araucaria forests while depicting Galletué as a destructive scourge of a national treasure and icon. They argued that they were responsible for preserving the forests from the depredations of logging companies since their legal and political struggles during the 1960s and 1970s had led to the prohibition on logging araucarias.[43] In November 1990, community members blocked roads in and out of the Lonquimay valley to prevent trucks loaded with araucaria logs from leaving Quinquén.[44] In addition, community members continued to collect piñones and pasture their livestock in Quinquén's forests, according to their long-standing tradition. As the community's lonko observed, evoking Mapuche understandings of property rights rooted in use rather than possession, this was not a land invasion but a use of the araucaria forests that stretched back for generations: "we must underline that this is not a toma of the estate. This is about collecting piñones and pasturing our livestock, an activity we do every year on land that belongs to us . . . a traditional labor that our community members call veranada."[45]

The community found support in environmental organizations such as CODEFF, as well in the government agency created to implement policy toward Chile's indigenous communities, the Comisión Especial de Pueblos Indígenas (Special Commission for Indigenous Peoples; CEPI), directed by José Bengoa and Víctor Hugo Painemal.[46] The community also received the backing of an increasingly mobilized and militant Mapuche movement. The Mapuche groups Ad-Mapu and Aukiñ Wallmapu Ngulam/Consejo de Todas las Tierras (Council of All Lands), for example, worked with both the Quinquén and Calfuqueo communities in their fight with Galletué.[47]

The support of these organizations made Quinquén both a national environmental issue, involving the fate of the country's remaining stands of ancient araucarias, and a broader issue of Mapuche land rights. Ad-Mapu was founded in 1978, when a group of Mapuche leaders, with the support of the Catholic Church in Temuco, organized opposition to a decree-law issued in 1978 by the dictatorship designed to subdivide and privatize Mapuche reducciones.[48] The Council of All Lands was a radical offshoot of Ad-Mapu, organized in 1988 when members of Ad-Mapu opposed negotiated deals with the Concertación to provide support for the coalition in exchange for a new indigenous law that would protect Mapuche land rights. The council articulated a new vision of Mapuche ethnic and territorial autonomy, including a more militant strategy of "recuperating" usurped land through land occupations.[49]

The Concertación government immediately sought to resolve the conflict in Lonquimay by seeking a negotiated settlement involving the state's purchase of Quinquén—or, at least, the part of the estate occupied by the Meliñirs—for its eventual redistribution.[50] A few months into the conflict, the Concertación, in some ways echoing the strategy adopted by the UP two decades earlier, took up the project of making the Quinquén valley into a national park. But whereas the UP had been able to employ forest and agrarian reform laws that codified its power to expropriate land in the name of conserving endangered species and managing "badly managed" properties or properties embroiled in social conflicts, in the 1990s the government was bound by laws handed down by the dictatorship to protect private property rights. In addition, the Aylwin government was unwilling to unleash a conflict that would recall the tumultuous years of agrarian reform during the first months of the delicate process of democratization.[51] Indeed, as soon as the government declared Quinquén's forests to be a national nature preserve, the business trade association the Sociedad de Fomento Fabril (Industrial Development Society; SOFOFA) denounced the measure as a violation of property rights that was as dangerous as the agrarian reform, and as a threat to the stability of the free-market model that would provoke a lack of confidence among businesses and private investors.[52] If Quinquén were going to be made into a nature reserve, the government would have to obey the laws of the market and pay the asking price. Unfortunately, in this case the market price was driven up by the political pressures to resolve the conflict as quickly as possible. Galletué, having already won its case against the state in court in 1987, now asked for the exorbitant price of $10 million for the Quinquén estate.[53]

The government responded by preparing a decree to declare Quinquén a national park, Park of the Araucarias, and to expropriate Quiquén while indemnifying the Sociedad Galletué.[54] In May, 1991, the Ministry of Agriculture issued a decree declaring the Quinquén valley a nature reserve and prohibiting the exploitation of its flora and fauna. The creation of the new 6,000-hectare Lago Galletué National Reserve did not expropriate the Sociedad Galletué's estate or resolve the property rights of the Quinquén community. The decree did place pressure on the company to change its position and negotiate a fair price by once again prohibiting logging in the araucaria forests. In response, Galletué resorted to the laws of the market, arguing that it was reluctant to put its land up for sale, since the decree banning logging araucaria forests had reduced its value.[55] The company continued to ask for a price that, as CEPI noted, had no relation to its market value.[56] Indeed, as Bengoa demonstrated, the Quinquén property had actually expanded illegitimately over the years: whereas the original title specified 6,789 hectares, by the 1990s Galletué's estate held 16,681 hectares.[57]

In the end, the conflict was resolved when the government agreed to purchase the Quinquén and Galletué estates (about 30,000 hectares) for $6.15 million, leaving 11,000 hectares in the company's hands. The government established a 22,000-hectare nature reserve on the land, distributing the property to four Mapuche-Pehuenche communities: Quinquén, Huenucal Ivante, Huallen Mapu, and Pedro Calfuqueo. In 1997, the Quinquén community was accorded legal status as an indigenous community with rights to more than 6,000 hectares owned by the Quinquén estate, as well as to part of the larger Galletué estate. A committee formed by members of the four Mapuche-Pehuenche communities was to administer the government reserve, and, as had occurred during the UP government, the communities were given usufruct rights, with the exception of prohibitions on exploiting araucaria.[58]

In 2007, after a decade of delays, the Quinquén community, having organized studies of the land and its resources with the support of the indigenous rights organization Observatorio de los Derechos de los Pueblos Indígenas (Observatory of the Rights of Indigenous Peoples), headed by José Aylwin, and the environmentalist group World Wildlife Fund (WWF), received title to the Quinquén valley.[59] The following year, the WWF began to work with the community to develop the land and forests as the Parque Pehuenche de Quinquén, oriented toward ecotourism. Conservation and stewardship of the native forests in Quinquén would be tied to developing a basic tourism infrastructure. This project also received support from the regional government; CORFO; the new Corporación Nacional de Desarrollo Indígena (National Indigenous Development Corporation; CONADI); and the Servicio Nacional de Turismo (National Tourism Service; SERNATUR).[60] The Quinquén ecotourism project belonged to a broader WWF and CODEFF program to develop sustainable community management of native forests.[61]

The solution to the Quinquén conflict followed closely the parameters of new legislation on indigenous rights introduced by the Aylwin government. A new Indigenous Law in 1993 made it the state's duty to respect, protect, and promote indigenous rights and culture and to safeguard indigenous lands. The law created CONADI to implement the indigenous law, as well as a Land and Water Fund "to subsidize the purchase of additional land for communities affected by land scarcity and to finance mechanisms to permit the solution of land conflicts and the provision of water." The law also established "areas of indigenous development," dedicating government funds to promoting the growth of regions with significant indigenous populations.[62]

The Concertación government sought to make the Quinquén settlement part of a broader program of development for the region around Lonquimay. With funds from the United Nations Food and Agriculture Organization, it developed Project Lonquimay to provide basic infrastructure for the area's indigenous communities, including building housing, schools, and cultural development programs that would emphasize Pehuenche folklore and traditional medicine, as well as the preservation of the region's biodiversity.[63] While in Quinquén the program focused on ecotourism and the preservation of the araucaria forests, in other areas of Lonquimay the government promoted reforestation with exotic species and logging. The National Forestry Corporation and a government development fund (the Fund for Solidarity and Social Investment), entered into agreements with sixty-seven families in Lonquimay, most of them Mapuche-Pehuenches, to forest with Ponderosa pine and Douglas fir. The head of CONAF underlined that forestation and forestry were essential to the development of the region.[64] In addition, CONAF provided forest management plans to indigenous and nonindigenous smallholders in Lonquimay through its Program for Forestation on Campesino and Indigenous Land. The National Forestry Corporation provided subsidies to campesinos so they could "recuperate the native forest, manage it technically, and order it." Its agents trained members of Mapuche communities in forestry techniques and supervised forest exploitation.[65]

Despite the benefits forestry training and subsidies provided to campesinos, for many members of Mapuche communities the norms and oversight imposed by CONAF constituted a reduction of their autonomy, limiting their access to native forests and pasture they had occupied for generations.[66] The case of the Bernardo Ñanco community in Lonquimay, not far from Quinquén, illustrates the tension between Mapuche-Pehuenche communities and CONAF. Under the UP, the community had benefited

9.2 Upper Bío Bío: Forest to Plantation (Courtesy of Felipe Orrego)

from the expropriation of the Chilpaco estate, a measure reversed by the Pinochet dictatorship. In 1984, however, in response to the military regime's new law obligating the subdivision and privatization of indigenous communities, the Ñanco community was divided into individual plots. In this case, the Ñanco community's subdivision actually worked in its favor: the community's president was able to increase Bernardo Ñanco's land-holding from 6,500 hectares to 16,000 hectares, employing cadastral maps produced during the process of reducción.[67]

With the restoration of democracy, the members of the subdivided Ñanco community underwent a second process of change. During the 1990s, members of the community pursued a strategy of petitioning CONADI for land allotments to form new communities under the Indigenous Law of 1993. These small communities, mostly families headed by young men, reflected Bernardo Ñanco's demographic growth and land scarcity, even after the expansion in 1984. Most community members scraped by, pasturing livestock on their small plots and extracting firewood and piñones from the forests. As in the past, the new, smaller communities, carved out of land once owned by neighboring estates, either sold firewood from their small patches of forests to merchants in Lonquimay and Curacautín or leased logging rights to local forestry companies. The forestry companies, in turn, often provided loans and credits to the communities when they needed money during the winter, then paid below-market prices for the wood as repayment of the loans. In many cases, grazing livestock and selling wood led to severely degraded land on the steep mountain slopes where many new communities were located, continuing a process that had lasted a century. Subdivision also dispersed the community, since many settled on land purchased by CONADI near the city of Victoria, a considerable distance from Bernardo Ñanco's original location.[68]

Some members of the now divided Ñanco community viewed CONAF's efforts to develop supervised forest management plans with apprehension. On the one hand, they received new land to settle and subsidies from CONAF through the government's Campesino Forestry Program for building fences and planting trees. Community members themselves recognized the importance of regulating the extraction of wood from their remaining patches of roble, lenga, and araucaria. On the other hand, they bridled at CONAF's supervision of their management of the forests. They resented restrictions placed on extracting firewood and pasturing cattle. They underlined that this contrasted with CONAF's regulation of large landowners in the region, noting that forestry regulation and supervision fell disproportionately on their shoulders, since large landowners easily fenced off their land and locked the gates to their estates, preventing CONAF officials from inspecting their forestry or livestock operations. In addition, forestry regulations threatened the pillars of indigenous and nonindigenous peasants' subsistence economy. As one member of the Bernardo Ñanco community argued, what resources did Ñanco's members have if they could not exploit their remaining forests? In this case, community members continued to rely on stands of high-altitude araucaria for piñones and as locations to hold religious ceremonies (the *nguillatún*) but collected wood and pastured cattle in degraded stands of beech, primarily lenga.[69]

Tension between CONAF's forestry programs and Mapuche communities extended farther south in the Andes cordillera, partly because of the ecological impact of foresting with exotic species such as eucalyptus. In the region around Panguipulli in the 1990s, for example, CONAF used the new reforms of DL 701 to direct subsidies and credits for forestation to campesino smallholders, many of them from Mapuche communities. As it had in Lonquimay, CONAF gave members of Mapuche communities Oregon and Monterey pine and eucalyptus saplings, wire to protect their plantations, monetary subsidies, and training. Technical assistance from CONAF often included instructions on how to clear invasive weeds and native vegetation to plant pine and eucalyptus. After a number of years, however, many members of Panguipulli's Mapuche communities noted the negative effects of the plantations, even on small parcels of land. One woman observed that her mother had planted three-quarters of a hectare of her land with eucalyptus with CONAF's support, but the trees' deep roots had sucked dry a spring on her property.[70]

Other members of Mapuche communities in Panguipulli also noted that

throughout the region, plantations developed by CONAF had led to desiccated soil and erosion once the trees were harvested. They also pointed
out that the plantations often replaced native vegetation used by Mapuche
communities for a wide variety of purposes. For example, CONAF foresters viewed maqui, a small tree found in the Valdivian evergreen forests, as
a competitor with eucalyptus and pine that required clearing. Yet Mapuche communities used maqui in many ways. Its berries produce jam and
the fruit liquor called chicha; they also have a number of medicinal uses.
Other invasive vegetation, such as quila and colihue bamboo, provided
fodder for Mapuches' livestock. Young colihue and huilo plants offered a
tasty vegetable akin to asparagus. For foresters, this vegetation had to be
cleared to plant pine and eucalyptus. As a result, by the first decade of the
twenty-first century, Mapuche communities around Lanco, near Panguipulli, had agreed not to forest their communities with pine or eucalyptus.
Many communities, represented by the Mapuche organization Parlamento
de Coz Coz, lobbied CONAF for programs to reforest with native species
instead. In interviews, members of the Parlamento de Coz Coz contended
that the exotic species had a negative impact on their soil and served as
nothing more than another subsidy for forestry companies to whom they
had to sell their wood.[71]

Conservation, in the form of private nature reserves on large estates,
also provoked tension with campesinos, as in the case of ecotourism and
forest conservation projects on some of the estates that formerly belonged
to the Complejo Maderero Panguipulli. Víctor Petermann, the owner of
the Huilo-Huilo fundo, purchased Neltume-Carranco in the late 1990s and
has kept it in operation, but his stated goal was to replace forestry with
ecotourism. In stark contrast to Quinquén's Parque Pehuenche, with its
emphasis on community use and management of the forests, the local population of the former complex, including neighboring indigenous communities, has played no role in the administration of Huilo-Huilo's tourism
projects or the estates' forests. While part of the tourist experience of the
Parque Pehuenche is the indigenous community and its traditional uses
of the araucaria forests, there is little room in Huilo-Huilo for either local
Mapuche communities or the remaining population of forestry workers
who are notably absent from the attractions Huilo-Huilo offers tourists.
Huilo-Huilo remakes Panguipulli's landscape once again, but now under
the initiative of private entrepreneurs wielding their own vision of a virgin
landscape free of human inhabitants.

On other estates that once belonged to the Complejo Maderero Panguipulli, new owners also halted logging, in part because of the exhaustion of
the forests' valuable species, and constituted the remaining forests as protected areas. For example, Andrónico Luksic, a wealthy millionaire who,

like Petermann, made his money in mining, purchased two properties—
Chan-Chan and Enco—to establish a private nature reserve. Luksic fenced
off his estates, inciting the ire of former workers in the region, who noted
that the his property held valuable reserves of raulí, coigüe, and Monterey
pine that had been planted by CONAF during the socialist administration
of the Complejo Maderero Panguipulli and during the 1976–81 period,
when the complex was run by the military regime. For the complex's former workers, a sustainable exploitation of the forests owned by Luksic,
rather than their preservation, would generate jobs for the region's impoverished labor force. Workers in the region also accused Luksic of creating
reserves on the estates and fencing off the plantations' native species to obtain carbon or pollution credits, which he could then sell. They voiced the
suspicion that Luksic was holding plantations of native species only until
the trees matured and market prices for native woods rose. And they noted
that fashioning nature reserves in the Valdivian forests allowed Luksic, an
owner of mining enterprises who has often faced charges of environmental
contamination, to create a "green" image.[72]

A similar tension between conservationist regulations and campesinos' use of native forests and mountain pasture continued to percolate in
older forest reserves and national parks. In Villarrica National Park, where
campesinos had claimed and squatted on land for decades, Mapuche communities engaged in a series of land occupations during the late 1990s,
demanding access to veranadas for their livestock.[73] In 1999, for example,
sixty members of the Pocura and Juan Chañapi communities from the
Panguipulli region demanded 3,000 hectares of park land on the border
of the Araucanía and Lakes regions. The communities occupied a small,
remote sector of the park, Pocura Alto, and requested that the land be
distributed to them, arguing that the park had a negative impact on their
territorial, cultural, economic, and religious rights.[74] Similarly, in Curarrehue and Pucón, Mapuche communities demanded the restoration of land
within the Hualafquén Forest Reserve and Villarrica National Park. In
1999, ninety members of the local communities were arrested following a
series of land invasions of the reserve and the park.[75]

A member of the Parlamento de Coz Coz noted that the Villarrica volcano holds deep symbolic importance for a number of Mapuche communities that have claimed land within the park's borders. Communities from
the region around Licán Ray, Choshuenco, and Coñaripe had ascended
the slopes of the volcano to collect piñones for generations. Their títulos
de merced, she observed, included land that reached to the very edge of
the imposing volcano.[76] The ongoing conflict exploded in 2006 when the
Concertación government opened the park to private bids for ecotourism
projects without consulting Mapuche communities who had occupied and

used the parks' forests and veranadas for generations. In addition, as the communities pointed out, not only had SERNATUR, and CONAF never sought their participation in designing ecotourism projects, but the auctions of the concessions ignored their long-standing land claims within the park. For Mapuche communities from Panguipulli and Calafquén, the ecotourism leases were redolent of earlier logging leases and constituted "a new method of consolidating usurpation and exclusion."[77]

The conflicts in Alto Bío Bío, Lonquimay, and Villarrica revolved around Mapuche communities' demands for access to high-altitude forests and veranadas in the Andes cordillera. In 1997, Mapuche communities in Lumaco and Traiguén, a region dominated by the coastal Nahuelbuta cordillera and piedmont of Malleco and Arauco provinces, initiated a series of land recuperations, invading estates owned by the large forestry companies Bosques Arauco and Forestal Mininco. Over the following years, a significant number of communities in Malleco, Arauco, Bío Bío, and Cautín provinces, many of them affiliated with either the Council of All Lands or a more militant Mapuche organization, the Coordinadora de Comunidades en Conflicto Arauco-Malleco (Coordinator of Communities in Conflict Arauco-Malleco), demanded land that had been usurped, challenged the validity of the forestry companies' land titles, and elaborated a searing critique of the ecological impact of pine plantations.[78]

By the end of the twentieth century, it had become quite clear to many Mapuche activists that the small plots of eroded soil on reducciones could do little to sustain the growing Mapuche population, especially in zones almost entirely cleared of their native vegetation. In addition, as monocultural pine plantations further eroded Mapuche communities' access to native forests, the communities increasingly defined the biodiversity of native forest ecosystems as fundamental to their survival. Communities in the coastal cordillera region and the undulating hills of the central valley relied on the income generated from extracting firewood and forest products, as well as hunting small game in their remaining forest stands. By the 1980s, however, many communities' small patches of native forests had become more and more fragmented and isolated within a sea of pine. They could no longer sustain the vegetation, birds, and animals many communities had depended on for survival. In addition, pine plantations had diminished supplies of clean water; reduced access to forest products and game; and contaminated soil and livestock with chemical pesticides, herbicides, and fungicides.[79] As a result, a number of Mapuche organizations and communities began to articulate demands for land in the language of modern environmentalism, attacking pine plantations' impact on the ecology of southern Chile's soil, water, and temperate forests.[80]

In late 1997, the valleys and coastal regions of Arauco and Malleco

9.3 *After the Pine Harvest (Courtesy of Felipe Orrego)*

9.4 *Arauco after the Pine Harvest (Courtesy of Felipe Orrego)*

9.5 *Arauco Bay Paper Mill (Courtesy of Felipe Orrego)*

provinces—notably, the communes of Lumaco and Traiguén—grabbed national headlines when three logging trucks owned by Celulosa Arauco were set on fire and twelve members of the Pichiloncoyán and Pililmapu communities were arrested under the Internal Security of the State Law.[81] The following month, one hundred Mapuches from the Loncoyán Grande community occupied a piece of Forestal Mininco's El Rincón estate and began to fell trees. El Rincón had been one of the many estates expropriated under Allende and then reforested by the Mapuche asentamiento in an agreement with CONAF.[82] A representative of CONADI noted that the community members did not want to abandon the estate "because during agrarian reform, they received the land where they planted the pine and eucalyptus that they are now cutting. They [the trees] grew, and now the Mapuches are asking Mininco to give them their trees."[83] As the El Rincón conflict wore on, Mapuche communities throughout the south engaged in numerous occupations of land claimed by forestry companies, in many cases organized with the support of the Council of All Lands, the Coordinadora Arauco-Malleco, and the Asociación Ñancucheo de Lumaco.[84] It was not uncommon during these years to see private security guards and carabineros guarding the forestry companies' pine plantations and escorting caravans of logging trucks throughout Malleco, Arauco, and Bío Bío.[85]

As Florencia Mallon has argued, in almost all cases it is possible to trace these tomas to histories of usurpation that go back to the first decades of the twentieth century or to the early 1930s. In addition, many tomas occurred in places where there had been tomas and expropriations during agrarian reform. In a number of cases, Mapuche communities demanded land they had won from the Corporación de la Reforma Agraria (Agrarian Reform Corporation) in the early 1970s.[86] In addition, many asentamientos organized during the agrarian reform had been reforested by CONAF and the campesinos. When the estates were returned to their owners during the military's counter-agrarian reform, land that was once eroded and infertile was now valuable because of the trees planted by the asentamiento members and CONAF. In many instances, estate owners eventually sold their reforested estates to forestry companies such as Forestal Mininco and Bosques Arauco, which assumed the responsibility for paying CONAF for the trees they acquired. In almost every case, the companies continued to forest the estates, taking advantage of the subsidies in DL 701 and sparking conflicts with Mapuche communities, who viewed the plantations initiated when the estates were asentamientos as their own. It is notable that in cases such as El Rincón, Mapuche communities made explicit demands not only for usurped land and access to native forests but for tree plantations they had planted during the agrarian reform.

The ecological dimensions of the explosive conflict between Mapuche communities and forestry companies were most clear in the much publicized case of the Temulemu community and the Santa Rosa de Colpi estate outside Traiguén.[87] The Temulemu community demanded the restoration of a piece of land that had been usurped over the years. The community's lonko, Pascual Pichún, noted that its land grant title had given the community 992 hectares, but that it occupied only 760 hectares in 1998. The missing land, he argued, had been folded into the estate now owned by Forestal Mininco.[88] In 1930, the community had won the rights to 58.4 hectares included in its land grant title but usurped by the estate in a decision by the Juzgado de Indios (Indian Courts) that was never enforced.[89] In addition, community members recalled that, with the process of reducción, they lost access to water, fields, and forests they had occupied "since time immemorial" and that lay outside of the boundaries defined by their title.[90]

Two dynamics exacerbated the conflict between the Temulemu community and the forestry companies. First, forestation with pine reduced the communities' customary access to neighboring estates. Before the estates were sold to the forestry companies, members of the community had worked seasonally on them to buy food, tools, and livestock, supplementing the income derived from their depleted plots. The community had also entered into arrangements with the former landowners to pasture their livestock on estate land for a small fee. When the forestry companies arrived, Temulemu's lonko recalled, they did not permit community members to collect firewood or pasture cattle on the estates, reducing the size of the herds they could maintain. A lonko of the neighboring El Pantano community recalled that "the forestry companies arrived with great arrogance from the beginning, not respecting our rights. For example, before we used to move freely around this land, but now they block our way." In addition, the estates, now covered with pine, no longer provided jobs, since subcontractors brought in workers from outside the region.[91] The mayor of Lumaco said that, while large estates traditionally had hired members of the region's reducciones as seasonal or temporary labor, "The forestry companies are not a source of employment for the commune. . . . [I]nstead of employing the local labor force . . . , they bring machinery that requires only a couple of workers."[92] Pascual Pichún similarly noted that "the forestry companies discriminate against hiring workers [from this region]. . . . [H]ere in the community we realized that the Mapuches aren't the only ones who are suffering because of the forestry companies. The poor of Traiguén are also suffering. These people are totally discriminated against. There is not one poor person in the entire commune who has a job in [Forestal] Mininco's camps."[93]

After a few years, the pine plantations took an environmental toll that also reduced Temulemu's subsistence base. As a member of the community described, "After four years, they sprayed, and that's when our animals and livestock died. . . . They were poisoned. . . . [L]ater we were left without water. We didn't even have water for the animals, [and] we could no longer have gardens." Temulemu's *machi* (healer) said, "To make medicine, I use the matico, palo santo, canelo, and other trees. All of these species existed in this zone, but all were removed by the forestry company. But our biggest problem is the lack of water. What falls is absorbed by the forestry company's trees." Another member of the community underlined the detrimental effects of aerial spraying: "this consumes the water, contaminating it and contaminating the pasture, which has a strange smell. This red liquid eliminates all the bushes and allows only pine to grow. . . . I don't know what they use now, but before they used pentachlorophenol [a highly toxic wood preservative and pesticide]."[94]

The government pursued a two-pronged approach to Mapuche communities' conflicts with forestry companies. CONADI purchased land for distribution to Mapuche communities through the Land and Water Fund. By 2002, CONADI had purchased 50,000 hectares from estate owners and distributed the land to 4,617 Mapuche families, including land on Forestal Mininco's El Rincón, Alaska, Pidenco, Tranquepe, and Santa Rosa de Colpi estates. In Temulemu, the government purchased the 58.4 hectares corresponding to the community's original land grant title. Still, most estimates were that Mapuche communities needed at least 150,000 more hectares to support basic subsistence.[95] The Temulemu community continued to invade land held by Forestal Mininco because the recuperation of the 58.4 hectares did not resolve the status of the land claimed by the community that was not included in the original title, but that the community viewed as its own because of generations of occupation.

In addition, when CONADI acquired land to resolve conflicts between estates and Mapuche communities, it often settled members of communities not on the usurped land they demanded but on land purchased elsewhere, distant from the original community. These *"tierras alternativas"* (alternative lands) were often the basis for new communities, frequently founded by young men in land-starved communities, as was the case with Bernardo Ñanco. In Panguipulli, the process of settling community members on new land provoked the rejection of the Parlamento de Coz Coz, whose members argued that the new land grants separated community members from places that held deep importance and placed them on land that did not hold the same generations-old cultural meaning. In addition, they observed that the process of settling community members on alternative land provoked divisions within communities between those who wanted to remain on their historical allotments and those who sought to settle in the new places outside the communities. The Parlamento de Coz Coz members argued that by fracturing community identities in terms of relations with an original place, dividing communities over strategies for recuperating land, and literally separating communities' families by settling some on the alternative lands, CONADI was weakening the bonds that held Mapuche communities together. One Parlamento de Coz Coz member lamented that in Panguipulli, the number of Mapuche communities had grown exponentially since the 1990s as they divided and settled on the new lands.[96] Similarly, in the case of Quinquén, in 1990 the community had rejected the government's proposals to resolve the bitter conflict with the Sociedad Galletué by resettling members of the community on alternative land in Malalcahuello. The Quinquén community underlined its longstanding occupation of the Quinquén valley, its historical use rights to meadows and forests, and its relationship to Quinquén as foundational to its identity.[97]

The Concertación combined its policy of purchasing land for Mapuche communities with harsh police responses to land invasions. In 2002, the government of President Ricardo Lagos Escobar, of the Socialist Party, attempted to crack down on the waves of land occupations in southern Chile by intensifying the use of antiterrorist legislation handed down by the dictatorship. Echoing a strategy adopted by the forestry companies, which in a number of cases had refused to sell land to the government to end tomas, the government made it clear that those who "used violence or occupied land illegally" would be excluded as beneficiaries of CONADI's Land and Water Fund program. As a report by Human Rights Watch and the Chilean Observatory of the Rights of Indigenous Peoples noted in 2004, after 2002 seven Mapuches and a pro-Mapuche activist were charged and convicted under a modified version of an antiterrorism law left in place by the Pinochet regime. They received sentences of up to ten years for arson or threats of arson against landowners and forestry companies. In addition, sixteen Mapuches were put on trial for belonging to the "terrorist" Coordinadora Arauco-Malleco and for "illicit terrorist association."[98]

While the Frei government used the ordinary criminal code, and three prosecutions against Mapuche activists under the Internal Security of the State Law of 1958, Lagos's government employed Pinochet's Antiterrorism Law to confront the escalating conflict between Mapuche communities and forestry companies.[99] The Antiterrorism Law had actually been expanded during the Aylwin administration to include arson, defined broadly as setting fires to uninhabited buildings, woods, fences, or fields.[100] As the Human Rights Watch report noted, in the case of the Mapuches detained for assaults on forestry company property, "The crimes committed in most cases are crimes against property and do not fit the characterization of terrorism contained in international treaties, including the Inter-American Convention against Terrorism, which requires grave violations against persons."[101] In July 2003, Rodolfo Stavenhagen, the United Nations Special Rapporteur on the situation of human rights and fundamental freedoms of indigenous peoples, similarly issued a report on detentions of Mapuche activists. He observed that "under no circumstances should legitimate protest activities or social demands by indigenous organizations and communities be outlawed or penalized.... [C]harges for offenses in other contexts [e.g., terrorist threat, criminal association] should not be applied to acts related to the social struggle for land and legitimate indigenous complaints." Stavenhagen called on the Chilean government to declare a general amnesty for those on trial "for social and/or political activities in the defense of indigenous lands."[102]

On the one hand, the Concertación governments employed tactics of repression redolent of the era of dictatorship, transforming zones in southern Chile covered with pine plantations into armed camps patrolled by carabineros. Mapuche communities in Malleco and Arauco, particularly, remained surrounded by both pine plantations and private forest guards, supported by carabineros, who protected Forestal Mininco's and Bosques Arauco's logging operations.

The Human Rights Watch report described the systematic violence executed by forestry guards and carabineros in zones of conflict, primarily in Arauco and Malleco. The report noted that Mapuches were frequently the victims of physical abuse and degrading treatment by the police "during operations to evict occupiers of disputed land and during raids into communities to capture suspects and seize evidence." In addition, "A disturbing feature of these incidents has been the ill treatment of women and old people, especially lonkos and machis.... [S]ome of the worst examples occurred in 1999 in Temulemu and in 2000 in the neighboring community of Temucuicui. Beatings during arrests, disproportionate and indiscriminate use of riot control weapons such as shotguns, racist insults, and destruction or theft of domestic articles are still common occurrences during such operations."[103]

On the other hand, the Concertación governments used CONADI's Land and Water Fund to purchase tens of thousands of hectares, often from the forestry companies themselves, to be distributed to land-starved Mapuche communities. In addition, the Concertación organized development and agricultural modernization programs, often to subsidize forestation with pine and eucalyptus, among Mapuche communities, as in the case of Project Lonquimay, often with the financial assistance of international development organizations and nongovernmental organizations. In the

end, the Concertación sought to defeat an increasingly radicalized Mapuche movement led by groups like the Coordinadora Arauco-Malleco and
Council of All Lands, while meeting the demands of individual communities within the context of its neoliberal economic policies. The center-left
governments incrementally restored land to communities to resolve their
generations-long conflicts with large estates without introducing reforms
that would challenge forestry companies' control of land and labor or reorganize the pine-based industrial forestry economy.

The case of Temulemu best exemplifies the Concertación's approach to
the militant Mapuche movement. In 2002, Pascual Pichún, along with a
number of lonkos from neighboring communities, was convicted of "terrorist threats" against forestry companies. Nonetheless, in 2009, after Pichún
had left jail, the Concertación government of the Socialist Party's Michelle
Bachelet negotiated a solution to Temulemu's claims to the Santa Rosa
de Colpi estate. With the financial support of the International Development Bank, the government purchased Santa Rosa de Colpi from Forestal
Mininco. As a condition for the sale, the Temulemu community signed a
contract with Forestal Mininco in which it agreed to sell it all the wood it
harvested from Santa Rosa de Colpi's pine plantations. The International
Development Bank agreed to provide Temulemu credits worth half the
value of the harvests to fund development projects within the community.
As José Antonio Viera-Gallo, the coordinator for indigenous affairs and
cabinet member in Bachelet's government, put it, the agreement was "tied
to a project of forestry development. The communities are very interested.
We are trying to create an innovative model of development with Mininco.
We hope that with this agreement we can begin the business association
between the forestry companies and the communities."[104]

The agreement recalled in many ways the contracts signed by Mapuche
communities with forestry companies during the 1940s and 1950s. Social
conflicts in the forests would be resolved by incorporating Mapuche peasants into the industrial forestry economy. Forestry development could
prove profitable for all parties. The communities would receive land, jobs
in forestry, and income from pine plantations. The forestry companies,
while agreeing to sell their land to the government, would be assured a
steady supply of inexpensive primary material for their cellulose plants.
The government would receive the social peace necessary for the continued growth of a dynamic export industry. The contract did little to challenge Mininco's dominance of the forestry economy. By purchasing the
land and its plantations, while guaranteeing that Mininco would maintain
its access to the pine logged on the estate, the government helped to subsidize both the solution to the conflict with Temulemu and the company's
industrial activity.

Indeed, since the late 1990s forestry industry executives had proposed that the solution to the "Mapuche conflict" lay in replacing impoverished soil no longer suitable for agriculture on reducciones with profitable tree plantations that would also expand the companies' supply of raw materials. In 1998, Jorge Serón, a regional director of the timber industry's trade association, argued in response to the conflicts in Traiguén, that "we have made CONADI see that it is feasible to generate a forestry development plan . . . that consists of the communities' basically dedicating their soil to planting with pine and eucalyptus. . . . [I]f the soil is capable of generating wealth for the forestry companies, why can't it create wealth in the hands of Mapuche communities?"[105] This proposal in many ways was consistent with CONAF's programs to subsidize reforestation with pine and eucalyptus within Mapuche communities in places as diverse as Lonquimay, Lumaco, and Panguipulli. In Temulemu, as in Lonquimay and Quinquén, government funds ultimately underwrote both the solution to a bitter, generations-old land conflict and a broader regional development project. And, as in Lonquimay, international development assistance helped a government committed to fiscal "responsibility" fund both the land purchase and the forestry development program.

The Concertación's approach to the Temulemu community's conflict with Forestal Mininco produced another significant result: it led Temulemu to negotiate with the company and the government as an individual community. The negotiated deal reaffirmed the Concertación's commitment to defining the causes of the Mapuche conflict as poverty and lack of development. By negotiating land purchases and development assistance with Mapuche communities individually, the Concertación sidestepped the larger issues raised by the violence in the southern countryside during the late 1990s: the social and ecological sustainability of its neoliberal model in the forests, the demands of Mapuche and non-Mapuche laborers and smallholders for reorganization of the forestry industry, and the militant demands of Mapuche organizations for collective ethnic political rights and territorial autonomy.[106]

Incorporating communities such as Temulemu into the industrial forestry economy did little to address the Mapuche movement's broader challenge to the social and ecological impact of pine plantations. As interviews with members of Mapuche communities in Panguipulli and the experiences of communities in places such as Temulemu made clear, biodiversity rooted in the conservation of native forests was central to Mapuche communities' moral and political critiques of forestry companies, as well as to their ethnic identity and cultural practices. Growing pine to supply the two big forestry conglomerates did nothing to address their central demand for a different kind of community-directed forestry, in which tree plantations were mixed with small plots dedicated to peasant agriculture, the garden crops of the chacra, and pasture, as well as small patches of remnant and second-growth native forest. The strongest argument for planting pine and eucalyptus on Mapuche land, especially in the cleared and eroded soil of regions such as Lumaco and Loncoche, where communities have effectively eliminated their forests over the years, is that both species grow well on cleared land and the slopes of hills and the foothills of the cordilleras where much of Mapuches' marginal agricultural land is located.[107] Yet as Roger A. Clapp notes, neither Monterey pine nor eucalyptus improves the fertility of soil. And as the experience of Mapuche communities in the Panguipulli region illustrates, in some cases tree plantations, even on small plots, degrade soil and dry up streams and watersheds. Pine and eucalyptus provide inexpensive, fast-growing prime material for the forestry industry. But they do not always fit well with a model of mixed agricultural and forestry production that would meet the demands of Mapuche communities.[108] Nor are they an adequate substitute for the patches of heterogeneous native vegetation on which many communities continue to depend for medicinal herbs, pasture, forest products such as mushrooms, and wild game.[109]

The environmentalist and Mapuche movements mounted the most serious challenge to the Concertación's neoliberal economic policies during the two decades that the center-left governments remained in power. While both produced incremental changes and small victories, neither was able to oppose effectively the organization of industrial forestry production or force a radical shift in the management of Chile's remaining temperate forests. Even while the transition to democracy during the 1990s created political openings for Mapuche communities and environmentalist groups to mobilize opposition to the authority of logging companies over land and natural resources, the democratic governments, like the dictatorship before them, remained wedded to a free-market economic model and bifurcated forestry policy in which the imperatives of economic growth based on industrial forestry, combined with efforts to preserve temperate forests viewed as a magnet for ecotourism, eclipsed the demands of local communities. Conservationist projects tied to ecotourism drew on traditional understandings of organizing national parks and nature preserves that excluded local populations, ignoring their historical occupation of the forests and local knowledge and practices of forest exploitation. Forestry science continued to focus on tree plantations as the engine of growth, maintaining a bifurcated vision of tree plantations and industrial development, on the one hand, and enclosed, pristine native forests subject to conservationist regulation, on the other. Neither case left room for local campesino communities. The Concertación did introduce environmental regulations, but like the Indigenous Law and reforms of the Labor Code, they did little to transform the organization of industrial forestry production.

During the transition to democracy the Chilean state did nothing to challenge the patterns of deep inequality in land and labor relations in southern Chile. Instead, in response to the waves of Mapuche land occupations during the 1990s, it addressed short-term issues of rural poverty by purchasing land and providing limited development assistance for Mapuche communities and non-Mapuche smallholders. For those who worked in the forests as laborers or depended on the forests for survival, the transition to democracy only intensified their dispossession, as forestry development accelerated with Chile's increasing integration into the global economy. Pine-fueled forestry on estates held by large financial conglomerates continued to squeeze peasants off their land. National parks and private forest reserves continued to exclude local populations from Chile's remaining stands of temperate forest, now in the name of growth motored by ecotourism. When the Concertación governments finally lost power in 2010, they left behind a forestry economy little reformed since the days of dictatorship.

═══ *Conclusion*

In the course of doing research for this book, I interviewed a number of forestry workers, union activists, and members of Mapuche communities throughout southern Chile. During these conversations over the years, I was struck by the fact that, across the board, the people I spoke with emphasized the ecological degradation and loss of biodiversity that has accompanied forestry development in Chile. Many attributed this to the Pinochet dictatorship and Chile's current neoliberal economic model, often folding together the tremendous social dislocations provoked by free-market reform; the trauma of state terror, which had been directed disproportionately at rural laborers; and the ecological changes produced by tree plantations. Campesinos concurred in viewing Monterey pine as a force responsible for the alienation of their labor and loss of their land, contrasting tree plantations unfavorably with native forests. To a significant degree, modern environmentalism has shaped a broader critique of forestry development and free-market economics, providing a language for understanding social, as well as environmental, injustice. Talking about the degradation of native forests was one way campesinos from all walks of life expressed a critique of the dispossession that has been produced by free-market restructuring since 1973.

The salience of environmentalist language in oral histories reflects a significant change in campesino politics throughout southern Chile. As this book shows, for most of the twentieth century, southern campesinos viewed environmentalism, in the form of early twentieth-century conservation, as responsible for excluding them from native forests on which they relied for survival. In addition, in Chile, conservationist regulations developed over the course of the twentieth century hand in hand with state-directed industrial forestry. Rather than implement managed or sustained-yield timbering in native forests, as in other parts of the world, the Chilean state promoted a bifurcated approach to forests and forestry. The commercial forestry industry focused on exploiting plantations of exotic species, often cultivated to replace logged and burned-over native forests both on privately held property and within state-run forest reserves, while remote stands of temperate forest were preserved in national parks in the mountains. In both cases, southern campesinos, evicted from forest reserves and national parks, as well as from estates that turned to planting pine, encountered growing restrictions on their access to forest resources.

By the first decade of the twentieth century, campesinos' loss of land and forests during colonization was exacerbated by ecological disasters in the frontier provinces. Soil erosion, climatic changes, drought, and flooding provoked by deforestation undermined the pillars of the southern peasant economy. Even as they sought to regulate the destruction of the forests on which they relied for myriad products, campesinos participated in this twentieth-century history of ecological degradation. Restricted to small plots of land and with land tenure increasingly insecure because of the constant threat of usurpation by large estates, campesinos extracted whatever value they could from the forests, selling firewood, manufacturing sleepers for the railroad, leasing logging rights both to small sawmills and to large timber companies, and clearing forests for livestock and crops. Mapuche communities and non-Mapuche ocupantes and colonos relied increasingly on exploiting the forests as soil erosion limited their cereal harvests and reduced their pasture. Necessity and insecurity forced land-starved southern campesinos into an increasingly destructive relationship with the forests.

In addition, forest laws alienated campesinos from the forests themselves. Conservationist regulations designed to stimulate the modernization of forestry enclosed public land on the frontier, inflicting ever more scarcity on poor campesinos. Deforestation and soil erosion on their small plots reinforced a perception, wielded by state land and colonization officials, and echoed by large landowners and forestry companies, that campesinos were irresponsible stewards of nature. Campesinos who claimed rights to frontier land came to be perceived as a threat to the forests, a view apparently reinforced by their frequent occupation of land within the borders of national parks and forest reserves. Chilean land and colonization officials employed a conservationist narrative of campesinos' detrimental impact on nature to justify handing over enormous tracts of land and frontier forest to large estates and logging companies, which ostensibly would enjoy the technical expertise, capital, and economies of scale that would permit rational exploitation of the forests according to the dictates of forestry science.[1]

By the late twentieth century, however, campesinos had sundered environmentalism from industrial forestry. During the 1980s and 1990s, forestry workers in southern Chile rebuilt unions decimated by military repression, challenging the authority of the large financial conglomerates that controlled vast expanses of tree plantations, cellulose plants, and stands of frontier forest. Forestry workers' unions set forth a broad set of demands that amounted to an alternative to tree-plantation-based, export-oriented industrial forestry. They produced a vision of a more just system of labor and land relations rooted in the biodiversity of Chile's native temperate forests, tying pine plantations to an agroindustrial economy, that alienated both workers' labor and nature. The unions' goal was to return to a moment in which campesino smallholders worked stands of native forest, mixing forestry production with peasant agriculture, and in which landless laborers retained rights to plots of land to pasture cattle and cultivate garden crops, as well as access to key forest resources. In addition, they proposed sustainable logging of native forests and the development of industries based on the processing of native woods that would add value to the logs extracted from the forests as an alternative to the pine and cellulose model of forestry development. Although far from preservationist, forestry workers' understanding of the value of nature, of the ecology of native forests, served as a vehicle for judging unjust the profoundly unequal social arrangements produced by free-market forestry.

During the same period, Mapuche communities throughout southern Chile organized a powerful movement to recuperate land usurped by large estates. The Mapuche movement also elaborated a powerful critique of industrial forestry. Mapuches drew clear connections between the ecological impact of pine-fueled forestry and their loss of land, articulating an alternative vision of southern Chile's social and environmental landscapes organized around the sustainable management of temperate forests. Like the forestry workers' unions, many Mapuche communities demanded not just the return of usurped land but also the restoration of an ecological order that mixed campesino agriculture with the customary use of the forests, from pasturing cattle to collecting firewood and forest products and hunting wild game. As in the case of forestry workers, this view of the forests was not preservationist. Mapuche communities sought to defend their multiple uses of the native forests, from gathering forest products to pasturing livestock and logging. Nonetheless, in both cases, Mapuche communities and forestry workers (and many forestry workers belonged to Mapuche communities) linked their demands to the contemporary environmentalist movement's concern with the ongoing destruction of southern Chile's remaining temperate forests.

I hope this book establishes a similar bridge between the concerns of social and environmental history. The two fields, especially the subfield of labor history, share a focus on the development of capitalism—the changes wrought by the expansion of the capitalist world market and new organizations of production based on wage labor, as well as the related histories of European imperial expansion.[2] Yet, as Gunther Peck notes, environmental and social historians have rarely connected parallel narratives focused on the degradation and alienation of nature, in the case of the former, and the degradation and alienation of labor, in the case of the latter.[3] For social historians, nature often serves as a backdrop to the central struggles over land and labor, not a key part of the production of commodities and value. Even in the context of peasant movements, nature has often been neglected, and workers' struggles have been defined in purely social materialist terms—against proletarianization, the penetration of market relations, and increasingly commodified forms of exchange and production—without considering the role of ecological change in these processes and the role of the environment and environmental ideas in peasant politics.[4]

At the same time, while pioneering environmental historians have written splendid accounts of the ecological impact of capitalist agriculture and agroindustry, the spread of market relations, and the changes wrought by European colonialism and settlement, their historical renditions of capitalism often elide the contradictions shaped by class relations and the tensions embedded in the organization of production, focusing on a capitalist "ethos" or an "agro-ecological system," in the words of Donald Wooster, largely undifferentiated by class, ethnicity, or gender.[5] In the case of William Cronon's pathbreaking *Nature's Metropolis*, the spread of capitalist markets produces the transformation of "first nature" into commodities whose value derives as much from "nature's wealth" as from labor. While labor and productive relations drive historical processes in the Marxist traditions that inform much social history, for Cronon labor is as much about the consumption and destruction of nature as about production. As a number of critics have argued, Cronon's brilliant rendition of environmental history as the extraction of value from nature through the process of commodification and market exchange ignores the social process of production, with all of its contradictions, tensions, and divisions.[6] As Cronon and Richard White have underlined, environmental historians continue to grapple with the question of how to make the history of nature part of human history, incorporating social historians' concerns with forms of social power based on class, gender, ethnicity, and race.[7]

In this book, I take up the challenge of bringing together social and environmental history.[8] In doing so, I revise traditional historical narratives of the frontier, Chilean agrarian history, the history of Chile's modern state, and the social history of campesino movements in several ways. Most social histories of rural Chile view campesino history largely in terms of class or, in the case of Mapuche communities, in terms of ethnicity, at best placing the environment as mere stage scenery for the major historical dramas of social conflict and transformation and ignoring the ways in which the environment has shaped rural class formations and campesino politics. Even the history of the colonization of the frontier during the late nineteenth century, following the defeat of formerly independent Mapuche groups, tends to neglect the environment, despite the great ecological changes introduced after the 1850s throughout southern Chile. More generally, very little has been written about Chile's environmental history, a striking lacuna, given the importance of contemporary environmental conflicts over degraded temperate forests; collapsed marine ecosystems; mining pollution; dangerously high levels of urban smog in Santiago; and the rampant use of pesticides, fungicides, and herbicides in the fruit and forestry industries.[9]

One contribution of this book is to show how both ecological conditions and environmental thought shaped colonization in southern Chile's frontier forests. Nineteenth-century ideas about forests, climate, and rainfall dictated massive burning of the southern forests during the second half of that century. The goal was to clear land for agriculture and settlement, but also to produce changes in climate that were more propitious to cultivating cereals. Nonetheless, the temperate forests shaped colonization in key ways. Their burning allowed landowners an easy way to extract rents—in this case, forest rents—by providing a cheap fertilizer that produced extraordinary yields during the first years of cultivation. Their logging offered landowners another quick and inexpensive way to make money without significant investments in production. This aided the speculative formation of large estates throughout southern Chile. Landowners acquired extensive stretches of land, often through fraud and violence; extracted quick profits through burning and logging forests; and then waited for their property to acquire greater value as the railroad extended its reach throughout the south.

To forest rents and the money made in land speculation were added the rents extracted from nature in workers' bodies. Inquilinos, medieros, and arrendatarios provided inexpensive labor to large estates predicated on the availability of forests to burn. Inquilinaje throughout Chile involved the exchange of labor for wages and regalías, including small plots of land, tools, livestock, and housing. In the south, laborers' regalías included forests because burning forest was the only way to acquire a good yield. The forest rent landowners extracted through logging and burning forests was also included in their forms of payment for labor. This produced a highly impermanent rural workforce that moved around southern Chile in search of stands of forest that could be burned and planted. In addition, labor and land relations were shaped by the ecological catastrophes created by this system of agricultural production. Within a generation following settlement, provinces such as Arauco, Malleco, and Cautín began to confront drought, flooding, and soil erosion. Large landowners could withstand growing ecological crises in the frontier because of their access to large stretches of land and forest, but smallholders, who worked their small plots intensively, faced ruin. By the first decade of the twentieth century, and until the agrarian reform of the 1960s, smallholders were increasingly forced by the changing conditions of climate and soil to sell their plots to large estates. They came to compose part of a swelling landless or, in the case of Mapuche communities, land-starved migrant labor force. As this book makes clear, ecological degradation drove the creation of this itinerant rural labor force and an unequal system of property relations organized around increasingly extensive haciendas.

While most histories of Chile locate narratives of state formation, the stage of national politics, in Santiago and central Chile, my research shows that a number of attributes of the modern state—especially the state's authority to regulate property relations and the exploitation of natural resources—developed in the frontier territory. Just as European conservationist, or "environmentalist," thinking about ecological catastrophe emerged out of the colonial experience, especially the laboratory of "tropical island Edens," in the words of Richard Grove, in Chile, the analogous colonial process in the frontier provinces produced a new concern with the environment in a region where easily observable ecological change was rapid and caused immediate devastation.[10] Ecological crises on the frontier produced a growing environmental consciousness and accelerated the state's governance of the southern forests. The state's capacity to assert its authority over land and natural resources was strengthened by the frontier's status as public land.

Drawing on European and North American forestry science, forest and land officials sought to remedy the ills of deforestation by promoting tree plantations. In addition, they worked to simplify nature, substituting monocultural pine plantations for the heterogeneous temperate forests. State-fomented industrial forestry also remade southern social formations, transforming large estates into modern forestry enterprises and itinerant and often rebellious campesino laborers into a proletarianized, if incompletely disciplined, workforce. This book demonstrates that beginning in the 1930s, the state worked to exercise governance by employing forestry science to engineer new landscapes and reorder people's relationship to the natural world, often in response to the ecological consequences of deforestation that its own colonization policies had produced. State subsidies helped drive forestry growth, even during the most doctrinaire moments of free-market restructuring during the 1970s, by making inexpensive land and tree plantations available to large forestry companies and then providing incentives to forest land often acquired from impoverished campesinos or the state itself.

My research shows that forestry science shaped social and environmental policy in the frontier territory during much of the twentieth century and that the ecological and social changes introduced under its industrial

forestry model were dialectically intertwined. While ecological degradation caused by deforestation drove the deracination of campesinos, exac-
erbating the fraud and violence that accompanied colonization at the beginning of the twentieth century, the ecological impact of tree plantations
constituted a second blow to the viability of southern Chile's campesino
economy at the end of the century. State-promoted forestation expelled
campesinos from their land both within and on the margins of estates
throughout southern Chile; by the 1990s, tree plantations had not only replaced a large population of former estate laborers, but also undermined
the ecological sustainability of forestry estates' neighboring campesino
smallholders. To the propitious conditions of southern Chile's soil and climate was added inexpensive labor, allowing pine plantations to flourish
and extend their reach throughout the frontier. State terror only served as
an additional subsidy to the major financial conglomerates that controlled
the forestry industry after 1973 by fracturing southern campesinos' organizational capacity to defend a peasant economy against the encroachments
of tree plantations and guaranteeing the forestry industry a permanent
supply of cheap labor in the growing population of sporadically employed
southern laborers.

Finally, this book bridges the gap between social and environmental
history by examining the ecological dimensions of campesino movements
in Chile since the beginning of the twentieth century. For much of the
twentieth century, campesinos in the southern frontier region engaged in
movements to take back from large estates land that they perceived as public or theirs by rights of occupation and use, a history of rural unrest that
has gone largely unnoticed by historians of modern Chile. My research
demonstrates that throughout the southern frontier territory, campesinos often defined their struggles in terms of a commons found in nature.
While conservation presented itself to southern campesinos as an ideology
justifying their exclusion from the frontier's forest resources, they articulated their own understanding of the just uses of nature. Campesinos frequently worded their moral critique of land fraud and the dominance of
large estates in terms of landowners' and loggers' wasteful destruction of
forests. They contrasted their reliance on the native forests first with logging, and then with the replacement of native forests by privately held pine
plantations. By the 1990s, they had come to oppose not only the privatization of land they viewed as public—or owned communally, in the case
of Mapuche communities—but also the reduction of complex ecosystems
to monocultural plantations worked by a proletarianized labor force up-
rooted from the land and expelled from the forests. Understanding forests
as a commons rather than private property allowed campesinos to critique

the increasing commodification of nature through logging and forestation with exotic species, as well as the commodification of their own labor.[11]

There is a sometimes unremitting pessimism to historical narratives of ecological degradation. Yet combining social and environmental history allows us to see moments of possibility that provide some relief from a sense of the inexorable alienation of labor and nature during southern Chile's increasingly intense integration into the global forestry economy. Campesinos' understanding of nature as a commons, which is now frequently articulated in the rhetoric of modern environmentalism, and their understanding of the importance of biodiversity to social, as well as ecological, sustainability provides an alternative to the widely held view that privatizing natural resources, allowing the market to be the final arbiter of the fate of the forests and the people who rely on them, is the only way to produce economic modernization and development and, ultimately, to complete the incorporation of the frontier and its forests into the nation. In part, this historical perspective allows us to approach the problems of land, labor, and nature from the viewpoint of local communities themselves. Instead of one dominant solution—for example, the centralizing and homogenizing prescriptions of forestry science for pine-driven development—we might locate different historical moments and places where campesinos came together to define alternative paths history might take.

Campesinos' proposals for the forests rarely reflected a preservationist sensibility. Nonetheless, their engagement with the southern frontier's shifting landscapes produced collective efforts to impose restrictions on ecological degradation. We can find examples in the outraged denunciations of extractive logging and clear-cutting in raulí forests in the Andes cordillera in Cunco and Panguipulli and in the alerce forests in Valdivia's Sarao coastal cordillera by ocupantes, colonos, and Mapuche communities who demanded the expropriation of large estates from the 1920s to the 1950s. Similarly, when squatters occupied land in the Malleco Forest Reserve in 1955, they phrased their demands for land as a critique of the government's policy of leasing the reserve's forests to logging companies who felled the native forests and then reforested with Monterey pine. Despite accusations that they were a threat to both pine plantations and the reserve's remaining native forests, the campesinos underlined that, along with cultivating crops, pasturing cattle, and collecting forest products, they had also protected stands of native forest from fire and logging. Likewise, when Concepción's Junta de Beneficencia (Welfare Society) leased the araucaria forests on estates it owned in the Nahuelbuta Cordillera to the Oelckers logging company during the 1940s, campesinos, who relied on the araucaria forests for extracting firewood and piñones, contended that they enjoyed rights to the land based on years of use and occupation. They buttressed their claims by pointing to Oelckers's indiscriminate destruction of the araucaria forests, which they contrasted with their own uses of the forests. Like the campesinos in the Malleco Forest Reserve, they put forward one of the earliest critiques of a government policy of leasing native forests to big logging companies and then overseeing their reforestation with Monterey pine, a policy of forest conversion that endured into the 1990s.

The brief experiment with socialist forestry in the Complejo Maderero Panguipulli possessed many of the limitations of the reigning industrial forestry model. However, it also constituted an important effort to balance the demands of development with the need to manage the Valdivian forests sustainably while meeting the needs of the region's large landless labor force and many Mapuche communities. Especially significant were efforts by the its foresters to end clear-cutting and extractive logging while initiating reforestation in degraded forests with native species and providing stable jobs for the region's many unemployed landless workers. In other zones already cleared of native vegetation and suffering severe soil erosion, such as Lumaco, planting pine made sense for campesinos and the members of Mapuche communities under the conditions provided by the agrarian reform. As in the Complejo Maderero Panguipulli, industrial forestry organized around pine plantations would provide jobs, recover degraded soil, and compose part of a peasant economy, combining tree plantations, stands of remnant and second-growth native forest, pasture for livestock, and plots for cultivating crops. In both Panguipulli and Lumaco, the agrarian reform made possible a mixed economy of forestry, pastoralism, and agriculture that might have allowed a significant degree of social and ecological sustainability. In part, this was so because during the agrarian reform, campesinos, empowered by the radical political movements sweeping through Chile, insisted that forestry and forestation include space for their garden crops and livestock, as well as for their small patches of native forest.

Farther north, one can trace the efforts made by Mapuche-Pehuenche communities to halt or restrict logging in araucaria forests in Alto Bío Bío and Lonquimay even as they contracted logging rights to their forests during the 1940s and 1950s. Campesinos proposed a similar mixed or multi-use approach to the forests based on extracting firewood and some lumber, as well as myriad forest products; pasturing livestock; and cultivating a mixture of crops, from maize, beans, potatoes, and squash to wheat and barley. They achieved momentary successes during the agrarian reform when they halted logging in the araucaria forests, pushed the socialist government to establish nature reserves, and won use rights to collect piñones and pasture their livestock in the forests. More recently, in Quinquén, one sees projects to organize locally directed forms of multi-use management of the forests, combining Mapuche-Pehuenche communities' historic uses of araucaria forests with ecotourism in a nature reserve administered by the communities themselves (in stark contrast to ecotourism projects initiated by the state in national parks and large landowners on their estates without the participation of local communities). This type of locally directed "community forestry" offers one method of creating both ecological and social sustainability while restoring a certain degree of local autonomy to campesino communities.[12]

If there was a common denominator in campesinos' struggles to exercise some governance over southern Chile's soil and forests across decades and diverse regions, it was that forestry be combined with a range of other land uses attached to a variety of landscapes, including patches of remnant and second-growth native forest for hunting game and collecting forest products, pasture for livestock, and plots for crops. Throughout the twentieth century, campesinos in southern Chile labored to produce the kind of mixed-use community forestry that can build the conditions for the sustainable reproduction of both their peasant economy and the native forests. In certain regions already cleared of native vegetation, such as Lumaco during the agrarian reform, this mixed-use agroforestry and silviculture also included tree plantations. As Roger A. Clapp argues, under the right conditions tree plantations can be part of a sustainable campesino economy. Systems of inter-cropping that include tree plantations may also promote the sustainability of the plantations and the soil in which they grow by protecting cultivated trees from diseases to which monocultures are particularly vulnerable, preventing soil erosion after trees are harvested, and restoring key nutrients to the soil. This community forestry model requires local control of remaining native vegetation, tree plantations, agricultural land, and pasture. Such a diversified "mosaic" landscape would supply a cash crop (pine) and soil for pasture and garden crops, as well as environments amenable to wild game, herbs, and other products of the native forests, which historically have been an important element of southern campesinos' subsistence.[13] As David Tecklin and Rodrigo Catalán also note, this kind of community forestry also requires flexible dialogue between the technical knowledge supplied by agronomy and forestry science, especially in regions where tree plantations occupy already cleared land and degraded soil, and the local knowledge and uses of the native forests.[14]

However, as the example of Quinquén suggests, the success of locally directed forest management requires significant state support and, most important, agrarian reform. Profound inequalities in landholding, as the

US Forest Service pointed out in 1946, place severe obstacles in the way of sustainable forest management. Given the complexity and reach of southern Chile's forest ecosystems and the economies of scale required for tree plantations, individual communities acting alone are at disadvantage because of limited landholdings and resources. A more collective and territorial approach to managing forests and tree plantations, combined with agricultural and pastoral activities, is required. In addition, without the kinds of state technical assistance and economic support that accompanied the agrarian reform, and that are supplied in smaller doses today by government agencies such as the Corporación Nacional Forestal (National Forestry Corporation), campesinos are at a significant disadvantage when it comes to managing native forests or planting pine. Campesinos' militant movements to take back land from large estates during the 1960s and early 1970s forced agrarian reform officials, who were as enthusiastic about forestry science and tree plantations as Federico Albert was in 1911, to restore spaces to pasture livestock and cultivate garden crops, as well as to restrict logging in araucaria forests in the Andes cordillera. Perhaps campesinos' growing understanding that social justice must also be environmental justice will force similar modifications of the current neoliberal forestry model, dedicated as it is to vast monocultural pine plantations and the production of cellulose for world markets, and place the reform of Chile's deeply unequal system of landholding on the table once again.[15]

Notes

Introduction

1. For a superb description of Chile's forest types, see Ken Wilcox, *Chile's Native Forests: A Conservation Legacy* (Redway, CA: Ancient Forest International, 1996), esp. 11–29. See also Claudio Donoso Zegers, *Bosques templados de Chile y Argentina: Variación, estructura y dinámica* (Santiago: Editorial Universitaria, 1993); Eduardo Neira, Hernán Verscheure, Carmen Revenga, Eduardo Neira, Carmen Revenga, *Chile's Frontier Forests: Conserving a Global Treasure* (Valdivia: World Resources Institute, Comité Nacional Pro Defensa de la Fauna y Flora, Universidad Austral de Chile, 2002), 6.

2. Joaquin Lavín, *Chile: Revolución silenciosa* (Santiago: Zig-Zag, 1987). For critical historical interpretations of Chile's vaunted economic miracle that focus on both the social and environmental costs of free-market development, see the essays in Peter Winn, ed., *Victims of the Chilean Miracle: Workers and Neoliberalism in the Pinochet Era, 1973–2002* (Durham, NC: Duke University Press, 2004), and Rayén Quiroga Martínez, Saar van Hauwermeiren, Jorge Berghammer Vega, Patricio Del Real Jaramillo, Marlene Gimpel Madariaga, eds., *El tigre sin selva: Consecuencias ambientales de la transformación económica de Chile, 1974–1993* (Santiago: Instituto de Ecología Política, 1994).

3. For works on the conflict between forestry companies and Mapuche communities, see Diane Haughney, *Neoliberal Economics, Democratic Transition, and Mapuche Demands for Rights in Chile* (Gainesville: University Press of Florida, 2006); Sara McFall, ed., *Territorio mapuche y expansión forestal* (Temuco: Ediciones Escaparate, 2001).

4. Karl Jacoby, *Crimes against Nature: Squatters, Poachers, Thieves, and the Hidden History of American Conservation* (Berkeley: University of California Press, 2003), 3. For an overview of forestry development in Chile, see Pablo Camus Gayán, *Ambiente, bosques y gestión forestal en Chile, 1541–2005* (Santiago: LOM Ediciones and Dirección de Bibliotecas, Archivos y Museos, 2006). For an environmentalist critique of forestry development, see also Defensores del Bosque Chileno, ed., *La tragedia del bosque chileno* (Santiago: Ocho Libros Editores, 1998).

5. See also Roger Alex Clapp, "Creating Competitive Advantage: Forest Policy as Industrial Policy in Chile," *Economic Geography* 71, no. 3 (July 1995): 273–96; Camus Gayán, *Ambiente, bosques y gestión forestal en Chile.*

6. Jacoby, *Crimes Against Nature*, 3.

7. For a case of South Asian peasants and their changing relationship to forests and environmentalism, see Arun Agrawal, *Environmentality: Technologies of Government and the Making of Subjects* (Durham, NC: Duke University Press, 2005).

8. I have followed Roger Alex Clapp, "The Unnatural History of the Monterey Pine," *Geographical Review* 85, no. 1 (January 1995): 1–19, here. A classic argument in favor of the ecological sustainability of Monterey pine plantations is Fernando Hartwig Carte, *Chile, desarrollo forestal sustentable: Ensayo de política forestal* (Santiago: Editorial Los Andes, 1991). For excellent summaries of the negative environmental and social impact of pine plantations, see René Montalba Navarro, Noelia Carrasco Henríquez, and José Araya Cornejo, *The Economic and Social Context of Monocultural Tree Plantations in Chile: The Case of the Commune of Lumaco* (Montevideo, Uruguay: World Rainforest Movement, 2006); Antonio Lara and Thomas T. Veblen, "Forest Plantations in Chile: A Successful Model?" in *Afforestation: Policies, Planning, and Progress*, ed. Alexander S. Mather (London: Belhaven, 1993), 118–139.

9. Clapp, "The Unnatural History of the Monterey Pine"; Lara and Veblen, "Forest Plantations in Chile"; Rafael Ros Vera, "Plantaciones: Sus efectos sobre el ambiente," *Actas: Jornadas Forestales* 14 (1992): 21–37.

10. Clapp, "The Unnatural History of the Monterey Pine"; Marlene Gimpel Madariaga, "El sector forestal ante la apertura económica: Exportaciones y medio ambiente," in Quiroga Martínez et al., *El tigre sin selva*; Ros Vera, "Plantaciones"; Rayén Quiroga Martínez and Saar van Hauwermeiren, *Globalización e insustentabilidad: Una mirada desde la economía ecológica* (Santiago: Instituto de Ecología Política, 1996).

11. Luis Otero Durán, *El problema social detrás de los bosques* (Concepción: Vicaría de la Pastoral Obrera, 1984); Grupo de Investigaciones Agrarias (GIA), "Región forestal: Empresas y trabajadores" (Santiago: Grupo de Investigaciones Agrarias, Academia de Humanismo Cristiano, n.d.); Rigoberta Rivera A. and María Elena Cruz, *Pobladores rurales* (Santiago: Grupo de Investigaciones Agrarias and Academia de Humanismo Cristiano, 1984); Pedro García Elizalde and José Ignacio Leyton, *El desarrollo frutícola y forestal en Chile y sus derivaciones sociales* (Santiago: Comisión Económica para América Latina y el Caribe, 1986).

12. See, e.g., E. P. Thompson, *Whigs and Hunters: The Origins of the Black Act* (New York: Pantheon Books, 1975); Peter Sahlins, *Forest Rites: The War of the Demoiselles in Nineteenth-Century France* (Cambridge, MA: Harvard University Press, 1994); Peter Linebaugh, "Karl Marx, The Theft of Wood, and Working Class Composition: A Contribution to the Current Debate," *Crime and Social Justice* 6 (Fall-Winter 1976): 5–16.

13. Human Rights Watch and Observatorio de Derechos de los Pueblos Indígenas, "Undue Process: Terrorism Trials, Military Courts and the Mapuche in Southern Chile," *Human Rights Watch* 16, no. 5 (October 2004): 1–63.

14. Thompson, *Whigs and Hunters*; John M. Merrriman, "The Demoiselles of the Ariège, 1829–1831," in *1830 in France*, ed. John M. Merriman (New York: New Viewpoints, 1975), 87–118; Sahlins, *Forest Rites*.

15. Ramachandra Guha, *The Unquiet Woods: Ecological Change and Peasant Resistance in the Himalaya* (Berkeley: University of California Press, 2000).

16. Thompson, *Whigs and Hunters*.

17. For a vivid description of the police presence in the southern forestry zones, see José A. Mariman, "Lumaco y el movimiento mapuche," in McFall, *Territorio mapuche y expansión forestal*, 99–113.

18. Among the many important works on the frontier, see Sergio Villalobos, Carlos Aldunate, Horacio Zapater, Luz María Méndez, and Carlos Bascuñan, *Relaciones fronterizas en la Araucanía* (Santiago: Ediciones Universidad Católica de Chile, 1982); Leonardo León, Patricio Herrera, Luis Carlos Parentini, Sergio Villalobos, *Araucanía: La frontera mestiza, siglo XIX* (Santiago: LOM Ediciones, 2003); Jorge Pinto Rodríguez, *La formación del estado y la nación, y el pueblo mapuche: De la inclusión a la exclusión* (Santiago: Dirección de Bibliotecas, Archivos y Museos and Centro de Investigaciones Diego Barros Arana, 2003).

19. Pinto Rodríguez, *La formación del estado y la nación*.

20. *Obras completas de Vicuña Mackenna, Volumen 12: Discursos parlamentarios*, Sesión 44 Ordinaria, 9 August 1868 (Santiago: Universidad de Chile, 1936), 409–10.

21. Juan J. Armesto, Pedro León Lobos, and Mary Kalin Arroyo, "Los bosques templados del sur de Chile y Argentina: Una isla biogeográfica," in Juan J. Armesti, Carolina Villagrán M., Mary Kali Arroyo, *Ecología de los bosques nativos de Chile* (Santiago: Editorial Universitaria, 1995), 25.

22. Rodrigo Catalán Labarías and Ruperto Ramos Antiqueo, *Pueblo mapuche, bosque nativo y plantaciones forestales: Las causas de la deforestatión en el sur de Chile* (Temuco: Universidad Católica de Temuco, 1999), 4, 20. See also Claudio Donoso Zegers, "Los bosques nativos de Chile: Patrimonio de la tierra," in Defensores del Bosque Chileno, *La tragedia del bosque chileno*, 83–87.

23. Ignacio Domeyko, *Araucanía y sus habitantes* (Santiago: Imprenta Chilena, 1846), 25–26.

24. Rudolph Philippi, "Valdivia en 1852," *La Revista de Chile*, no. 74 (May 1901), 330.

25. See Otto Berninger, "Bosque y tierra despejada en el sur de Chile desde la conquista española" (Thesis, Universidad de Chile, Santiago, 1966), 126–27.

26. Catalán Labarías and Ramos Antiqueo, *Pueblo mapuche, bosque nativo y plantaciones forestales*, 4, 20.

27. Luis Otero Durán, *La huella del fuego: Historia de los bosques nativos, poblamiento, y cambios en el paisaje del sur de Chile* (Santiago: Corporación Nacional Forestal and Pehuén, 2006), 24, 42–48.

28. Durán, *La huella del fuego*, 32–33.

29. Sergio Villalobos, *Vida fronteriza en la Araucanía: El mito de la Guerra de Arauco* (Santiago: Editorial Andrés Bello, 1995); Villalobos et al., *Relaciones fronterizas en la Araucanía*; León et al., *Araucanía*; Andrea Ruiz-Esquide, *Los indios amigos en la frontera araucana* (Santiago: Dirección de Bibliotecas, Archivos y Museos and Centro de Investigaciones Diego Barrios Arana, 1993). Jorge Pinto Rodríguez, "Redes indígenas y redes capitalistas: La Araucanía y las pampas en

el siglo XIX" in *Los pueblos campesinos de las Américas. Etnicidad, cultura e historia en el siglo XIX*, Heraclio Bonilla and Amado A. Guerrero, eds. (Bucaramanga, Colombia: Universidad Industrial de Santander, Escuela de Historia, 1996): 137–144; Leonardo León, *Maloqueros y conchavadores en la Araucanía y las pampas, 1700–1800* (Temuco: Ediciones Universidad de la Frontera, 1991).

30. Alfred W. Crosby, *Ecological Imperialism: The Biological Expansion of Europe, 900–1900* (Cambridge: Cambridge University Press, 2004).

31. José Bengoa, *Historia del pueblo mapuche, siglo XIX y XX* (Santiago: LOM Ediciones, 2000), 48, 54–56.

32. "Informe sobre el territorio de Arauco y la población indígena, 1868–1869," in Iván Inostroza Córdova, ed., *Etnografía mapuche del siglo XIX* (Santiago: Dirección de Bibliotecas, Archivos y Museos, 1998), 122–23; Ernesto Wilhelm de Mosbach, *Botánica indígena de Chile* (Santiago: Editorial Andrés Bello, 1992). See especially the prologue by Carlos Aldunate and Carolina Villagran.

33. Bengoa, *Historia del pueblo mapuche*, 62.

34. "Informe sobre el territorio de Arauco y la población indígena," 122.

35. See William Cronon's argument for New England in *Changes in the Land: Indians, Colonists, and the Ecology of New England* (New York: Hill and Wang, 1983), 53, and the analogous argument for agro-pastoralism in precolonial South Africa in Nancy J. Jacobs, *Environment, Power, and Injustice: A South African History* (Cambridge: Cambridge University Press, 2003), 49.

36. "Informe sobre el territorio de Arauco y la población indígena," 123. Edmond Reuel Smith, *The Araucanians; or Notes of a Tour among the Indian Tribes of Southern Chili* (New York: Harper and Brothers, 1855), 179. Domeyko, *Araucanía y sus habitantes*, 26.

37. David L. Aagesen, "Burning Monkey-Puzzle: Native Fire Ecology and Forest Management in Northern Patagonia," *Agriculture and Human Values* 21 (2004): 232–244; David L. Aagesen, "Indigenous Resource Rights and Conservation of the Monkey-Puzzle Tree (*Araucaria araucana, Araucariaceae*): A Case Study from Southern Chile," *Economic Botany* 52, no. 2 (1998): 146–160; David L. Aagesen, "On the Northern Fringe of the South American Temperate Forest: The History and Conservation of the Monkey-Puzzle Tree," *Environmental History* 3, no. 1 (January 1998): 64–85; Mauro E. González, Thomas T. Veblen, and Jason S. Sibold, "Fire History of *Araucaria-Nothofagus* Forests in Villarrica National Park, Chile," *Journal of Biogeography* 32 (2005): 1187–202.

38. Patricia Nelson Limerick, "Turnerians All: The Dreams of a Helpful History in an Unintelligible World," *American Historical Review* 100 (1995): 697–716.

39. Bengoa, *Historia del pueblo mapuche*, 51–52.

40. Eduard Poeppig, *Un testigo en la alborada de Chile (1826–1829)* (Santiago: Zig-Zag, 1960), 413.

41. See, e.g., Domeyko, *Araucanía y sus habitantes*; Smith, *The Araucanians*; Poeppig, *Un testigo en la alborada de Chile*.

42. Consuelo Figueroa Garavagno, "Geografías en disputa: La construcción del Chile territorial," *Revista 180* (Universidad Diego Portales), no. 27 (2011): 10–13; Consuelo Figueroa Garavagno, "De rastros y extravíos: Guerras en exhibición,

Chile, 1880s–1930s," in *Entre el humo y la niebla: Guerra y cultura en América Latina,* ed. Felipe Martínez P. and Javier Uriarte (Pittsburgh, PA: University of Pittsburgh Press, forthcoming).

43. Pinto Rodríguez, *La formación del estado y la nación.* Bengoa, *Historia del pueblo mapuche;* Bengoa, *Historia de un conflicto: El estado y los mapuches en el siglo* XX (Santiago: Planeta, 1999). Florencia E. Mallon, *Courage Tastes of Blood: The Mapuche Community of Nicolás Ailío and the Chilean State, 1906–2001* (Durham, NC: Duke University Press, 2005). For an alternative approach to the frontier's history that emphasizes trade and cultural interaction, see León et al., *Araucanía;* Ruiz-Esquide, *Los indios amigos en la frontera araucana;* Villalobos, *Vida fronteriza en la Araucanía;* Villalobos et al., *Relaciones fronterizas en la Araucanía.*

44. For an important account of a similar process of primitive accumulation in the Patagonia region of Chile and Argentina, see Albert Harambour Ross, "Borderland Sovereignties: Postcolonial Colonialism and State Making in Patagonia, Argentina and Chile, 1840s–1992" (PhD diss., Stony Brook University, Stony Brook, NY, 2012).

45. William Cronon, *Nature's Metropolis: Chicago and the Great West* (New York: W. W. Norton, 1992). See also the superb discussion of the role of nature in the creation of value in Marxist theory in Fernando Coroníl, *The Magical State: Nature, Money, and Modernity in Venezuela* (Chicago: University of Chicago Press, 1997), 45–48.

46. David Harvey, *The New Imperialism* (Oxford: Oxford University Press, 2003).

47. Brian Balogh, "Scientific Forestry and the Roots of the Modern American State: Gifford Pinchot's Path to Progressive Reform," *Environmental History* 7, no. 2 (April 2002): 198–225; Char Miller, *Gifford Pinchot and the Making of Modern Environmentalism* (Washington, DC: Island Press, 2001); Patricia Nelson Limerick, *The Legacy of Conquest: The Unbroken Past of the American West* (New York: Norton, 1987), 293–303.

48. See the discussion of forestry in James C. Scott, *Seeing like a State: How Certain Schemes to Improve the Human Condition Have Failed* (New Haven, CT: Yale University Press, 1998).

49. Alberto Peña, "Reflexiones en torno a la extensión forestal en Chile," in *Bosques y comunidades del sur de Chile,* ed. Rodrigo Catalán, Petra Wilken, Angelika Kandzior, David Tecklin, and Heinrich Burschel (Santiago: Editorial Universitaria, 2005), 67–72.

50. David Tecklin and Rodrigo Catalán, "La gestión comunitaraia de los bosques nativos en el sur de Chile: Situación actual y temas en discusión" in Catalán et al., *Bosques y comunidades del sur de Chile,* 19.

51. Ranajit Guha, "The Prose of Counter-Insurgency," in *Selected Subaltern Studies,* ed. Ranajit Guha and Gayatri Chakravorty Spivak (Oxford: Oxford University Press, 1988), 45–88.

52. For an important ethnographically informed case study of the history of one Mapuche community, see Mallon, *Courage Tastes of Blood.* For a larger discussion of indigenous history and ethnography that emphasizes the important role

of dialogue and collaboration between "native" intellectuals and non-indigenous historians and social scientists, see Florencia E. Mallon, ed., *Decolonizing Native Histories: Collaboration, Knowledge, and Language in the Americas* (Durham, NC: Duke University Press, 2011).

1 State Sovereignty on the Frontier

1. Congreso Nacional, *Comisión Parlamentaria de Colonización: Informe, proyectos de ley, actas de las sesiones, y otros antecedentes* (Santiago: Sociedad "Imprenta y Litografía Universo," 1912), lvii.

2. Congreso Nacional, *Comisión Parlamentaria de Colonización*, lviii.

3. Carl E. Solberg, "A Discriminatory Land Policy: Chile, 1870–1914," *The Americas* 26, no. 2 (October 1969), 131–32.

4. William Cronon makes a similar argument for colonial New England in *Changes in the Land: Indians, Colonists, and the Ecology of New England* (New York: Hill and Wang, 1983).

5. For useful reviews of colonization laws, see Solberg, "A Discriminatory Land Policy"; Julio Zenteno Barros, *Recopilación de leyes y decretos supremos sobre colonización* (Santiago: Imprenta Nacional, 1892). For histories of the colonization of the frontier, see also José Bengoa, *Haciendas y campesinos: Historia social de la agricultura chilena*, vol. 2 (Santiago: Ediciones Sur, 1990); José Bengoa, *Historia del pueblo mapuche, siglo XIX y XX* (Santiago: LOM Ediciones, 2000).

6. *Sesta Memoria del Director de la Oficina de la Mensura de Tierras* (Santiago, 1913), BCN, 178–83. For a discussion of the 1866 law see Darío Ulloa H., *La cuestión indígena ante la legislación nacional* (Santiago: Imprenta i Encuadernación Chile, 1917), 31–34. For Mapuche loss of land during colonization, see José Bengoa, *Historia de un conflicto: El estado y los mapuches en el Siglo XX* (Santiago: Planeta, 1999), 50–55.

7. *Sesta Memoria del Director de la Oficina de la Mensura de Tierras* (Santiago, 1913), 180. All translations in this chapter and throughout the volume are mine.

8. For a history of this process, see Bengoa, *Historia de un conflicto*, 354–60; Bengoa, *Historia del pueblo mapuche*, 60–61.

9. *Memoria de la Inspección Jeneral de Tierras i Colonización* (Santiago, 1906), BN, 174.

10. *Memoria de la Inspección Jeneral de Tierras i Colonización* (Santiago, 1903), BN, 30.

11. "Informe anual del presidente de la Comisión de Títulos de Merced," Temuco, 1 May 1905, in *Memoria de la Inspección Jeneral de Tierras i Colonización* (Santiago, 1905), BN, 163.

12. Solicitante Caciques i Reducciones de Cautín, Materia Piden Mejora Situación Indígenas, 14 December 1906, Terminada 22 May 1907, MRREE Sección Colonización, Solicitudes Particulares Mandadas Archivar en 1907, Letras A–LL, ARNAD.

13. "Informe del Protector de Indígenas de Malleco," in Congreso Nacional, *Comisión Parlamentaria de Colonización*, 113, 115.

14. *Memoria de la Inspección Jeneral de Colonización e Inmigración* (Santiago, 1917–18), BN, 247.

15. "Memoria de los trabajos llevados a cabo por el injeniero de la Comisión Radicadora de Indígenas que suscribe, durante el año 1901," *Memoria de la Inspección Jeneral de Tierras i Colonización* (Santiago, 1905), 229–30.

16. "Memoria del Protector de Indígenas de Valdivia i Llanquihue," *Memoria de la Inspección Jeneral de Colonización e Inmigración* (Santiago, 1908), BN, 211–12.

17. "Memoria del Protector de Indígenas de Valdivia i Llanquihue," *Memoria de la Inspección Jeneral de Colonización e Inmigración* (Santiago, 1908), BN, 212; see also 235–41, 250–51.

18. *Memoria de la Inspección Jeneral de Colonización e Inmigración* (Santiago, 1906), 36.

19. "Memoria del Protector de Indíjenas de Valdivia," *Memoria de la Inspección Jeneral de Colonización e Inmigración* (Santiago, 1912), BN, 652.

20. Oficina del Protectorado de Indíjenas to Inspector Jeneral de Tierras i Colonización, *Memoria de la Inspección Jeneral de Tierras i Colonización* (Santiago, 1903), 168.

21. Oficina del Protectorado de Indíjenas to Inspector Jeneral de Tierras i Colonización, 165–66.

22. Oficina del Protectorado de Indíjenas to Inspector Jeneral de Tierras i Colonización, 164.

23. *Memoria de la Inspección Jeneral de Tierras i Colonización* (Santiago, 1903), 33.

24. *Memoria de la Inspección Jeneral de Tierras i Colonización* (Santiago, 1903), 42–43, 173–74.

25. Oficina del Protectorado de Indíjenas to Inspector Jeneral de Tierras i Colonización, 181–82.

26. "Memoria del Protector de Indíjenas de Valdivia i Llanquihue," 227.

27. Oficina del Protectorado de Indíjenas to Inspector Jeneral de Tierras i Colonización, 167.

28. "Relación de los trabajos efectuados por el injeniero de la Comisión Radicadora de Indígenas," 31 December 1904, in *Memoria de la Inspección Jeneral de Tierras i Colonización* (Santiago, 1905), 181. See also "Informe del Protector de Indígenas de Malleco," 111; Oficina del Protectorado de Indíjenas to Inspector Jeneral de Tierras i Colonización, 166.

29. Solicitante, Domingo Caniu i otros indígenas. Materia: Piden lanzamiento de Otto Reusch i otros, i que se le niegue lugar a una solicitud, MRREE Sección Colonización, Solicitudes Particulares Mandadas Archivar en 1907, Letras A–Ll, vol. 1334, ARNAD.

30. For a discussion of the organization of Mapuche society around individual households and the imposition of new community legal identities and land rights during reducción, see Bengoa, *Historia del pueblo mapuche*, 360–61.

31. Oficina del Protectorado de Indíjenas to Inspector Jeneral de Tierras i Colonización, 167.

32. Oficina del Protectorado de Indíjenas to Inspector Jeneral de Tierras i Colonización, 163.

33. Oficina del Protectorado de Indíjenas to Inspector Jeneral de Tierras i Colonización, 151.

34. Congreso Nacional, *Comisión Parlamentaria de Colonización*, xii.

35. *Memoria de la Inspección Jeneral de Tierras i Colonización* (Santiago, 1903), 29.

36. *Memoria de la Inspección Jeneral de Tierras i Colonización* (Santiago, 1903), 33.

37. See Ministerio de Fomento, *Memoria de Fomento* (Santiago, 1927), BCN, 124–27.

38. *Memoria del Ministro de Colonización i Culto* (Santiago, 1899), BCN, 9–10.

39. *Memoria de la Inspección Jeneral de Tierras i Colonización* (Santiago, 1901), BN, 11.

40. *Memoria de la Inspección Jeneral de Tierras i Colonización* (Santiago, 1905), 8.

41. *Memoria del Ministerio de Colonización i Culto* (Santiago, 1899), 228.

42. Cornelio Saavedra, "Resumén i apreciación jeneral," in Cornelio Saavedra, *Documentos relacionados a la ocupación de Arauco* (Santiago: Imprenta de la Libertad, 1870), 256.

43. Cornelio Saavedra, "Comunicaciónes a favor del avance de nuestras fronteras en el territorio indíjena, i del establecimiento de una nueva linea sobre el Bío Malleco," 11 October 1861, in *Documentos relacionados a la ocupación de Arauco*, 10.

44. Saavedra, "Comunicaciónes a favor del avance de nuestras fronteras en el territorio indígena," 21.

45. Saavedra, "Resumén i apreciación jeneral," 252.

46. Bengoa, *Historia del pueblo mapuche*, 158–159; Bengoa, *Haciendas y campesinos*, 151–52. See also Andrea Ruiz-Esquide, "Migration, Colonization, and Land Policy in the Former Mapuche Frontier, Malleco, 1850–1900" (PhD thesis, Columbia University, New York, 2000), 185–91.

47. Quoted in Bengoa, *Historia del pueblo mapuche*, 159.

48. Claudio Gay, "Viaje a la Araucanía en 1863," in Iván Inostroza Córdova, *Etnografía mapuche del siglo xx* (Santiago: Dirección de Bibliotecas, Archivos y Museos, 1998), 94.

49. Bengoa, *Haciendas y campesinos*, 154. See also Ricardo Ferrando Keun, *Y así nació la frontera* (Santiago: Editorial Antartica, 1986), 504–509.

50. *Sesta Memoria del Director de la Oficina de la Mensura de Tierras*, 142; Nicolás Palacios, *La raza chilena: Libro escrito por un chileno y para los chilenos* (Valparaíso: Imprenta y Litografia Alemana, 1904), 594–95; *Memoria de la Inspección Jeneral de Tierras y Colonización* (Santiago, 1902), 41. For an excellent discussion of land auctions, see Ruiz-Esquide, "Migration, Colonization, and Land Policy in the Former Mapuche Frontier," 237–85.

51. For a list of thousands of delinquent payments of land purchased at auction, see Inspección Jeneral de Tierras y Colonización, *Nómina de deudores morosos*, (Santiago, 1903), BCN.

52. Nicolás Palacios, "Algunos efectos de la colonización extranjera," in Congreso Nacional, *Comisión Parlamentaria de Colonización*, 384.

53. Petition to the Ministro de Colonización, 15 February 1895, in Congreso Nacional, Cámara de Diputados, Diario de Sesiones, Sesión Ordinaria, 13 July 1911, BCN.

54. Petition to the Ministro de Colonización, 15 February 1895. See various documents relating to this case in Zenteno Barros, *Recopilación de leyes i decretos supremos sobre colonización,* 1290–96.

55. *Sesta Memoria del Director de la Oficina de la Mensura de Tierras,* 148–49.

56. Argentina, Ministerio de Relaciones Exteriores, Informes Consulares, 1905, *Boletín,* vols.7–8, 11.

57. Congreso Nacional, Cámara de Diputados, Diario de Sesiones, Sesión Extraordinaria, 16 January 1914, BCN.

58. Sindicato Pellahuén, "Demanda sobre posesión de este fundo," Imprenta El Colono, 1906, BN.

59. Congreso Nacional, Cámara de Diputados, Diario de Sesiones, Sesión Extraordinaria, 16 January 1914, BCN; *El Sur,* 15–27 October 1913.

60. Congreso Nacional, *Comisión Parlamentaria de Colonización,* xvi.

61. *Memoria del Ministro de Colonización i Culto* (Santiago, 1899), 19.

62. *Sesta memoria del Director de la Oficina de la Mensura de Tierras,* 157.

63. "Memoria del Interventor de Colonias Correspondiente al año 1911," in *Memoria de la Inspección Jeneral de Colonización e Inmigración* (Santiago, 1912), 59.

64. *Sesta memoria del Director de la Oficina de la Mensura de Tierras,* 158.

65. *Memoria de la Inspección Jeneral de Colonización e Inmigración* (Santiago, 1910), BN, 6–7.

66. See *Sesta memoria del Director de la Oficina de la Mensura de Tierras,* 167.

67. See *Setima memoria del Director de la Oficina de Mensura de Tierras* (Santiago, 1914), 63.

68. Jaime Flores Chávez, "Europeos en la Araucanía: Los colonos del Budi a principios del siglo XX," in *Emigración centroeuropea a América Latina,* ed. Josef Optrny (Prague: Editorial Karolinum, 2000), 319.

69. *Memoria de la Inspección Jeneral de Colonización e Inmigración* (1917–18), 227–28.

70. For a discussion of colonization concessions, see also Bengoa, *Haciendas y campesinos,* 166–74.

71. *Memoria de la Inspección Jeneral de Tierras i Colonización* (Santiago, 1908), 6–9.

72. *Memoria de la Inspección Jeneral de Tierras i Colonización* (Santiago, 1905), 32.

73. "Memoria de la Colonia 'Nueva Italia' año 1906," in *Memoria de la Inspección Jeneral de Tierras i Colonización* (Santiago, 1906), 200–201.

74. *Memoria de la Inspección Jeneral de Tierras i Colonización* (Santiago, 1903), 13–14.

75. *Memoria de la Inspección Jeneral de Tierras i Colonización* (1902), BN, 16–17.

76. Administrador de las Colonias de la Frontera, 1 May 1908, *Memoria de la Inspección Jeneral de Colonización e Inmigración* (Santiago, 1908), 110.

77. *Memoria de la Inspección Jeneral de Colonización e Inmigración* (Santiago, 1910), BN, 58.

78. *Memoria de la Inspección Jeneral de Tierras i Colonización* (Santiago, 1906), 25–26.

79. I. Calderon Ruiz to Inspector Jeneral de Tierras, 1 May 1905, *Memoria de la Inspección Jeneral de Tierras i Colonización,* (Santiago, 1905), 80.

80. *Primera Memoria del Director de la Oficina de la Mensura de Tierras* (Santiago, 1908), BCN, 94–99. For a detailed history of the Sociedad Agrícola Budi, see Florencia E. Mallon, *Courage Tastes of Blood: The Mapuche Community of Nicolás Ailío and the Chilean State, 1906–2001* (Durham, NC: Duke University Press, 2005), 52–54.

81. Solicitante: Comerciantes del Departamento de Nueva Imperial, Materia: Reclamo contra empresa colonizadora del Lago Budi por malos tratos a colonos nacionales, MRREE, Sección Colonización, Solicitudes Particulares, 1907–10, vol. 1336, ARNAD; Bengoa, *Haciendas y campesinos*, 169–71.

82. Solicitante, Manuel Calfuala por si i por indígenas Valdivia, Materia; Contra "Sociedad Quele," 1906, MRREE, Sección Colonización, Solicitudes Particulares, A–V, vol. 1252, ARNAD. See another description of this case in Aureliao Díaz Meza, *En la Araucanía: Breve relación del último parlamento araucano de Coz-Coz en 18 de enero de 1907* (Santiago: Imprenta "El Diario Ilustrado," 1907), 91–92.

83. *Segunda Memoria del Director de la Oficina de Mensura de Tierras* (Santiago, 1909), BCN, 67–68.

84. *Tercera Memoria del Director de la Oficina de Mensura de Tierras* (Santiago, 1910), BCN, 177–78.

85. *Sesta Memoria del Director de la Oficina de la Mensura de Tierras*, 67; *Memoria de la Inspección Jeneral de Colonización e Inmigración* (1917–18), 235.

86. "Memoria de las Colonias del Departamento de Temuco," in *Memoria de la Inspección Jeneral de Colonización e Inmigración* (Santiago, 1910), 62.

87. "Memoria del Interventor Fiscal de Colonias," in *Memoria de la Inspección Jeneral de Colonización e Inmigración* (Santiago, 1908), 41–42.

88. *Memoria de la Inspección Jeneral de Colonización e Inmigración* (Santiago, 1908), 11.

89. *Memoria de la Inspección Jeneral de Colonización e Inmigración* (Santiago, 1908), 10.

90. *Memoria de la Inspección Jeneral de Colonización e Inmigración* (Santiago, 1908), 12.

91. See, e.g., the speeches by the Democratic Party Deputy Malaquías Concha in Congreso Nacional, *Comisión Parlamentaria de Colonización,* and by the Conservative Party Deputy José Ramón Gutiérrez in Congreso Nacional, Cámara de Diputados, Sesión Ordinaria, 7 January 1911.

92. *Memoria del Ministro de Colonización i Culto* (Santiago, 1899), 12.

93. See the report from Víctor Aquiles Bianchi to MTC, 10 January 1907, in Congreso Nacional, Cámara de Diputados, Diario de Sessiones Sesión Ordinaria, 7 January 1911, BCN. Bengoa also discusses this report in *Haciendas y campesinos,* 177–78.

94. Congreso Nacional, *Comisión Parlamentaria de Colonización,* xix.

95. On the failures of national colonization, see Bengoa, *Haciendas y campesinos,* 176–78.

96. Letter from Agustín Edwards, 10 October 1903, in Congreso Nacional, Cámara de Diputados, Diario de Sesiones, Sesión Ordinaria, 8 July 1911, BCN.

97. *Sesta Memoria del Director de la Oficina de la Mensura de Tierras,* 171.

98. *Memoria de la Inspección Jeneral de Colonización e Inmigración* (Santiago, 1903), 12.

99. Luis Urrutia Ibáñez, "Constitución de la propiedad raiz en la zona austral," *Boletín de Bosques, Pesca i Caza*, vol. 1, no. 8, February 1913, 557–58.

100. Congreso Nacional, *Comisión Parlamentaria de Colonización*, ix.

101. Congreso Nacional, *Comisión Parlamentaria de Colonización*, xiii.

102. "Memoria de Intendencia de Llanquihue," 8 April 1927, Ministerio del Interior, Providencias, 1928, vol. 7062, ARNAD.

103. "Memoria de Intendencia de Arauco," 5 August 1927, Ministerio del Interior, Providencias, 1928, vol. 7062, ARNAD.

104. "Memoria de Gobernación de Cañete," 1 June 1927, Ministerio del Interior, Providencias, vol. 7061, 1928, ARNAD.

105. Congreso Nacional, *Comisión Parlamentaria de Colonización*, xxxiv.

106. *Sesta Memoria del Director de la Oficina de la Mensura de Tierras*, 185.

107. *Memoria de la Inspección Jeneral de Colonización e Inmigración* (1917–18), 249.

108. Ministerio de Fomento, *Memoria de Fomento*, 112–23.

2 Natural Disorder

1. Letter from Lorenzo Anadón to José Ramón Gutiérrez, in Congreso Nacional, *Comisión Parlamentaria de Colonización: Informe, proyectos de ley, actas de las sesiones, y otros antecedentes* (Santiago: Sociedad "Imprenta y Litografía Universo," 1912), 402.

2. Letter from Lorenzo Anadón to José Ramón Guitiérrez, 402.

3. Federico Albert, "Informe del Jefe de la Sección de Aguas y Bosques" in Congreso Nacional, *Comisión Parlamentaria de Colonización*, 397–400.

4. Vicente Pérez Rosales, *Memoria sobre colonización de la provincia de Valdivia* (Valparaíso: Imprenta del Diario, 1852), 11.

5. Vicente Pérez Rosales, *Times Gone By: Memoirs of a Man of Action*, trans. John H. R. Polt (Oxford: Oxford University Press, 2003), 323–24.

6. Rudolph Philippi, "Valdivia en 1852," *La Revista de Chile*, no. 74 (May 1901): 332.

7. Rafael Elizalde Mac-Clure, *La sobrevivencia de Chile: La conservación de sus recursos naturales renovables* (Santiago: Ministerio de Agricultura, 1970), 24.

8. Lorenzo Anadón to José Ramón Gutiérrez, 403.

9. Quoted in Otto Berniger, "Bosque y tierra despejada en el sur de Chile desde la conquista española" (Thesis, Universidad de Chile, Santiago, 1966), 71.

10. Teodoro Schneider, *La agricultura en Chile en los últimos cincuenta años* (Santiago: Imprenta Barcelona, 1904), 50.

11. Quoted in Victorino Rojas Magallanes, *Informe sobre bosques presentado al Ministerio de Industria i Obras Públicas* (Santiago: Imprenta Litografía y Encuadernación Barcelona, 1904), 15.

12. *Corte de bosques: Informe de la comisión nombrada para dictaminar esta materia i reglamento dictado por el Presidente de la República* (Santiago: Imprenta Nacional, 1873), 7–8, 43.

13. *Corte de bosques,* 45–46. See also Pablo Camus Gayán, *Ambiente, bosques y gestión forestal en Chile, 1541–2005* (Santiago: LOM Ediciones and Dirección de Bibliotecas, Archivos y Museos, 2006), 118–19.

14. Roberto Opazo, *Desarrollo agrícola de los territorios situados al sur del río Bio-Bio* (Temuco: Sociedad Cooperativa y de Fomento Agrícola, 1920), 25.

15. George McCutchen McBride, *Chile: Land and Society* (New York: American Geographical Society, 1936), 22.

16. *El Agricultor,* no. 16–17, January–February 1940, 17.

17. Juan Jirkal H., "Generalidades sobre el problema de la erosion del suelo en Chile" in *Primera Asamblea Forestal Nacional,* ed. Sociedad Amigos del Árbol (Santiago: n.p., 1943), 292.

18. Eduardo Baezas, "Monografía del fundo Poco a Poco de Temuco" (Thesis, Instituto Agronómico, Universidad de Chile, Santiago, 1910), 6, 14, 23. José Bengoa also discusses this thesis in a very useful analysis of southern labor relations in *Haciendas y campesinos: Historia social de la agricultura chilena,* vol. 2 (Santiago: Ediciones Sur, 1990), 155–58.

19. Baezas, "Monografía del fundo Poco a Poco de Temuco," 29.

20. Baezas, "Monografía del fundo Poco a Poco de Temuco," 23.

21. Prudencio Tardío, "Protección a los bosques," *Pacífico Magazine,* no. 2 (February 1913): 181. One cuadra is equal to 1.572 hectares.

22. Baezas, "Monografía del fundo Poco a Poco de Temuco," 6, 14, 23.

23. "Informe anual del Presidente de la Comisión de Títulos de Merced," 1 May 1905, in *Memoria de la Inspección Jeneral de Tierras i Colonización* (Santiago, 1905), BN, 190–92.

24. *El Diario Austral,* 22 March 1916.

25. See the discussion of the same process in colonial New England in William Cronon, *Changes in the Land: Indians, Colonists, and the Ecology of New England* (New York: Hill and Wang, 1983), 143–47. For a discussion of Mapuches' transition from pastoralists to campesino wheat producers, see José Bengoa, *Historia del pueblo mapuche, siglo XIX y XX* (Santiago: LOM Ediciones, 2000), 364–65; José Bengoa, *Historia de un conflicto: El estado y los mapuches en el siglo XX* (Santiago: Planeta, 1999), 82–94.

26. *Memoria de la Inspección Jeneral de Colonización e Inmigración* (Santiago, 1912), 155; Carta de la Sociedad Caupolicán, Temuco, 19 October 1914, Ministerio de Relaciones Exteriores, Providencias, 1915, Sección Colonización, vol. 1928, ARNAD.

27. Opazo, *Desarrollo agrícola de los territorios situados al sur del río Bio-Bio,* 32.

28. Tardío, "Protección a los bosques," 182.

29. Congreso Nacional, Cámara de Diputados, Diario de Sesiones, Sesión Extraordinaria, 27 November 1913, BCN.

30. Elizalde Mac-Clure, *La sobrevivencia de Chile,* 22.

31. Federico Albert, "El problema forestal en Chile," *Boletín de Bosques, Pesca i Caza,* vol. 1, no. 10, April 1913, 681, 687.

32. Arnold Bauer, *Chilean Rural Society from the Spanish Conquest to 1930* (Cambridge: Cambridge University Press, 1975), 76.

33. Opazo, *Desarrollo agrícola de los territorios situados al sur del río Bio-Bio,* 20–25.

34. Opazo, *Desarrollo agrícola*, 22; Albert, "El problema forestal en Chile," 681, 687.

35. "Relación de los trabajos efectuados por el injeniero de la Comisión Radicadora de Indígenas," 31 December 1904, in *Memoria de la Inspección Jeneral de Tierras i Colonización* (Santiago, 1905), 192.

36. *Memoria de la Inspección Jeneral de Tierras i Colonización* (Santiago, 1903), BN, 40–41.

37. *Memoria de la Inspección Jeneral de Tierras i Colonización* (Santiago, 1906), BN, 174.

38. Raphael Zon and William N. Sparhawk, *Forest Resources of the World* (New York: McGraw-Hill, 1923), 739.

39. Zon and Sparhawk, *Forest Resources*, 747.

40. Opazo, *Desarrollo agrícola de los territorios situados al sur del río Bio-Bio*, 5, 15–16.

41. Intendencia de Malleco, "Informa sobre visita efectuada al Dpto. de Collipulli," 10 January 1928, Ministerio del Interior, Providencias, vol. 7061, 1928, ARNAD.

42. Opazo, *Desarrollo agrícola de los territorios situados al sur del río Bio-Bio*, 5.

43. "Memoria de Intendencia de Cautín, Temuco," 1 August 1927, Ministerio del Interior, Providencias, vol. 7062, 1928, ARNAD.

44. Rojas Magallanes, *Informe sobre bosques presentado al Ministerio de Industria i Obras Públicas*, 22–23.

45. Gustave Verniory, *Diez años en Araucanía, 1889–1899* (Santiago: Pehuén, 2001), 246.

46. Paul Treutler, *Andanzas de un alemán en Chile, 1851–1863* (Santiago: Editorial del Pacífico, 1958), 290–92.

47. Federico Albert, "El agotamiento de los recursos naturales de bosques, pesca i caza," *Boletín de Bosques, Pesca i Caza*, vol. 1, no. 4 (1912), 217–88.

48. Verniory, *Diez años en Araucanía*, 234. For railroad consumption and shipments of wood, see Camus Gayán, *Ambiente, bosques y gestión forestal en Chile*, 132.

49. Albert, "El problema forestal en Chile."

50. Tardío, "Protección a los bosques," 177.

51. Oficina Mensura de Tierras, 18 January 1911, MRREE, Sección Colonización, Oficios de la Oficina Mensura de Tierras, 1911, ARNAD.

52. Inspección General de Tierras y Colonización, 8 January 1907, MRREE, Sección Colonización, Oficios Dirigidos por la Inspección General de Tierras y Colonización, 1907, ARNAD.

53. "Irregularities in the Reserva Forestal Villarrica," 30 April 1917, MRREE, Sección Colonización, Oficios de la Inspección General de Colonización, vol. 2058, 1917, ARNAD.

54. See A. Ramírez to Sub-Inspector de Tierras y Colonización de Cautín i Valdivia, Colonia Transvaal, 25 April 1905, in *Memoria de la Inspección Jeneral de Tierras i Colonización* (Santiago, 1905), 39.

55. *Memoria de la Inspección Jeneral de Tierras i Colonización* (Santiago, 1905), 13.

56. *Memoria de la Inspección Jeneral de Tierras i Colonización* (Santiago, 1906), 219–21.

57. *Memoria de la Inspección Jeneral de Colonización e Inmigración* (Santiago, 1910), BN, 59–60.

58. Renato Bussi Soto, "Monografía económico-cultural del fundo 'Puello,' Provincia de Cautín, Departamento de Temuco" (Thesis, Universidad de Chile, Santiago, 1941). Bengoa also discusses the Puello estate using this thesis in *Haciendas y campesinos*, 154–55.

59. *Memoria de la Inspección Jeneral de Tierras i Colonización* (Santiago, 1905), 100.

60. Inspección General de Tierras y Colonización, 30 October 1912, MRREE, Sección Colonización, Inspección de Colonización, vol. 1769, 1913, ARNAD.

61. Consejo de Defensa Fiscal to Ministerio de Tierras y Colonización, Informe, 10 December 1910, MRREE, Sección Colonización, Providencias, 1910–11, ARNAD.

62. *Memoria de la Inspección Jeneral de Tierras i Colonización* (Santiago, 1906), 213.

63. *Memoria de la Inspección Jeneral de Tierras i Colonización* (Santiago, 1906), 217.

64. *Memoria de la Inspección Jeneral de Tierras i Colonización* (Santiago, 1906), 203.

65. *Memoria de la Inspección Jeneral de Tierras i Colonización* (Santiago, 1906), 205.

66. Solicitante, Eleuterio Domínguez, Materia, se Entrega a Soc. Budi Tres Aserraderos, 1906, MRREE, Sección Colonización, Solicitudes Particulares, A–V, 1906, ARNAD. Bengoa notes that Budi had ten sawmills logging the forests in 1911: Bengoa, *Haciendas y campesinos*, 169. See also Bengoa, *Historia de un conflicto*, 66–67.

67. Opazo, *Desarrollo agrícola de los territorios situados al sur del río Bio-Bio*, 6.

68. Opazo, *Desarrollo agrícola*, 14.

69. Albert, "El problema forestal en Chile," 675; Opazo, *Desarrollo agrícola*; Zon and Sparhawk, *Forest Resources of the World*, 747. See also the very useful descriptions of the problems confronting wood producers in Congreso Nacional, Cámara de Diputados, Diario de Sesiones, Sesión Ordinaria, 14 June, 28 June 1913, BCN.

70. Zon and Sparhawk, *Forest Resources of the World*, 747.

71. Congreso Nacional, Cámara de Diputados, Diario de Sesiones, Sesión Ordinaria, 14 June, 28 June 1913, BCN.

72. Congreso Nacional, Cámara de Diputados, Diario de Sesiones, Sesión Ordinaria, 11 August 1913.

73. Zon and Sparhawk, *Forest Resources of the World*, 749.

74. Irving Theodore Haig, *Forest Resources of Chile as a Basis for Industrial Expansion* (Forest Service, U.S. Department of Agriculture, 1946), 88–89.

75. Mark Jefferson, *Recent Colonization in Chile* (New York: Oxford University Press, 1921), 15.

76. Lilian E. Elliott, *Chile: Today and Tomorrow* (New York: Macmillan, 1922), 213.

77. Elliott, *Chile*, 220.

78. Elliott, *Chile*, 221–25.

79. Rojas Magallanes, *Informe sobre bosques*, 1.

80. Magallanes, *Informe sobre bosques*, 9–10.

81. Magallanes, *Informe sobre bosques*, 18.

82. Magallanes, *Informe sobre bosques*, 31.

83. Schneider, *La agricultura en Chile en los últimos cincuenta años*, 53–55.

84. Schneider, *La agricultura*, 54.

85. Here I have followed Fernando Hartwig Carte, *Federico Albert, pionero del desarrollo forestal en Chile* (Talca: Editorial Universidad de Talca, 1999), 11–14. See also Pablo Camus Gayán, "Federico Albert: Artífice de la gestión de los bosques de Chile," *Revista de Geografía Norte Grande*, no. 30 (2003): 55–63; Camus Gayán, *Ambiente, bosques y gestión forestall en Chile*, 153–65.

86. Schneider, *La agricultura en Chile en los últimos cincuenta años*, 54.

87. Brian Balogh, "Scientific Forestry and the Roots of the Modern American State: Gifford Pinchot's Path to Progressive Reform," *Environmental History* 7, no. 2 (April 2002); Char Miller, *Gifford Pinchot and the Making of Modern Environmentalism* (Washington, D.C.: Island Press, 2001).

88. Albert, "Informe del Jefe de la Sección de Aguas y Bosques," 397–400.

89. Albert, "El problema forestal en Chile," 649.

90. Albert, "Informe del Jefe de la Sección de Aguas y Bosques," 398–99.

91. Albert, "El problema forestal en Chile," 655.

92. Albert, "El problema forestal en Chile," 662.

93. Albert, "El problema forestal en Chile," 701.

94. Albert, "El problema forestal en Chile," 670–71.

95. Federico Albert, "Los bosques: Su conservación, esplotación i fomento," *Boletín de Bosques, Pesca i Caza*, vol.2, no.1, July 1913, 4–46.

96. Albert, "El problema forestal en Chile," 700–703.

97. See, e.g., "Plantaciones de árboles" *Boletín de la Sociedad Agrícola del Sur*, May 1912.

98. "Plantaciones forestales," *Boletín de la Sociedad Agrícola del Sur*, May 1912.

99. *El Diario Austral*, 9 May 1916.

100. *El Sur*, 16 April 1911, 23 April 1911.

101. El Sur, 22 July 1911, 24 July 1911, 25 July 1911.

102. See Troels Bay-Schmith, "Algunas observaciones sobre ensayos de especies forestales en la provincia de Arauco," *Seminario Forestal*, no. 10 (July 1965): 4–5.

103. Congreso Nacional, Cámara de Senadores, Diario de Sesiones, Sesión Extraordinaria, 28 May 1911, BCN.

104. *Primera Memoria del Director de la Oficina de la Mensura de Tierras* (Santiago, 1908), BCN, 87.

105. *Sesta Memoria del Director de la Oficina de la Mensura de Tierras* (Santiago, 1913), BCN, 62–63.

106. Angel Cabeza Monteira, *Aspectos históricos de la legislación forestal vinculada a la conservación, la evolución de areas silvestres protegidas de la zona de Villarrica y la creación del primer parque nacional de Chile* (Santiago: Corporación Nacional Forestal, 1988), 21, 32.

107. Juan Schilling, *Industria de papel: Ante-proyecto de instalación de una fábrica de papel i celulosa en Chile* (Santiago: Sociedad de Fomento Fabríl and Imprenta Universo, 1911), 5.

108. Opazo, *Desarrollo agrícola de los territorios situados al sur del río Bio-Bio*, 6–7.

109. Luis Urrutia Ibáñez, "La propiedad raiz en la zona austral," *Boletín de Bosques, Pesca i Caza*, vol. 1, no. 9, March 1913, 612–13.

3 Peasant Protest

1. For accounts of the events, see *El Sur*, 1–22 July 1934; *El Malleco*, 30 June–10 July 1934; *El Mercurio*, 1–4 July 1934; *La Opinión*, 26 March, 2 April, 9 May, 19 May, 13 June, 30 June–4 July 1934. Sebastián Leiva, "El partido comunista de Chile y levantamiento de Ránquil," *Cyber Humanitas*, no. 28 (2003), available at http://www.cyberhumanitatis.uchile.cl/index.php/RCH/article/viewArticle/5711/5579; Eduardo Téllez Lúgaro, Cristian Arancibia, Juan Canales, Larisa de Rut, Rodrigo Quinteros, and Yuri Quintuppirray, "El levantamiento del Alto Bíobio y el Soviet y la República Araucana de 1934," *Annales de la Universidad de Chile*, sixth series, no. 13 (August 2001), available at http://www.revistas.uchile.cl/index.php/ANUC/article/viewArticle/2530/2448; Olga Uliánova, "Levantamiento Campesino de Lonquimay y la Internacional Comunista," *Estudios Públicos*, no. 89 (Summer 2003), available at http://www.cepchile.cl/1_3185/doc/levantamiento_campesino_de_lonquimay_y_la_internacional_comunista.html#.UfZ-jPY3VCjo. There are also a number of memoirs written by observers or tangential participants. See Harry Fahrenkrog Reinhold, *La verdad sobre la revuelta de Ránquil: memorias de Harry Fahrenkrog Reinhold* (Santiago: Editorial Universitaria, 1985); Arturo Huenchullán, *Los sucesos del Alto Bío-Bío y el diputado Huenchullán* (Santiago: Selectam, 1934). A couple of important works on peasants in twentieth-century Chile deal briefly with the Ránquil uprising: see Brian Loveman, *Struggle in the Countryside: Politics and Rural Labor in Chile, 1919–1973* (Bloomington: Indiana University Press, 1976); José Bengoa, *Haciendas y campesinos: Historia social de la agricultura chilena*, vol. 2 (Santiago: Ediciones Sur, 1990). The best two works on Ránquil are Jaime Flores Chávez, "Un episodio en la historia social de Chile: Ránquil, una revuelta campesina" (master's thesis, Universidad de Santiago de Chile, 1993); Germán Palacios, *Ránquil: La violencia en la expansión de la propiedad agrícola* (Santiago: Instituto de Ciencias Alejandro Lipschutz, 1992).

2. *El Malleco*, 5 July 1934.

3. *El Mercurio*, 4 July 1934.

4. *El Diario Austral*, 29 June 1934, 1.

5. Elizabeth Lira and Brian Loveman, *Las ardientes cenizas del olvido: Via chilena de reconciliación política, 1932–1994* (Santiago: LOM Ediciones, 2000), 40. Lira and Loveman refer to the victims of the Ránquil massacre as Chile's first detenidos-desaparecidos, even though, as they point out, the term was not used in 1934. Some details of the repression can be found in the military judicial processes of the Juzgado Militar de Concepción: see esp. the testimony in the case against

carabineros initiated by the FOCH on 18 July 1934, Causa 195–34, Juzgado Militar de Concepción, ARNAD.

6. See the debate about inquilino contracts in the Congreso Nacional, Cámara de Diputados, Diario de Sesiones, Sesión Ordinaria, 25 July 1944, BCN.

7. Fahrenkrog Reinhold, *La verdad sobre la revuelta de Ránquil*, 22–23.

8. George McCutchen McBride, *Chile: Land and Society* (New York: American Geographical Society, 1936), 294.

9. Fahrenkrog Reinhold, *La verdad sobre la revuelta de Ránquil*.

10. McBride, *Chile*, 294.

11. Mark Jefferson, *Recent Colonization in Chile* (New York: Oxford University Press, 1921), 5.

12. Some insight into life on a fundo dedicated to logging and forestry is provided by Neltume, in Valdivia Province, near Lake Panguipulli. Before 1938, it was a privately owned estate that belonged to the Echavarri family. Then it was transformed into a public timber company that exploited the native forests in the Andes cordillera and produced plywood in one of the country's largest wood factories. For important histories of Neltume, see Claudio Barrientos, "Emblems and Narratives of the Past: The Cultural Construction of Memories and Violence in Peasant Communities of Southern Chile, 1970–2000 (PhD diss., University of Wisconsin, Madison, 2003); Victor Espinoza Cuevas, Paz Rojas Baeza, María Luisa Ortíz Rojas, *Derechos humanos: Sus huellas en el tiempo: Una experiencia de trabajo en derechos humanos y salud mental en una zona rural del sur de Chile* (Santiago: Comité de Defensa de los Derechos del Pueblo, 1999), 31–36. Ida Sepúlveda, interview by author, Valdivia, March, 2002.

13. Palmera Maldonado v. De Mansilla, Carelmanu, Agricultura, December 1925, Ministerio de Agricultura, Industria, i Obras Públicas, Sección Colonización, decretos 133–96, May 1926, ARNAD.

14. Cayetano Risco C., Presidente del Sindicato Profesional Agrícola de Pichi-Ropulli, Depto. de la Unión, "Reiterated Demands and Denunciations of Abuses," and report of Dirección General de Tierras y Colonización, Oficina Osorno, November, 1934, MTC, Providencias, vol. 1076, 1935, ARNAD.

15. "Ley sobre constitución de la propiedad austral. Anomolias y omisiones en los expedientes que se tramitan," Dirección General de Tierras y Colonización to Ministerio de Tierras y Colonización, November 1939, MTC, Providencias, 1939, vol. 1814, ARNAD. The use of inquilinaje contracts to usurp smallholders' land also affected indigenous communities: see, e.g., the case of communities in Cañete in Juan Bautista Marileo Paillaray, Reducción Juan Segundo Marileo, Contulmo, Cañete, Arauco to MTC, 28 November 1946, and Report of Juzgado de Indios, Victoria, MTC, 21 December 1946, MTC, Providencias, 1947, vol. 3114, ARNAD.

16. *El Comercio*, 20 October 1929.

17. See Ministerio de Fomento, *Memoria de Fomento* (Santiago, 1927), 112–13.

18. Ministerio de Fomento, *Memoria de Fomento* (Santiago, 1928), 194.

19. Ministerio de Fomento, *Memoria de Fomento*, 11–13.

20. MTC to Ministerio del Interior, 4 November 1925, MTC, Oficios 1–341, 1925, ARNAD.

21. See an account of this more general process in *El Diario Austral,* 14 October 1916.

22. Ministerio de Propiedad Austral, Oficina de Osorno, 15 March 1930, MTC, Providencias, vol. 167, 1930, ARNAD.

23. Luis Silva Rivas to Dirección General de Tierras y Colonización, 12 January 1926, Ministerio de Agricultura, Industria, i Obras Públicas, decretos 197–226, April 1926; MTC to Ministro del Interior, 31 December 1925, MTC, Oficios 1–341, 1925; Consejo de Defensa Fiscal, Antecedentes, May 1926, MTC, Oficios Dirigidos y Antecedentes, 301–484, 1925–26; letter to Don Carlos Ibáñez del Campo, 9 April 1927, Ministerio de Agricultura, Industria y Colonización, Sección Colonización, Providencias, July–September 1927, ARNAD.

24. Letter to Don Carlos Ibáñez del Campo (9 April 1927). See the similar case of conflict between ocupantes and the Nueva Italia concession in MTC to Ministro del Interior, 22 May, 23 November, 16 December, 30 December 1925, MTC, Oficios 1–341, 1925; MTC, Representative of Jorge Ricci, Sociedad Nueva Italia, 30 April 1925, report, 24 June 1925, MTC, Selección Colonización, decrees 199–300, May–June 1925, ARNAD.

25. Circular from Ministerio de Fomento to Intendente, 29 April 1929, Intendencia de Cautín, vol. 281, 1929, AN.

26. Communique from Intendente to Prefecto de Carabineros, 26 August 1933, Intendencia de Cautín, vol. 414, 1933, AN.

27. Telegram from Pedro Soto B. to Intendente, 27 December 1929, Intendencia de Cautín, vol. 298, 1930, AN. See the similar case of the Colilco estate described in a telegram from Alfredo Latour to Intendente, 29 July 1929 and "Informe del teniente de carabineros enviado a investigar informa," 30 August 1929, Intendencia de Cautín, vol. 278, 1929, AN.

28. Communique from Secretario de Bienestar to Intendencia, 7 April 1930; "Informe de carabineros a la Prefectura de Cautín," 20 April 1930, Intendencia de Cautin, vol. 305, 1930, AN.

29. Petition to Intendente from Pedro Ugalde Barrios, Agricultor, 13 October 1926, Intendencia de Cautín, vol. 261, 1928, AN.

30. Petition to Intendente from Eduardo Slano Illanes y Santiago Viñuela, Agricultores, 13 October 1928, Intendencia de Cautín, vol. 261, 1928, AN.

31. Communique from Ministerio del Interior to Intendente, 23 July 1930, Intendencia de Cautin, vol. 306, 1930, AN.

32. Ministerio de Fomento, *Memoria de Fomento* (1927), 117.

33. Ministerio de Fomento, *Memoria de Fomento* (1928), 207.

34. Ministerio de Fomento, *Memoria de Fomento* (1928), 119, 207.

35. In 1929, the Vicuña Mackenna National Park and Villarrica Forest Reserve were collapsed into one large forest reserve. In 1940, a national park was reestablished in Villarrica. The Villarrica National Park was on land that was near, but not superimposed on, the original national park created in 1925. In 1967, a new national park, Parque Nacional Huerquehue, was located within the borders of the old and short-lived Parque Nacional Vicuña Mackenna: Dirección General de Tierras, Oficio no. 844, 17 June 1925, in Angel Cabeza Monteiro, *Aspectos históricos de la leg-*

islación forestal vinculada a la conservación, la evolución de areas silvestres protegidas de la zona de Villarrica y la creación del primer parque nacional de Chile (Santiago: Corporación Nacional Forestal, 1988), 41–42, 52.

36. "Memoria que presenta el Silvicultor de la Reserva Forestal de Villarrica," 1929, Intendencia de Cautín, vol. 284, 1929, AN.

37. "Memoria anual," Gobernador de Villarrica, 27 February 1929, Intendencia de Cautín, vol. 285, 1929, AN.

38. Copies of telegrams received from the Prefectura de Carabineros de Pucón to Intendente, 27 August 1932, Intendencia de Cautín, vol. 382, 1932, AN.

39. Confidential communique from Intendente Suplente to Prefecto de Carabineros, 10 April 1931, Intendencia de Cautín, vol. 376, 1931, AN.

40. Communique from Silvicultor de la Reserva Forestal to Intendente, 15 October 1932, Intendencia de Cautín, vol. 383, 1932, AN.

41. Communique from Intendente sobre intromisión de gente extraña en terrenos de la reserva en el Río Pedregoso, 12 November 1932, Intendencia de Cautín, vol. 384, 1932, AN.

42. Oficio al Ministro de Fomento, 25 October 1928, Intendencia de Cautín, vol. 269, 1928, AN.

43. Congreso Nacional, Cámara de Diputados, Diario de Sesiones, Sesión Extraordinaria, 1 March 1932, BCN.

44. A *pulgada* is a measurement of volume for wood that is one inch deep, ten inches wide, and twelve feet long.

45. Congreso Nacional, Cámara de Diputados, Diario de Sesiones, Sesión Extraordinaria, 1 March 1932, BCN. Similar cases of conflicts between campesinos and loggers in Cunco are described in "Comunicación de particulares a la Intendencia," 20 April 1937, Intendencia de Cautín, vol. 467, 1938, AN.

46. Communique from Subdelegado de la Comuna de Cunco to Intendente, 15 October 1930, Intendencia de Cautín, vol. 346, 1931, AN.

47. *El Comercio,* 12 October 1930.

48. "Solicitan investigaciones sobre la actuación del señor Abdón Rodríguez, Silvicultor de Malalcahuello," 28 October 1931, Intendencia de Cautín, vol. 380, 1932, AN.

49. *El Comercio,* 28 December 1930.

50. Communique from Gobernador de Victoria to Intendente, 20 January 1933, Intendencia de Cautín, vol. 400, 1933, AN.

51. Fahrenkrog Reinhold, *La verdad sobre la revuelta de Ránquil; El Sur,* 22 July 1934.

52. Communique from Gobernador de Victoria Salomón Ascui G. to Intendente, 14 October 1929, Intendencia de Cautín, vol. 279, 1929, AN.

53. For a brief discussion of this history, see Fahrenkrog Reinhold, *La verdad sobre la revuelta de Ránquil.* See also a detailed description of the property titles in *Boletín de Tierras y Colonización,* decreto no. 265, 27 March 1930, published in its entirety in Congreso Nacional, Cámara de Diputados, Sesión Ordinaria, 9 July 1934, BCN. See also the excellent reviews of this history in Jaime Flores Chávez, "Tierra y territorio Mapuche-Pehuenche, 1860–1997," in *Estudios introductorios en relaciones interétnicas,* ed. Belén Lorente Molina and Carlos Vladimir Zambrano

(Bogotá: Corporación Colombiana de Investigaciónes Humanísticas, 1992), 79–104; Flores Chávez, "Un episodio en la historia social de Chile"; Téllez Lugaro et al., "El levantamiento del Alto Bíobio y el Soviet y la República Araucana de 1934"; Leiva, "El partido comunista de Chile y levantamiento de Ránquil."

54. Communique from Sindicato Agrícola de Lonquimay to Intendente, 29 September 1929, Intendencia de Cautín, vol. 278, 1929, AN.

55. See the full text of the interview between Ibáñez and Leiva in *El Comercio*, 9 December 1928.

56. Telegram from Carlos Ibáñez del Campo, 11 November 1929, Intendencia de Cautín, vol. 279, 1929, AN.

57. See, e.g., *El Comercio*, 16 June 1929.

58. Congreso Nacional, Cámara de Diputados, Diario de Sesiones, Sesión Ordinaria, 9 July 1934, BCN.

59. Communique from Gobernador de Villarrica to Ministerio del Interior, 10 November 1932, Intendencia de Cautín, vol. 403, 1933, AN.

60. Communique from Ministerio del Interior to Intendente, 6 April 1933, Intendencia de Cautín, vol. 342, 1931–33, AN.

61. "Informe sobre colonización en Alto Bío-Bío del Gobernador de Victoria al Ministro de Tierras y Colonización," 5 January 1932, Intendencia de Cautín, vol. 378, 1932, AN.

62. See Congreso Nacional, Cámara de Diputados, Diario de Sesiones, Sesión Extraordinaria, 4 February 1932, BCN; *La Opinión*, 30 June 1934.

63. *El Comercio*, 5 June 1932.

64. *El Comercio*, 5 November 1933.

65. *El Sur*, 22 July 1934.

66. Fahrenkrog Reinhold, *La verdad sobre la revuelta de Ránquil*, 43–44.

67. *La Opinión*, 1 July 1934.

68. Fahrenkrog Reinhold, *La verdad sobre la revuelta de Ránquil*, 44–45.

69. *La Opinión*, 30 June 1934. Copies of petitions to Congress from the Sindicato Agrícola Lonquimay are also in Congreso Nacional, Cámara de Diputados, Sesión Extraordinaria, 4 February 1932, BCN.

70. "Damos cuenta al señor subdelegado de que somos víctimas de atropellos y amenazas de parte del Señor Gonzalo Bunster Gómez," 16 December 1932; "Informa reclamación de don Gonzalo Bunster contra Sindicato Agrícola Lonquimay," 5 January 1933, Intendencia de Cautín, vol. 400, 1933, AN.

71. *Justicia*, 10 February 1933, 2.

72. *El Sur*, 10 March 1933.

73. Telegram to Intendencia, 27 October 1932, Intendencia de Cautín, vol. 383, 1932, AN.

74. See, e.g., telegram from Ministro de Tierras to Intendente, 18 November 1932, Intendencia de Cautín, vol. 384, 1932, AN.

75. "Comunicación de la Compañía Agrícola y Ganadera de Toltén," 9 November 1932, Intendencia de Cautín, vol. 403, 1933, AN.

76. *El Sur*, 22 January, 27 January 1933.

77. *La Opinión*, 12 April 1934.

78. *La Opinión*, 19 May 1934.

79. *La Opinión*, 26 March 1934.

80. Segundo Juzgado de Letras, Provincia de Cautín, Departamento de Temuco, Causa Criminal no. 5680, "Usurpación," 13 December 1933, ARA.

81. Segundo Juzgado de Letras, Provincia de Cautín, no. 6363, "Incendio" (11 June 1935).

82. *El Diario Austral*, 14 November 1935.

83. Congreso Nacional, Cámara de Diputados, Sesión Ordinaria, 9 July 1934, BCN.

84. Fahrenkrog Reinhold, *La verdad sobre la revuelta de Ránquil*, 46–47.

85. *El Sur*, 2 July 1934, 6 July 1934; *El Mercurio*, 4 July 1934.

86. *El Sur*, 6 July 1934.

87. *El Sur*, 21 July 1934.

88. Juez Titular de Indios de Victoria, Sentencia, November 1942, Gustavo Risquertz Susarte, Juzgado de Indios, CONADI Archive, ARA. For an important account of this history, see Raúl Molina and Martín Correa, *Territorio y comunidades pehuenches del Alto Bío-Bío* (Santiago: Corporación Nacional de Desarrollo Indígena, 1996).

89. Ministerio de Relaciones Exteriores, Culto i Colonización, Sección: Colonización, Solicitante Ignacio Maripí y Otros, Materia: Piden que se les radique, Iniciada, 1912, in MRRE, vol. 1699, Sección Colonización, Solicitudes Particulares 1912, Letras A-LL, ARNAD.

90. Ministerio de Relaciones Exteriores, Culto i Colonización, Sección: Colonización, Solicitante Ignacio Maripí y Otros, Materia: Piden que se les radique, Iniciada, 1912.

91. Radicación no. 2099, 19 January 1939, Juzgado de Indios de Victoria, CONADI Archive, ARA.

92. Congreso Nacional, Cámara de Diputados, Diario de Sesiones, Sesión Extraordinaria, 4 February 1932, BCN.

93. There is an important, if brief, history of the Mapuche-Pehuenche communities in the region in Domingo Namuncura, *Ralco: Represa o pobreza?* (Santiago: LOM Ediciones, 1999). See also Molina and Correa, *Territorio y comunidades pehuenches del Alto Bío-Bío.*

94. Radicación no. 2099. This decision is also cited in Namuncura, *Ralco*, 260.

4 Changing Landscapes

1. Adolfo Ibáñez, "Reforestación en Concepción," *El Campesino*, vol. 75, no.4, February, 1943, 268–69.

2. Congreso Nacional, Cámara de Diputados, Diario de Sesiones, Sesión Extraordinaria, 2 December 1931, BCN.

3. George McCutchen McBride, *Chile: Land and Society* (New York: American Geographical Society, 1936), 283.

4. Caja Nacional de Empleados Públicos y Periodistas, Comunidad Yrarrázaval Mac-Clure, "Parcelas-bosques plantadas con pinos insignis: Fundos Siberia,

Los Litres, Cruz de Piedra y Las Mercedes (Provincia de Ñuble–Departamento de Yungay)," BN.

5. Luis Contreras, "La forestación en la hacienda 'Canteras,'" in *Primera Asamblea Forestal Nacional*, ed. Sociedad Amigos del Árbol (Santiago: n.p., 1943), 22, BN.

6. *El Sur*, 30 July 1937, 4.

7. Carlos Keller, *Revolución en la agricultura* (Santiago: Zig-Zag, 1956), 131.

8. *El Sur*, 23 November 1938, 10.

9. *El Sur*, 23 November 1938, 10.

10. *El Sur*, 14 October 1937, 3. See also José Maige A. and Ernesto Bernath, "Viveros y plantaciones forestales de la Caja de Colonización Agrícola," in Sociedad Amigos del Árbol, *Primera Asamblea Forestal Nacional*, 259–270, BN.

11. Corporación de Fomento de la Producción, *Esquema de diez años de labor, 1939–1949*, BN.

12. *El Sur*, 5 January 1946, 7.

13. Vicente Pastor P., "Fomento efectivo al la forestación del país realizado por la Corporación de Fomento a la Producción," in Sociedad Amigos del Árbol, *Primera Asamblea Forestal Nacional*, 246–248.

14. Luis Ortega Martínez, Cármen Norambuena Carrasco, Julio Pinto Vallejos, Guillermo Bravo Acevedo, *Corporación de Fomento de la Producción: 50 Años de Realizaciones, 1939–1989* (Santiago: CORFO, 1989), 94.

15. Alfonso Mora Venegas, "Los bosques en la economía chilena" (Licenciatura thesis, Universidad de Chile, Santiago, 1950), 90.

16. Venegas, "Los bosques en la economía chilena," 88.

17. Irving Theodore Haig, *Forest Resources of Chile as a Basis for Industrial Expansion* (Washington, DC: Forest Service, US Department of Agriculture, 1946), 93.

18. Horacio Recart, "La Corporación de Fomento a la Producción en su cooperación en actividades forestales y madereras del país," in Sociedad Amigos del Árbol, *Primera Asamblea Forestal Nacional*, 249–250; Ministerio de Economía y Comercio, "Plan de fomento y racionalización de las industrias forestales, 1946–1952," Archivo Alessandri, BN.

19. Corporación de Fomento de la Producción de Chile, *Esquema de diez años de labor*. For a description of CORFO's forestry initiatives, see also Pablo Camus Gayán, *Ambiente, bosques y gestión forestal en Chile, 1541–2005* (Santiago: LOM Ediciones and Dirección de Bibliotecas, Archivos y Museos, 2006), 174–75.

20. *El Sur*, 28 January 1940, 18.

21. *El Sur*, 17 May 1939, 12.

22. *El Sur*, 29 March 1940, 3; a useful company account summarizing this history is available at http://www.cmpc.cl/eng/our_company/history, accessed on 1 November 2005.

23. *El Sur*, 29 March 1940, 3.

24. "La madera de *pinus insignis* para la fabricación de pulpa," *El Campesino*, vol. 65, no. 11, November 1933, 590.

25. *El Agricultor*, nos. 13–15, January–March 1943, 1.

26. Dirección General de Tierras y Colonización, 21 December 1939, "Guillermo Otto solicita certificado," MTC, Providencias, vol. 1817, 1939, ARNAD.

27. "Alberto Heck solicita se registran nuevas plantaciones, Fundo 'Los Pinos,'" 6 December 1945, MTC, Providencias, vol. 2954, 1946, ARNAD.

28. "Comunidad Fundo Selva Negra solicita se registran nuevas plantaciones," 4 December 1945, MTC, Providencias, vol. 2954, 1946, ARNAD.

29. Letter to Jefe del Departamento de Bosques, 13 October 1945, Santiago, MTC, Providencias, vol. 2954, 1946, ARNAD. For a discussion of requests for forestry certification, see also Camus Gayán, *Ambiente, bosques y gestión forestal en Chile*, 172.

30. Ministerio de Agricultura, Oficios Recibidos, 1929–35, ARNAD.

31. Dirección General de Tierras y Colonización, Departamento de Bienes Nacionales, 6 May 1938. MTC, Providencias, vol. 1926, 1940, ARNAD.

32. Juan Moty to Dirección General de Tierras y Colonización, 17 September 1946, MTC, Providencias, vol. 3116, 1947, ARNAD.

33. Prospectuses for a number of public forest companies are held at the BN. See, e.g., Caja Nacional de Empleados Públicos y Periodistas, Comunidad Yrarrázaval Mac-Clure, "Parcelas-bosques plantadas con pinos insignis"; "Parcelas forestadas de inmediata explotación por cooperativas agrícolas madereras: Parcelación fundo Comao"; "Parcelas-bosques 'Los Pinares de Cauquenes'"; prospectus of the Sociedad Forestal "Shangri-La"; "Parcelas-bosques de pino Radio La Cooperativa Vitalicia"; prospectus of the Comunidad Forestal "Malloga"; prospectus of the Comunidad Forestal "El Avellano"; Compañía Forestal "Constitución"; and prospectuses of the Sociedad Anónima y Forestal los Pinos.

34. "Parcelas forestadas de inmediata explotación por cooperativas agrícolas madereras."

35. "Parcelas-bosques de pino Radio La Cooperativa Vitalicia."

36. Leoncio Chaparro Ruminot, *Colonias de medieros* (Santiago: Multitud, 1941).

37. Haig, *Forest Resources of Chile as a Basis for Industrial Expansion*, 73.

38. Haig, *Forest Resources of Chile*, 74.

39. Caja Nacional de Empleados Públicos y Periodistas, Comunidad Yrarrázaval Mac-Clure, "Parcelas-bosques plantadas con pinos insignis."

40. Quoted in "Primeros forestadores: Visionarios plantadores," *Lignum*, no. 29 (1997): 32–34.

41. Mora Venegas, "Los bosques en la economía chilena," 93. For a discussion of the "parcelas-bosques" projects, see Camus Gayán, *Ambiente, bosques y gestión forestal en Chile*, 173–74.

42. Caja Nacional de Empleados Públicos y Periodistas, Comunidad Yrarrázaval Mac-Clure, "Parcelas bosques plantadas con pinos insignis."

43. *El Agricultor*, no. 779, March–June 1951, 6–7.

44. *Agricultura Austral*, vol. 5, no. 55, August, 1937, 926.

45. Sociedad Ganadera, Agrícola y Forestal "Rio Laja," folleto preliminar, 1935, BN.

46. Belsario Neira Salas, "Una grande idea que se realiza mediante los ahorros de unos pocos amigos del árbol," in Sociedad Amigos del Árbol, *Primera Asamblea Forestal Nacional*, 251–52, BN.

47. Mora Venegas, "Los bosques en la economía chilena," 92.

48. Mora Venegas, "Los bosques en la economía chilena," 93.
49. Bosques e Industrias Maderas, *Séptima Memoria*, 1946, BN.
50. Mora Venegas, "Los bosques en la economía chilena," 90.
51. Bosques e Industrias Maderas, *Séptima Memoria*.
52. See, e.g., "El pino de Monterey o pino insignis (Pinus radiata): Árbol de enorme porvenir en Chile como productor de madera y para pasta de papel," reprinted in *El Agricultor*, no. 7, June 1937; 8 "Nuevamente el problema del trigo," *El Agricultor*, nos. 16–17, January–February 1940, 16–17.
53. José Maige A., *Plantaciones forestales en las colonias agrícolas* (Boletín del Ministerio de Agricultura, n.p., 1937), 6.
54. Chaparro Ruminot, *Colonias de medieros*.
55. Juan Jirkal H., "Generalidades sobre el problema de la erosión del suelo de Chile" in Sociedad Amigos del Árbol, *Primera Asamblea Forestal Nacional*, 296.
56. Haig, *Forest Resources of Chile as a Basis for Industrial Expansion*, 142.
57. *Chile Maderero*, vol. 4, no. 3, May 1954, 11–12.
58. Rafael Elizalde Mac-Clure, *La sobrevivencia de Chile: La conservación de sus recursos naturales renovables* (Santiago: Ministerio de Agricultura, 1970), 22–23.
59. See, e.g., "Nuevamente el problema del trigo," *El Agricultor*, nos. 16–17, January–February, 16–17. See also numerous articles in *El Campesino* throughout out the 1930s and 1940s. One good example is a speech made by Jaime Larraín, president of the Sociedad Nacional de Agricultura, that was published in *El Campesino*, "La asamblea de agricultores de Ñuble," vol. 66, no. 6, June 1934, 267. See also "Notas de Actualidad" and "El presidentede la Soc. N. DE A. da cuenta de la labor del año," *El Campesino*, vol. 72, no. 1, January, 1940, 1–2; "Un agricultor sureño haba para nuestra revista del 'problema del trigo,'" *El Campesino*, vol. 71, no. 1, January, 1939.
60. Arnold Bauer, *Chilean Rural Society from the Spanish Conquest to 1930* (Cambridge: Cambridge University Press, 1975).
61. Adolfo Mathei, *La agricultura en Chile y la política agraria chilena* (Santiago: Imprenta Nacimento, 1939), 40.
62. Rafael Elizalde Mac-Clure, "La muerte del bosque," *En Viaje*, no. 116 (June 1943).
63. Quoted in Ortega Martínez et al., *Corporación Fomento de la Producción*, 103.
64. Ortega Martínez et al., *Corporación Fomento*, 103–104.
65. *Guía del veraneante: Revista de turismo de los FFCC del E (Chile)*, 1941, 9, BN.
66. Departamento de Turismo, *Visite Ud. Chile: País del Pacífico Sur de clima suave bello y hospitalario*, 1943, BN.
67. Telegram from Ministerio del Interior to Intendente, 5 February 1937, Intendencia de Cautín, vol. 441, 1935, AN.
68. Telegram from Municipalidad de Pucón to Supremo Gobierno, 11 April 1949, Intendencia de Cautín, vol. 570, 1949, AN. See also communique from Club Andino de Cautin to Intendente, 22 February 1949, Intendencia de Cautín, vol. 575, 1949, AN.

69. Raúl González to Ministerio de Tierras y Colonización, MTC, Providencias, vol. 3183, 1947, ARNAD.

70. Sociedad Amigos del Árbol, *Primera Asamblea Forestal Nacional,* 22, 23–30, 90–97. For an extended discussion of the society, see Camus Gayán, *Ambiente, bosques y gestión forestal en Chile,* 195–97.

71. Haig, *Forest Resources of Chile as a Basis for Industrial Expansion.* Camus Gayán also discusses the report in *Ambiente, bosques y gestión forestal en Chile,* 198–204.

72. U.S. Department of Agriculture, *Forest Service Information Digest,* November 1950.

73. Haig, *Forest Resources of Chile as a Basis for Industrial Expansion,* 41.

74. Haig, *Forest Resources of Chile,* 29.

75. Haig, *Forest Resources of Chile,* 53.

76. Haig, *Forest Resources of Chile,* 45.

77. Haig, *Forest Resources of Chile,* 45.

78. Haig, *Forest Resources of Chile,* 144.

79. Haig, *Forest Resources of Chile,* 144.

80. Haig, *Forest Resources of Chile,* 126.

81. Haig, *Forest Resources of Chile,* 126.

82. Haig, *Forest Resources of Chile,* 94.

83. Haig, *Forest Resources of Chile,* 90.

84. *Chile Maderero,* vol. 4, no. 4, June–July 1954, 15, 17.

85. Haig, *Forest Resources of Chile as a Basis for Industrial Expansion,* 92.

86. Haig, *Forest Resources of Chile,* 143.

87. Haig, *Forest Resources of Chile,* 27.

88. Mora Venegas, "Los bosques en la economía chilena," 53.

89. For a discussion of these issues, see Brian Loveman, *Chile: The Legacy of Hispanic Capitalism* (Oxford: Oxford University Press, 2001), 215.

90. Ministerio de Economía y Comercio, "Plan de fomento y racionalización de las industrias forestales"; Mora Venegas, "Los bosques en la economía chilena," 88.

91. *Chile Maderero,* vol. 4, no. 3, May 1954, 3.

92. *Chile Maderero,* no. 4, June–July 1954, 2–3.

93. *Chile Maderero,* vol. 6, no. 4, July–August 1956, 7.

5 *Politics of Social Reform*

1. Juan Sánchez Guerrero, *Hijo de las piedras* (Santiago: Zig-Zag, 1962), 26–29. I thank Consuelo Figueroa Garavagno for pointing me to this collection of stories. She refers to it in her history of women in the coal-mining camps, *Revelación del subsole: Las mujeres en la sociedad minera del carbón, 1900–1930* (Santiago: Dirección de Bibliotecas, Archivos y Museos, 2009), 36, n. 30.

2. Sánchez Guerrero, *Hijo de las piedras,* 35.

3. Octavio Astorquiza and Oscar Galleguillos V., *Cien años del carbón de Lota* (Santiago: Compañía Carbonífera e Industrial de Lota, 1952), 261. For an interest-

ing technical study of the company's pine plantations, see Konrad Peters, *Estudio experimental sobre selvicultura en Chile* (Santiago: Imprenta Nascimento, 1938).

4. *El Sur,* 2 September 1937, 10.

5. Article reproduced in MTC, Providencias, vol. 1927, 1939, ARNAD.

6. *El Sur,* 5 December 1939, 6.

7. See interview with the Minister of Lands and Colonization in *El Sur,* 5 December 1939.

8. *El Sur,* 5 January 1940, 7.

9. Leoncio Chaparro Ruminot, *Colonias de medieros* (Santiago: Multitud, 1941).

10. Comunicación de la Agricultural Colonization Fund al Intendente, 11 June 1939, Intendencia de Cautín, vol. 478, 1939, AN.

11. *El Sur,* 5 December 1939, 5; Chaparro Ruminot, *Colonias de medieros.*

12. Chaparro Ruminot, *Colonias de medieros.*

13. Chaparro Ruminot, *Colonias de medieros.*

14. Luis Contreras, "Necesidad de formar el obrero forestal," in *Primera Asamblea Forestal Nacional,* ed. Sociedad Amigos del Árbol (Santiago: n.p., 1943), 200–201, BN.

15. Contreras, "Necesidad de formar el obrero forestal," in *Primera Asamblea,* 200.

16. Contreras, "Necesidad de formar el obrero forestal," in *Primera Asamblea,* 201.

17. *El Sur,* 25 June 1939, 16.

18. *El Sur,* 18 July 1939, 11.

19. *El Sur,* 27 October 1939, 13.

20. *El Sur,* 5 December 1939, 6.

21. Brian Loveman, *Chile: The Legacy of Hispanic Capitalism* (Oxford: Oxford University Press, 2001), 212–16; Brian Loveman, *Struggle in the Countryside: Politics and Rural Labor in Chile, 1919–1973* (Bloomington: Indiana University Press, 1976).

22. Dirección General de Tierras y Colonización, Colonias Forestales, Fundos Chanlil y Raguintulelfu, 3 April 1946, MTC, Providencias, vol. 3008, 1946, ARNAD.

23. Colonias Forestales, Fundos Chanlil y Raguintulelfu, 3 April 1946.

24. Colonias Forestales, Fundos Chanlil y Raguintulelfu, 3 April 1946.

25. Colonias Forestales, Fundos Chanlil y Raguintulelfu, 3 April 1946.

26. Dirección General de Tierras y Colonización, Mensura de Tierras, Osorno, 13 November 1946, MTC, Providencias, vol. 3114, 1947; Dirección General de Tierras y Colonización, Departamento Mensura de Tierras Puerto Montt, 13 November 1946, MTC, Providencias, vol. 3114, 1947; "Solicita avaluo fiscal predios que indica," Santiago, 14 November 1946, MTC, Providencias, vol. 3117, 1947, ARNAD.

27. Colonias Forestales, Fundos Chanlil y Raguintulelfu 3 April 1946.

28. Colonos of Raintulelfu [*sic*] to Ministro de Tierras y Colonización, 5 September 1947, MTC, Providencias, vol. 3183, 1947, ARNAD.

29. *El Siglo,* 4 June 1946, 8.

30. Gobernador de Maullín to Ministro del Interior, 19 March 1946, MTC, Providencias, vol. 3180, 1947, ARNAD.

31. Ministerio de Tierras y Colonización, report, 18 October 1954, MTC, Providencias, vol. 1, 1954, ARNAD.

32. "Reiteran solicitud de radicación en el fundo Trapul," hoy Traful, Valdivia (Licán Ray); Dirección General de Tierras y Colonización, Departamento Mensura de Tierras, Valdivia, 26 September 1947, MTC, Providencias, vol. 3178, 1947, ARNAD.

33. *El Sur,* 14 April 1939, 9.

34. Junta de Beneficencia de Santa Juana, Concepción, 28 July 1939, MTC, Providencias, vol. 1923, 1940, ARNAD.

35. *El Sur,* 12 May 1939, 10.

36. Memorandum, Dirección General de Tierras y Colonización, Santiago, 22 November 1939, "Comité de Colonos de Futa y Niebla Reclamos," MTC, Providencias, vol. 1818, 1939, ARNAD.

37. Memorandum, Dirección General de Tierras y Colonización, Santiago, 22 November 1939.

38. Francisco Vega Cáceres, Colono Fundo Fiscal "Vegas Blancas," Angol, Malleco, 1939; Informe, MTC, Departamento de Bienes Nacionales, Nahuelbuta, Vegas Blancas, 27 November 1939, MTC, Providencias, vol. 1818, 1939, ARNAD.

39. *El Siglo,* 3 April 1955, 9.

40. *El Siglo,* 20 September 1958, 11

41. "Autorización explotación maderas, reducción Bernardo Ñanco," 14 July 1947, no. 2713, Juzgado de Indios, Victoria; "Autorización para explotar maderas," 11 June 1947, no. 2726, Comunidad Bernardo Ñanco, Título de Merced (TM) 1075, Carpetas Administrativas, CONADI Archive, ARA.

42. "Autorización explotación maderas, reducción Bernardo Ñanco," 14 July 1947, no. 2713, Juzgado de Indios, Victoria; "Autorización para explotar maderas," 11 June 1947, no. 2726.

43. Don Aladino, Comunidad Bernardo Ñanco, interview by author, Lonquimay, August 2010.

44. "Cuaderno de medidas precautorias, Reducción Bernardo Ñanco," n. d., no. 24120A, Juzgado de Indios de Temuco, Victoria; letter to Director de Asuntos Indígenas from Juez Oscar R. Izarnotegui, Victoria, 9 August 1963; Hugo Ormeño Medet, Defensor de Indígenas, December 1963, 1966, 1970, Juzgado de Letras de Indios, Victoria, Comunidad Bernardo Ñanco, TM 1075, Carpetas Administrativas, CONADI Archive, ARA.

45. Petition to Tribunal de Indios from members of the Reserva of Cacique Ancanco, 15 December 1955; petition related to the remeasurement of the Galletué fundo from Aspirantes a Colonos, Icalma, Lonquimay, June 1955, Conunidad Pedro Calfuqueo, TM 998, Carpetas Administrativas, CONADI Archive, ARA.

46. See the discussion of this process in José Bengoa, *Quinqén: Cien años de historia pehuenche* (Santiago: Ediciones Chile América-CESOC, 1992), 49–62.

47. Paz Neira, Samuel Línker, and Irene Romero, *Memorias de Llaima: Historias de las comunidades mapuches de Melipeuco* (Santiago: Talleres Designofpasten, 2011).

48. Moises Durán, interview by the author, Panguipulli, June 2012.

49. Juzgado Civil de Mayor Cuantía de Panguipulli, 24 February 1956, no. 401, Victór Kunstmann v. Vicente Reiñehuel, Comunidad Vicente Reiñehuel, TM 2410, Carpetas Administrativas, CONADI Archive, ARA.

50. Alex Rudloff, interview by the author, Panguipulli, June, 2010.

51. Communique from Jefe de la V Zona Forestal to Intendencia, 29 April 1953, communique from Administrador de la Reserva Forestal de Villarrica to Intendencia, 15 April 1953, Intendencia de Cautín, vol. 646, 1954, AN. The government reports from 1932 spell the indigenous community's name "Huaquivir."

52. *El Siglo,* 24 July 1947, 4.

53. *El Siglo,* 12 October 1947, 8.

54. *El Siglo,* 5 August 1947, 4.

55. *El Agricultor,* no. 70, June–July 1948, 24–25.

56. Comité de Inquilinos Medieros del Fundo Pilpilco Alto, MTC, Providencias, vol. 4, 1953, ARNAD.

57. Congreso Nacional, Cámara de Diputados, Diario de Sesiones, Sesión Extraordinaria, 6 December 1944, BCN.

58. Congreso Nacional, Sesión Extraordinaria, 21 November 1944.

59. For a history of this process, see Loveman, *Struggle in the Countryside.*

60. *El Siglo,* 25 November, 4, 1 December, 12 1946.

61. *El Siglo,* 13 August 1943, 4.

62. *El Siglo,* 25 November, 4, 1 December, 12 1946.

63. *El Siglo,* 15 April 1947, 7.

64. Congreso Nacional, Cámara de Diputados, Diario de Sesiones, Sesión Ordinaria, 25 July 1944, BCN.

65. *El Siglo,* 26 May 1947, 4.

66. Petition from Sindicato Apolítico, Profesional, Agricultural, Fundos El Chacay and Pillén Pillén, MTC, Providencias, vol. 3180, 1947, ARNAD.

67. *Chile Maderero,* vol. 6, no. 4, July–August 1956, 54.

68. *Chile Maderero,* no. 1, May 1958, 69.

6 State-Directed Forestry Development

1. *El Diario Austral,* 24 February 1966, 9.

2. Congreso Nacional, Cámara de Diputados, *Informe de la Comisión de Agricultura y Colonización,* bulletin no. 10.485, 1965, vii, ix.

3. *Informe de la Comisión de Agricultura y Colonización,* xii–xiii.

4. *El Siglo,* 2 November 1957, 3.

5. *El Siglo,* 2 January 1958, 2.

6. *Chile Maderero,* no. 3, December 1957, 17.

7. *El Siglo,* 27 February 1953, 6.

8. Luis Rosales, interview by the author, Panguipulli, March, 2002.

9. *El Siglo,* 18 June 1957, 8.

10. *El Diario Austral,* 15 November 1966, 3, 17; November 1966, 10.

11. *El Diario Austral,* 16 November 1966, 12, 22; November 1966, 3.

12. *El Diario Austral,* 22 November 1966, 3.

13. *El Diario Austral,* 31 August 1966, 28.

14. *El Diario Austral,* 31 August 1966, 23.

15. *El Diario Austral,* 31 August 1966, 23.

16. *El Diario Austral,* 31 August 1966, 12.

17. *El Diario Austral,* 31 August 1966.

18. *Chile Maderero,* vol. 4, no. 3, May 1954, 2.

19. *El Diario Austral,* 9 August 1965, 3.

20. *El Diario Austral,* 4 December 1966, 9.

21. Arthur Cranmer Kenrick Bell to Vice-President, CORA, 28 January 1977, Fundo Chan-Chan, record no. 1124, CORA Archive, SAG.

22. *El Diario Austral,* 9 January 1967, 7.

23. *El Diario Austral,* 24 March 1967, 3.

24. *El Siglo,* 10 November 1958, 13.

25. *El Siglo,* 6 December 1964, 15.

26. *El Siglo,* 22 August 1959, 14.

27. "Urgencia y posibilidades de la industrialización maderera," *Panorama Económica,* no. 84, 28 August 1953, 555.

28. "La industria de celulosa y papel," *Panorama Económica,* no. 81, 28 August 1953, 557–559. For a discussion of the FAO and CEPAL, see Pablo Camus Gayán, *Ambiente, bosques y gestión forestal en Chile, 1541–2005* (Santiago: LOM Ediciones and Dirección de Bibliotecas, Archivos y Museos, 2006), 204–9.

29. "The Work of FAO: Forestry Education in Chile," *Unasylva,* vol. 7, no. 3, September 1953. See also Camus Gayán, *Ambiente, bosques y gestión forestal en Chile,* 214.

30. "The Work of FAO: Forestry Education in Chile."

31. "The Work of FAO: The Llancura Forest Research and Training Center," *Unasylva,* vol. 9, no. 2, June 1955.

32. *Chile Maderero,* no. 1, May 1958, 18.

33. *Chile Maderero,* vol. 4, no. 3, May 1954, 11. For a discussion of forestry education, see Camus Gayán, *Ambiente, bosques y gestión forestal en Chile,* 214–17.

34. C. W. Scott, "Radiata Pine as an Exotic," *Unasylva,* vol. 14, no. 1, 1960. See also "Latin-American Forestry and Forest Products Commission," *Unasylva,* vol. 3, no. 5, September–October 1949.

35. Cited in "Nuevas plantas de celulosa abrirán perspectivas a las plantaciones de pino insigne," *Chile Maderero,* no. 2, September 1957, 21.

36. "Nuevas plantas de celulosa abrirán perspectivas a las plantaciones de pino insigne"; "News of the World: Chile," *Unasylva,* vol. 11, no. 1, 1957.

37. For a discussion of the CMPC's cellulose plant project, as well as the ending of the Empresa Nacional de Celulosa project, see the account in *El Siglo,* 28 April 1970, 5.

38. See the CMPC's advertisement in *El Siglo,* 1 January 1958, 4, and a description of both the CMPC and CORFO cellulose plant projects in Eduardo Astorga Schneider, "Factores que han afectado la tasa de reforestación en Chile," *Actas: Jornadas Forestales* 7 (1971): 35–39. See also Camus Gayán, *Ambiente, bosques y gestión forestal en Chile,* 232–33.

39. For a discussion of the CMPC's monopoly and its role in driving down prices for pine, and thus in contributing to a decline in the cultivation of plantations between 1958 and 1964, see Astorga Schneider, "Factores que han afectado la tasa de reforestación en Chile."

40. Camus Gayán also notes the importance of the CMPC's monopoly in *Medio ambiente, bosques y gestión forestal en Chile*, 239. Similar arguments are made in Sergio Salcedo, "Las plantaciones forestales en Chile," *Pensamiento iberoamericano*, no. 12, 1987, 155–66.

41. Comité Interamericano de Desarrollo Agrícola, *Chile: Tenencia de la tierra y desarrollo socio-económico del sector agrícola* (Santiago: Comité Interamericano de Desarrollo Agrícola, 1966), 170–71.

42. For two useful discussions of the agrarian reform see Kyle Steenland, *Agrarian Reform under Allende: Peasant Revolution in the South* (Albuquerque: University of New Mexico Press, 1977); and Heidi Tinsman, *Partners in Conflict: The Politics of Gender, Sexuality, and Labor in the Chilean Agrarian Reform, 1950–1973* (Durham, NC: Duke University Press, 2002), 174–75.

43. Camus Gayán attributes the exemption of properties engaged in forestry from agrarian reform to the lobbying of the forestry industry's trade association in *Ambiente, bosques y gestión forestal en Chile*, 224–25.

44. See Steenland, *Agrarian Reform under Allende*, 8–9; Jacques Chonchol, *Sistemas agrarios en América Latina* (Mexico City: Fondo de Cultura Económica, 1994), 291–93.

45. Corporación de la Reforma Agraria, *Reforma agraria chilena* (Santiago: 1970).

46. Trafún, record no. 1117, CORA Archive, SAG; Germán Kunstmann, "Explotación de los bosques autóctonos del fundo Trafún desde el punto de vista de sus propietarios," *Seminario Forestal: Boletín Informativo*, no. 10 (July 1965), 20–22.

47. Corporación de la Reforma Agraria, "Informe sobre renuncia a la indemnización," 19 July 1974, record no. 2208, Galletué, CORA Archive, SAG.

48. Sociedad Agrícola Fressard Hermanos to Vice-Presidente, CORA, 13 May 1974, record no. 4478, Fundo Porvenir o Lolén, CORA Archive, SAG.

49. Petition from Ralco company to CORA, record no. 1081, Fundo Pitrilín, CORA Archive, SAG.

50. See also Camus Gayán, *Ambiente, bosques y gestión forestal en Chile*, 224.

51. See the analysis of the Agrarian Reform Law and Mapuche communities in Martín Correa, Raúl Molina, and Nancy Yáñez, *La reforma agraria y las tierras mapuches: Chile, 1962–1975* (Santiago: LOM Ediciones, 2005), 103–35.

52. Correa, Molina, and Yáñez, *La reforma agraria*, 131.

53. Correa, Molina, and Yáñez, *La reforma agraria*, 133.

54. Rafael Elizalde Mac-Clure, *La sobrevivencia de Chile: La conservación de sus recursos naturales renovables* (Santiago: Ministerio de Agricultura, 1970), 120–21.

55. "Informa Comisión de Problemas Indígenas existentes en fundos Queco y Ralco, Provincia de Bío-Bío, Oficina de Tierras y Bienes Nacionales de Concepción," 13 January 1969, Carpeta Administrativa Comunidad Antonio Canío, TM 2874, CONADI Archive, ARA.

56. Raúl Molina and Martín Correa, *Territorio y comunidades pehuenches del Alto Bío-Bío* (Santiago: Corporación Nacional de Desarrollo Indígena, 1996), 53–56. The Ralco-Lepoy community was created in 1954 when the original Ralco community discussed in chapter 3 split in two. The other community formed out of this division was called Quepuca-Ralco. See Domingo Namuncura, *Ralco: ¿represa o pobreza?* (Santiago: LOM Ediciones, 1999), 258–261.

57. This was the case of the Francisco Cayul community in Lonquimay, which claimed veranadas where community members had collected piñones for a small fee in an agreement with the father of Bruno Ackermann, the presumptive owner of the mountain forests. By the 1960s, Ackermann had prohibited the community from collecting piñones, abrogating the long-standing agreement: see surveyors' reports in Juzgado de Indios, Victoria, record no. 390, CONADI Archive, ARA.

58. "Informa Comisión de Problemas Indígenas existentes en fundos Queco y Ralco." See also Molina and Correa, *Territorio y comunidades pehuenches del Alto Bío-Bío*, 73–119.

59. "Informa Comisión de Problemas Indígenas existentes en fundos Queco y Ralco."

60. "Informa Comisión de Problemas Indígenas existentes en fundos Queco y Ralco."

61. "Informa problemas existentes en el fundo Trapa-Trapa y Queco. Provincia de Bío Bío," Oficio no. 51, Concepción, 26 January 1970, Oficina de Tierras y Bienes Nacionales de Concepción, Carpeta Administrativa Comunidad Antonio Canío, TM 2874, CONADI Archive, ARA.

62. "Informa problemas existentes en el fundo Trapa-Trapa y Queco."

63. "Al H. Consejo de la CORA," Maderas Ralco, record no. 1081, Fundo Pitrilán, CORA Archive, SAG.

64. *El Sur,* 13 February 1972, 18. The initial plan in 1966 was that CORFO would provide 55 percent and Parsons and Whitmore would provide 45 percent of the initial capital: *El Diario Austral,* 1 December 1966, 1–5.

65. "Plan de celulosa constitución," 24 July 1969, vol. 8367; Corporación de Fomento de la Producción de Chile, report no. 67, 26 August 1971, Gerencia de Finanzas y Control, vol. 4845; Corporación de Fomento de la Producción de Chile, resolution no. 50, 9 January 1970, vol. 4845, CORFO Archive, ARNAD. See also Camus Gayán, *Ambiente, bosques y gestión forestal en Chile,* 237–38.

66. *El Diario Austral,* 27 January 1965, 11–13.

67. Grupo de Investigaciones Agrarias, *Región forestal: Empresas y trabajadores* (Santiago: Academia de Humanismo Cristiano, 1984), 37–40.

68. *El Diario Austral,* 2 August 1965, 7; 7 August 1965, 1; 9 August 1965, 3.

69. *El Diario Austral,* 17 May 1966, 3; 23 May 1966, 8; 22 June 1966, 3; 19 June 1967, 10–11.

70. Ministerio de Agricultura, "Plan nacional de reforestación, periodo 1966–70," doc. no. 68, 11 October 1965, Biblioteca Nacional (BN).

71. Ministerio de Agricultura, "Plan nacional de reforestación."

72. *El Diario Austral,* 22 June 1966, 3.

73. Alex Rudloff, interview by the author, Valdivia, June 2010.

74. Jaime Tohá, "Plan nacional de reforestación," *Actas: Jornadas Forestales* 2 (1966): 32.

75. For the reforestation campaign, see also Camus Gayán, *Ambiente, bosques y gestión forestal en Chile,* 239–41.

76. *El Diario Austral,* 14 August 1966, 7.

77. *El Diario Austral,* 15 July 1965, 10.

78. *El Diario Austral,* 28 July 1965, 6.

79. *El Diario Austral,* 3 December 1966, 11.

80. *El Diario Austral,* 24 September 1964, 3; 1 October 1964, 6; 6 June 1965, 9.

81. *Punto Final,* 1–15 July 1967.

82. *El Diario Austral,* 1 October 1965 6; 14 May 1967, 8.

83. Tohá, "Plan nacional de reforestación," 36.

84. José Ignacio Leyton Vásquez, "Tenencia forestal en Chile," United Nations Food and Agriculture Organization, 2009, 6.

85. *El Diario Austral,* 31 October 1969, 1. See also Molina et al., *La reforma agraria y las tierras mapuches,* 129.

86. Forestación Nacional, Memorandum, 29 May 1974, vol. 228, CORFO Archive, ARNAD.

87. Instituto Forestal, *Inventario forestal de los fundos adquiridos por la CORFO a la Sociedad Agrícola y Forestal Colcura S.A.* (Santiago: INFOR, 1968), copy available at the INFOR library in Santiago.

88. "Plan de Celulosa Constitución"; Corporación de Fomento de la Producción de Chile, informe no. 67: Corporación de Fomento de la Producción de Chile, resolution no. 50; Corporación de Fomento de la Producción de Chile, Gerencia de Finanzas y Control, 3 December 1970, and Gerente General, Agrícola y Forestal Lebu, 12 June 1970, vol. 4882, CORFO Archive, ARNAD.

89. See *El Siglo,* 10 June, 6, 12; June, 6, 13; June, 6, 1967. Roger A. Clapp notes that the government also purchased Colcura's saw mill to rescue it from bankruptcy: Roger A. Clapp, "Creating Comparative Advantage: Forest Policy as Industrial Policy in Chile," *Economic Geography* 71, no. 3 (July 1995): 280.

90. Corporación de Fomento de la Producción de Chile, "Antecedentes," 16 June 1970, vol. 4846; Corporación de Fomento de la Producción de Chile, resolution no. 50; Corporación de Fomento de la Producción de Chile, "Forestal Pilpilco, Antecedentes Generales," 13 August 1970, vol. 4846, CORFO Archive, ARNAD.

91. Corporación de Fomento de la Producción de Chile, Gerencia de Finanzas y Control (3 December 1970); Gerente General, Agrícola y Forestal Lebu, 12 June 1970, vol. 4882, CORFO Archive, ARNAD.

92. Astorga Schneider, "Factores que han afectado la tasa de reforestación en Chile."

93. Salcedo, "Las plantaciones forestales en Chile."

94. *El Siglo,* 24 April 1955, 6; 29 May 1955, 7; 5 June 1955, 7.

95. *El Siglo,* 4 January 1964, 9; 20 January 1964 9; 9 March 1964, 9; 21 March 1964, 5.

96. This account is taken from the interview with Filadelfo Guzmán in *El Pino*

Insigne, August 1983. *El Pino Insigne* was a pamphlet produced and distributed by the Vicaría Pastoral Obrera in Concepción as part of its efforts to lend support to unions in the region during the early 1980s. This was a project of Luis Otero, now of the Universidad Austral de Chile, who, although anonymous in *El Pino Insigne,* should receive credit for the extraordinary fieldwork interviews in the pamphlet. I consulted copies held by the Vicaría Pastoral Obrera in Concepción.

97. "Primer taller forestal," *Boletín de Estudios Agrarios,* no. 22, May 1988.

98. See Brian Loveman, *Struggle in the Countryside: Politics and Rural Labor in Chile, 1919–1973* (Bloomington: Indiana University Press, 1976).

99. *Periódico Forestal,* no.1, October 1992, 3; "Primer taller forestal." For a history of labor law and rural unionization, see Brian Loveman, *Struggle in the Countryside.*

100. *Ránquil,* no. 7, February 1970, 15.

101. *El Siglo,* 23 April 1970, 12.

102. Acuerdos con Trabajadores Fundo Peñuelas, Concepción, 17 April 1973; Acta de Avenamiento, Monte Águila, 26 May 1972; Acuerdo, Monte Águila, Sociedad Agrícola Lebu y Trabajadores de Monte Águila and Hacienda Peñuelas, 26 May 1972, vol. 315, CORFO Archive, ARNAD.

103. José Ignacio Leyton, "El fomento de la actividad forestal y su impacto sobre el desarrollo rural en Chile," in Pedro García Elizalde and José Ignacio Leyton, *El desarrollo frutícola y forestal en Chile y sus derivaciones sociales* (Santiago: Comisión Económica para América Latina, 1986).

7 Reform Arrives in the Forests

1. For histories of the UP period, see Peter Winn, *Weavers of Revolution: The Yarur Workers and Chile's Road to Socialism* (Oxford: Oxford University Press, 1986); Julio Pinto Vallejos, ed., *Cuando hicimos historia: La experiencia de la Unidad Popular* (Santiago: LOM Ediciones, 2005); Florencia E. Mallon, *Courage Tastes of Blood: The Mapuche Community of Nicolás Ailío and the Chilean State, 1906–2001* (Durham, NC: Duke University Press, 2005); Heidi Tinsman, *Partners in Conflict: The Politics of Gender, Sexuality, and Labor in the Chilean Agrarian Reform, 1950–1973* (Durham, NC: Duke University Press).

2. Winn, *Weavers of Revolution.*

3. "Discurso del Ministro de Agricultura Sr. Jacques Chonchol," *Actas: Jornadas Forestales* 6 (1970).

4. Hugo Bianchi, "Política forestal frente al problema de tenencia y uso de la tierra," *Actas: Jornadas Forestales* 6 (1970): 33–38.

5. *El Correo de Valdivia,* 1 December 1970, 7; 12 December 1970, 1, 6; 24 December 1970, 9. Interviews with Luis Rosales, former director of the Neltume workers' union, and Ida Sepúlveda, March, 2002, provided important background on the history of Panguipulli's forestry estates from the perspective of their workers. For background on Neltume, see also Víctor Espinoza Cuevas, Paz Rojas Baeza, María Luisa Ortíz Rojas, *Derechos humanos: Sus huellas en el tiempo. Una experiencia de trabajo en derechos humanos y salud mental en una zona rural del sur de Chile* (San-

tiago: Comité de Defensa de los Derechos del Pueblo, 1999); Comité de Defensa de los Derechos del Pueblo, *Chile: recuerdos de la guerra: Valdivia, Neltume, Chihio, Liquiñe* (Santiago: Comité de Defensa de los Derechos del Pueblo, 1991).

6. Luis Rosales, interview by the author, Panguipulli, March, 2002. *Poder Campesino,* 15–30 January 1971, 6.

7. *Poder Campesino,* 15–30 January 1971, 6; *El Diario Austral,* 31 January 1971.

8. *Punto Final,* 8 March 1971.

9. *Punto Final,* 8 March 1971, 6.

10. *Punto Final,* 8 March 1971, 6.

11. *El Diario Austral,* 31 January 1971, 1, 6.

12. *El Correo de Valdivia,* 1 December 1970, 1, 6, 7.

13. *El Correo de Valdivia,* 1 December 1970.

14. *El Diario Austral,* 31 January, 1971, 1, 6.

15. *El Diario Austral,* 9 May 1966, 10.

16. Moises Durán, interview by the author, Panguipulli, June 2012.

17. *El Diario Austral,* 1 December, 1, 3 December 1970, 8.

18. *El Siglo,* 24 February 1971, 8.

19. *El Sur,* 29 March 1971, 7, 22.

20. *El Sur,* 24 February 1971, 12; *El Siglo,* 24 February 1971, 8, 12.

21. *El Sur,* 21 March 1971, 3.

22. *El Diario Austral,* 18 March 1971, 1; *El Correo de Valdivia,* 18 March 1971, 5; 20 March 1971, 18; 5 July 1971, 7.

23. *El Correo de Valdivia,* 20 March 1971, 18, 30; July 1971, 7; *El Diario Austral,* 5 December 1971, 11.

24. *El Diario Austral,* 12 January 1971, 1; 18 January 1971, 1; 3 March 1971, 9; 9 October, 9, 1971.

25. *El Sur,* 17 February 1971.

26. *El Correo de Valdivia,* 20 March 1971, 18.

27. Corporación de Fomento de la Producción de Chile, "Mensura forestal de la provincia de Valdivia," 1952, BN.

28. E. I. Kotok to F. T. Wahlen, ETAP, "Interim Report on Economic Aspects in Forestry," FAO, 30 October 1951, cited in Corporación de Fomento de la Producción de Chile, "Mensura forestal de la provincia de Valdivia," 1952, BN.

29. Corporación de Fomento de la Producción de Chile, "Mensura forestal de la provincia de Valdivia."

30. Trafún, record no. 1117, and Pirihueico, record no. 1123, CORA Archive, SAG.

31. Germán Kunstmann, "Explotación de los bosques autóctonos del fundo Trafún desde el punto de vista de sus propietarios," *Seminario Forestal,* no. 10 (July 1965), 19–22. See also Alex Rudloff, interview by the author, Valdivia, March, 2002; Rubén Peñaloza Wagenknecht, Federico Schlegel Sachs, and Claudio Donoso Zegers, "Importancia del recurso forestal del Complejo Forestal y Maderero Panguipulli," report, Facultad de Ciencias Forestales, Universidad Austral, Valdivia, n.d., 4.

32. Kunstmann, "Explotación de los bosques autóctonos del fundo Trafún desde el punto de vista de sus propietarios."

33. "Interesante experiencia de política forestal ofrece la Caja de EE.PP. y PP," *Panorama Económico,* no. 81, 28 August 1953, 565–66.

34. Friedrich Reinhold, "La transformación del bosque valdiviano vírgen a bosques económicos," *Chile Maderero,* no. 25, 1968, 14–15.

35. "Plan de desarrollo para el area del Complejo de Panguipulli," research project, Facultad de Ingenieria Forestal, Universidad Austral de Chile, Valdivia, n.d.

36. Gobierno Interior, Departamento de Panguipulli, *Memoria Anual, 1970– 1971,* Gobernación Panguipulli, copy in my possession.

37. "Plan de desarrollo para el area del Complejo de Panguipulli."

38. Alex Rudloff, interview by author, Valdivia, March 2002.

39. Rudloff, March, 2002 interview.

40. For a discussion of reforestation programs in the Complejo Maderero, see *El Correo de Valdivia,* 29 June 1971, 8. Interestingly, a program to reforest 6,000 hectares in 1971 was funded with a credit from the German government. That year, the Complejo Maderero projected reforestation of 11,000 hectares.

41. *El Correo de Valdivia,* 18 October 1971, 7.

42. *El Correo de Valdivia,* 18 October 1971.

43. *El Correo de Valdivia,* 7 August 1971.

44. *El Correo de Valdivia,* 5, 6 October 1971.

45. *El Correo de Valdivia,* 18 October 1971, 7; 2 November, 1971, 8.

46. *El Correo de Valdivia,* 24 December 1970, 9.

47. *El Diario Austral,* 14 May 1971, 1.

48. *El Diario Austral,* 14 July 1972, 8.

49. Rudloff, March, 2002, interview.

50. Pedro Cardyn and Humberto Manquel Maillañguer, interviews by the author, Panguipulli, June 2012.

51. Corporación de Fomento de la Producción de Chile, "Informe sobre algunos predios propiedad de la CORFO administrados por el Complejo Forestal Maderero Panguipulli," 24 August 1979, volume 281, CORFO Archive, ARNAD.

52. Gobierno Interior, Departamento de Panguipulli, *Memoria Anual, 1970– 1971.* I thank Gonzalo Toledo Martel for providing me with a copy of this report. Also see his excellent thesis, "La muerte y sus interpretaciones: Represión política en el Complejo Forestal y Maderero Panguipulli Ltda., 1973," Universidad Austral de Chile, Valdivia, 1994.

53. *El Diario Austral,* 3 February 1971, 7.

54. See Mallon, *Courage Tastes of Blood;* Kyle Steenland, *Agrarian Reform under Allende: Peasant Revolt in the South* (Albuquerque: University of New Mexico Press, 1977).

55. Martín Correa, Raúl Molina, and Nancy Yáñez, *La reforma agraria y las tierras mapuches: Chile, 1962–1975* (Santiago: LOM Ediciones, 2005), 208, 220.

56. *El Diario Austral,* 16 June 1969, 7.

57. *El Diario Austral,* 5 December 1970, 13.

58. Raúl Molina and Martín Correa, *Territorio y comunidades pehuenches del Alto Bío-Bío.* (Santiago: Corporación Nacional de Desarrollo Indígena, 1996), 50–71.

59. *El Diario Austral,* 5 December 1970, 13.

60. *El Sur,* 2 December 1970, 1, 14.

61. *El Sur,* 11 December 1970, 10.

62. *El Siglo,* 29 January 1971, 4.

63. Molina and Correa, *Territorio y comunidades pehuenches del Alto Bío-Bío,* 118; *El Sur,* 4 February 1971, 1, 12; *El Siglo,* 29 January 1971, 4.

64. Molina and Correa, *Territorio y comunidades pehuenches del Alto Bío-Bío.*

65. José Bengoa, *Quinquén: Cien años de historia pehuenche* (Santiago: Ediciones Chile América-cesoc, 1992).

66. See *El Diario Austral,* 19 May 1991, A16, as well as Bengoa, *Quinquén,* for an account of this history.

67. Hugo Ormeño Melet to Juzgado de Letras de Temuco, 14 August 1970, Comunidad Pedro Calfuqueo, tm no. 998, Carpetas Administrativas conadi Archive, ara.

68. Letter from Surveyor to Abogado Defensor de Indígenas, 5 July 1971; Ministerio de Tierras y Colonización, Dirección de Asuntos Indígenas, Oficina Regional de Victoria, 17 June 1971, Comunidad Pedro Calfuqueo, tm no. 998, Carpetas Administrivas conadi Archive, ara.

69. Corporación de Reforma Agraria, "Informe sobre renuncia a la indemnización," 1974, record 2208, Galletué, cora Archive, sag.

70. Corporación de Reforma Agraria, "Informe sobre expropiación, Fundo Galletué" and Corporación Nacional Forestal, "Informe de Parque Nacional Galletué," record 2208, Galletué, cora Archive, sag. See also Bengoa, *Quinquén.*

71. Corporación Nacional Forestal, "Informe de Parque Nacional Galletué."

72. *El Siglo,* 10 August 1972, 6; *El Diario Austral,* 12 August 1972, 1–5.

73. Letter from Fressard Hermanos to Alberto Araneda, Vice-President, cora, 13 May 1974, record no. 4478, Fundo Porvenir o Lolén, cora Archive, sag.

74. Ministerio de Tierras y Colonización, Dirección de Asuntos Indígenas, Comisión de Restitución, Angol, July 1972; "Informa sobre conflicto reducción Bernardo Ñanco y Colonos" to Jose Montecinos, Jefe de Gabinete, Ministerio del Interior, 3 October 1972; Departamento Asuntos Indígenas, Departamento Técnico, 21 June 1991, Comunidad Bernardo Ñanco, tm 1075, Carpetas Administrativas, conadi Archive, ara.

75. Ministerio de Tierras y Colonización, Dirección de Asuntos Indígenas, Comisión de Restitución, Angol (July 1972).

76. Correa et al., *La reforma agraria y las tierras mapuches,* 180.

77. Fressard Hermanos, Fundo Porvenir o Lolén, record no. 4478, cora Archive. sag. The expropriation was revoked in 1974.

78. Sergio Salcedo, "Las plantaciones forestales en Chile," *Pensamiento iberoamericano,* no. 12, 1987, 155–166.

79. *El Siglo,* 14 March 1971, 10–13.

80. *El Sur,* 24 February 1971, 1.

81. *El Sur,* 27 June 1971, 12.

82. *El Siglo,* 9 October 1971, 3; 21 March 1972, 6.

83. *El Sur,* 21 July 1971, 12.

84. *El Sur,* 30 September 1971, 12.

85. *El Sur,* 12 February 1972, 8.

86. *El Siglo,* 11 September 1972, 4.

87. *El Siglo,* 14 February 1972, 3.

88. "Inventario de las plantaciones forestales y fundos Llohué," vol. 1671, CORFO Archive, ARNAD.

89. Luis Otero Durán, *El problema social detrás de los bosques* (Concepción: Vicaría de la Pastoral Obrera, 1984), 16.

90. *El Sur,* 13 November 1971.

91. The indemnification was for 20 percent of the property's value; the value of the improvements was paid in cash, and the rest was paid in agrarian reform bonds. Sociedad Forestadora Nacional, Ñuble, El Guanaco, record no. 4458, CORA Archive, SAG. Another estate forested with pine and also expropriated in 1972 was the Granja Cosmito in Concepción: see Parte Granja Cosmito, record no. 2034, CORA Archive, SAG.

92. *El Diario Austral,* 11 April 1972, 11.

93. This is based on a review of the CORA Archive, SAG.

94. Comité Interamericano de Desarrollo Agrícola, *Chile: Tenencia de la tierra y desarrollo socio-económico del sector agrícola* (Santiago: Comité Interamericano de Desarrollo Agrícola, 1966), 82.

95. *El Diario Austral,* 14 January 1971, 6.

96. *El Diario Austral,* 8 January 1971, 1, 7.

97. Comité Interamericano de Desarrollo Agrícola, *Chile,* 84, 167.

98. *El Diario Austral,* 4 September 1965, 7.

99. *El Sur,* 7 January 1971, 3.

100. Comité Interamericano de Desarrollo Agrícola, *Chile,* 81–82.

101. Comité Interamericano de Desarrollo Agrícola, *Chile,* 313.

102. *El Siglo,* 24 December 1970, 12; *Poder Campesino,* 15–30 January 1971, 12.

103. Intendencia de Cautin, vol. 470, 1971, ARA.

104. *El Siglo,* 23 May 1971, 2.

105. Correa et al., *La reforma agraria y las tierras mapuches,* 144.

106. *Poder Campesino,* 15–30 January 1971, 12.

107. *El Diario Austral,* 15 November 1971, 11; *El Siglo,* 24 June 1971, 6.

108. *El Diario Austral,* 16 November 1971, 8; 15 December 1971, 11.

109. *El Diario Austral,* 14 December 1971, 9.

110. *El Diario Austral,* 12 February 1971, 9.

111. *El Siglo,* 24 December 1970, 12.

112. *El Siglo,* 30 January 1960, 11.

113. *El Siglo,* 24 August 1960, 4.

114. In 1960, for example, campesinos in Melipeuco near Temuco proposed that the government implement an extensive program for exploiting native forests in public reserves and then reforesting. The state would employ five hundred workers in this program and provide them modern saw mills. *El Siglo,* 24 August 1960, 4.

115. *El Correo de Valdivia,* 4 July 1972, 9.

116. *El Correo de Valdivia,* 4 December 1972, 7.

8 Free-Market Forestry

1. "Un millón de hectáreas forestales: ¿Quién sale ganando?," *Noticiero de la Realidad Agraria,* no. 15, January 1984, 2.

2. Instituto Forestal, "Exportaciones chilenas por sectores de la economía, 1960–1997," statistical bulletin no. 61, *Estadísticas Forestales 1997* (Santiago: INFOR, 1978), 73, INFOR Library, Santiago.

3. Grupo de Investigaciones Agrarias, "Chile: Un proyecto de país forestal," Santiago, 1980, 20.

4. Grupo de Investigaciones Agrarias, "Chile," 17.

5. Markos J. Mamalakis, *Historical Statistics of Chile: Forestry and Related Activities,* vol. 3 (Westport, CT: Greenwood, 1982), 261.

6. Joaquin Lavín, *Chile: Revolución silenciosa* (Santiago: Zig-Zag, 1987).

7. See Mamalakis, *Historical Statistics of Chile,* 262, 267.

8. In Chile during the late 1970s, foresting a hectare with pine cost $90, with 85 percent of this cost coming from labor. In the United States and Europe, the cost was more than $1,000: Grupo de Investigaciones Agrarias, "Chile," 22. See Roger A. Clapp, "Creating Comparative Advantage: Forestry Policy as Industrial Policy," *Economic Geography* 71, no. 3 (July 1995): 273–296. The social and ecological impact of forestry growth after 1973 are also discussed in Pablo Camus Gayán, *Ambiente, bosques y gestión forestal en Chile, 1541–2005* (Santiago: LOM Ediciones and Dirección de Bibliotecas, Archivos y Museos, 2006), 276–288; Pedro García Elizalde and José Ignacio Leyton, *El desarrollo frutícola y forestal y sus derivaciones sociales* (Santiago: Comisión Económica para América Latina, 1986).

9. Marlene Gimpel Madariaga, "El sector forestal ante la apertura económica: Exportaciones y medio ambiente," in *El tigre sin selva: Consecuencias ambientales de la transformación económica de Chile,* ed. Rayén Quiroga Martínez, Saar van Hauwermeiren, Jorge Berghammer Vega, Patricio Del Real Jaramillo, and Marlene Gimpel Madariaga (Santiago: Instituto de Ecología Política, 1994), 310.

10. Jorge Morales Gamboni, "El estado y el sector privado en la industria forestal: El caso de la región de Concepción," *Boletín de Estudios Agrarios,* no. 24, August 1989, 21–22.

11. Gamboni, "El estado y el sector privado en la industria forestal." See also Camus Gayán, *Ambiente, bosques y gestión forestal en Chile,* 264–69.

12. Francisco Reusch, "La política forestal del gobierno y la concentración económica en el sector forestal," *Boletín de Estudios Agrarios,* 7 March 1981, 35–36. For a description of DL 701, see also Camus Gayán, *Ambiente, bosques y gestión forestal en Chile,* 250–53.

13. Grupo de Investigaciones Agrarias, "Chile," 21.

14. Hugo Fazio R., *Mapa actual de la extrema riqueza en Chile* (Santiago: LOM Ediciones and Universidad de Arte y Ciencias Sociales, 1997), 173.

15. "Un millón de hectáreas forestales," 3.

16. Grupo de Investigaciones Agrarias, *Región forestal: Empresas y trabajadores* (Santiago: Academia de Humanismo Cristiano, n.d.), 19.

17. Morales Gamboni, "El estado y el sector privado en la industria forestal."

18. Patricio Escobar S. and Diego López F., *El sector forestal en Chile: Crecimiento y precarización del empleo* (Santiago: Programa de Economía del Trabajo, 1996), 80.

19. Instituto Forestal, "Monto de las exportaciones principales productos forestales, 1962–1997," statistical bulletin no. 61, *Estadísticas Forestales 1997*, 76, INFOR Library, Santiago; Escobar S. and López F., *El sector forestal en Chile*, 82.

20. Cruzat-Larraín controlled Forestal Arauco, Celulosa Arauco, Celulosa Constitutción, Forestal CELCO, and Forestal Chile. Matte-Alessandri owned CMPC, with plants in Puente Alto, Laja, Bío Bío, and Valdivia; Laja Crown; Forestal Mininco, with 150,000 hectares of plantations; Fábrica Chilena de Moldeados; Aserradero San Pedro; and Sociedad Recuperadora de Papel. The Vial group, through the Banco Hipotecario y de Fomento de Chile, owned INFORSA, with its pulp plant and two paper factories, the Forestales Rio Vergara, CRECEX, Georgia Pacific CRECEX, Maderas Nacimineto, Papelero Sud America, and Maderas y Paneles: Grupo de Investigaciones Agrarias, *Región forestal*, 38–45.

21. Luis Otero Durán, *El problema social detrás de los bosques* (Concepción: Vicaría de la Pastoral Obrera, 1984).

22. Union leader, Federación Forestal Liberación, interview by the author, Concepción, April 2000.

23. *El Pino Insigne*, October 1984.

24. Two very important histories of the events in Panguipulli preceding the military coup and of the violation of human rights, including detentions and disappearances, after the coup are Comité de Defensa de los Derechos del Pueblo, *Chile: Recuerdos de la guerra: Valdivia, Neltume, Chihuio, Liquiñe* (Santiago: Comité de Defensa de los Derechos del Pueblo, 1991), and Víctor Espinoza Cuevas, Paz Rojas Baeza, María Luisa Ortíz Rojas, *Derechos humanos: Sus huellas en el tiempo: Una experiencia de trabajo en derechos humanos y salud mental en una zona rural del sur de Chile* (Santiago: Comité de Defensa de los Derechos del Pueblo, 1999).

25. Comité de Defensa de los Derechos del Pueblo, *Chile*, 143.

26. Comité de Defensa de los Derechos del Pueblo, *Chile*, 3.

27. Ministerio del Interior, Intendencia Región Puerto Montt, 28 May 1981, vol. 281, CORFO Archive, ARNAD; Luis Rosales and Pedro Cardyn, interviews by the author, Panguipulli, March, 2002.

28. "Modificación cesión de crédito, Complejo Forestal y Maderero Panguipulli Limitada, Sociedad Reforestadora Collico Limitada, Corporación de la Reforma Agraria"; Corporación de la Reforma Agraria, Secretaria de Consejo, Santiago, 29 April 1977, Trafún, CORA Archive, SAG.

29. Alex Rudloff, interviews by the author, Valdivia, March, 2002 and June, 2010.

30. Ministerio del Interior, Intendencia Región Puerto Montt (28 May 1981). Complejo Forestal y Maderero Panguipulli, "Informe sobre acuerdo del directorio al traspaso al sector privado," 15 June 1981, vol. 281, CORFO Archive, ARNAD.

31. Corporación de Fomento de la Producción de Chile, Gerencia de Normalización, Subsecretaria de Comercialización, 17 June 1980; Dirección Nacional de Fronteras y Límites del Estado, 12 February 1980, vol. 1856, CORFO Archive, ARNAD.

32. Corporación de Fomento de la Producción de Chile, "Fundo Pilmaiquén," 26 August 1987; "Contrato venta y explotación Fundo Huilo-Huilo," 1 March 1989, vol. 91, CORFO Archive, ARNAD.

33. Rubén Peñaloza Wagenknecht, Federico Schlegel Sachs, Claudio Donoso Zegers, "Importancia del recurso forestal del Complejo Forestal y Maderero Panguipulli," Universidad Austral, Valdivia, 1985, 11, 13.

34. "Importancia del recurso forestal del Complejo Forestal y Maderero Panguipulli."

35. Foresters, Federación Nacional de Sindicatos de CONAF (FENASIC), interviews by the author, Santiago, April 2000.

36. Instituto Forestal, "La pequeña empresa madera de bosque nativo: Su importancia, perspectiva y una propuesta para su desarrollo," technical report no. 128, 1991, INFOR Library, Santiago.

37. Corporación de Fomento de la Producción de Chile, "Transacción juicios fundos Neltume y Carranco," 1 October 1987, vol. 1847, CORFO Archive, ARNAD.

38. Corporación de Fomento de la Producción de Chile, "Fundo Pilmaiquén," 26 August 1987, vol. 91, CORFO Archive, ARNAD.

39. "Contrato venta y explotación Fundo Huilo-Huilo," 1 March 1989, vol. 91, CORFO Archive, ARNAD.

40. Record no. 2183, Galletué, CORA Archive, SAG.

41. José Bengoa, *Quinquén: Cien años de historia pehuenche* (Santiago: Ediciones Chile América-CESOC, 1992), 71–84.

42. "Se opone a un contrato de explotación de maderas en reducción Pedregoso de la Comunidad Vicente Guaiquillán de Lonquimay," Comunidad Paulino Huaiquillan, TM no. 1056, Carpetas Administrativas, CONADI Archive, ARA.

43. "Bases para un contrato de explotación Maderera," Comunidad Paulino Huaiquillan, TM no. 1056, Carpetas Administrativas, CONADI Archive, ARA.

44. Ministerio de Agricultura, "Informa una solicitud sobre forma de hacer un pago, Temuco," 9 May 1975, Comunidad Paulino Huaiquillan, TM no. 1056, Carpetas Administrativas, CONADI Archive, ARA.

45. Ministerio de Agricultura, "Solicitan autorización que indica, Lonquimay," 10 March 1976, Comunidad Paulino Huaiquillan, TM no. 1056, Carpetas Administrativas, CONADI Archive, ARA.

46. "Informe reserva Forestal Madera Araucaria–Comunidad Bernardo Ñanco, Victoria," 2 January 1975, Comunidad Bernardo Ñanco, TM no. 1075, Carpetas Administrativas, CONADI Archive, ARA.

47. Letter to Raúl Arias, Abogado de INDAP, Temuco, Lonquimay, 4 December 1979, Comunidad Bernardo Ñanco, TM no. 1075, Carpetas Administrativas, CONADI Archive, ARA.

48. Segundo Carilao and Don Aladino, Comunidad Bernardo Ñanco, interviews by the author, Lonquimay, August 2010.

49. Foresters, FENASIC, interviews.

50. Sergio Gómez, "Forestación y campesinado," doc. no. 19, Instituto de Desarrollo Agropecuario, Santiago, January 1994.

51. Foresters, FENASIC, interviews.

52. Union leader, Confederación de Trabajadores Forestales (CTF) interview by the author, Concepción, April, 2000.

53. Foresters, FENASIC, interviews.

54. Agronomist, Vicaría Pastoral Obrera, interview by the author, Concepción, January 2009.

55. Rudloff interview, June, 2010.

56. Intendencia de Cautín, 1974–75, vol. 589, ARA.

57. Fundo "El Volcán," record no. 990, CORA Archive, SAG.

58. Rudloff interview, June, 2010.

59. Corporación de Reforma Agraria, "Encuesta," 1972, Fundo Villacura, record no. 4477, CORA Archive, SAG.

60. Bartolomé Medina Reyes, Agrónomo, Servicio Agrícola y Ganadero, Los Angeles, March 1974, record no. 4477, CORA Archive, SAG.

61. *El Pino Insigne*, n.d., no. 2.

62. *El Pino Insigne*, n.d., no. 2

63. *El Pino Insigne*, August 1983.

64. *El Pino Insigne*, n.d., no. 11.

65. *El Pino Insigne*, August 1983.

66. Union leader, Federación Forestal Liberación, interview.

67. García Elizalde and Leyton, *El desarrollo frutícola y forestal en Chile y sus derivaciones sociales*, 184.

68. *El Pino Insigne*, December 1983.

69. *El Mercurio*, 17 July 1983, quoted in *El Pino Insigne*, December 1983.

70. El Pino Insigne, December 1983.

71. For a discussion of the policy of replacing native forests with plantations, see Gimpel Madariaga, "El sector forestal ante la apertura económica."

72. Guillermo Castilleja, "Changing Trends in Forest Policy in Latin America: Chile, Nicaragua and Mexico," *Unasylva*, vol. 44, no. 175, 1993–94.

73. Rayén Quiroga Martínez and Saar van Hauwermeiren, *Globalización e insustentabilidad: Una mirada desde la economía ecológica* (Santiago: Instituto de Ecología Política, 1996), 67; Antonio Lara and Thomas T. Veblen, "Forest Plantations in Chile: A Successful Model?" in *Afforestation: Policies, Planning, and Progress*, ed. Alexander S. Mather (London: Belhaven, 1993), 129.

74. Quoted in "Mujeres forestales: Una vida sufrida," *Noticiero de la Realidad Agraria*, no. 19, June 1981, 22. See also Rafael Ros Vera, "Plantaciones: Sus efectos sobre el ambiente," *Actas: Jornadas Forestales* 14 (1992), 27.

75. Quiroga Martínez and van Hauwermeiven, *Globalización e insustentabilidad*, 69–70.

76. García Elizalde and Leyton, *El desarrollo frutícola y forestal en Chile y sus derivaciones sociales*, 184.

77. Rigoberta Rivera A. and María Elena Cruz, *Pobladores rurales* (Santiago: Grupo de Investigaciones Agrarias and Academia de Humanismo Cristiano, 1984), 131–38.

78. Union leader, Federación Forestal Liberación, interview.

79. Union leader, Confederación de Trabajadores Forestales, interview by the author, Concepción, April 2000.

80. Felipe Pozo, "Las penas de la Octava Región," *Análisis*, January 1983, 19–21.

81. *Páginas Sindicales,* no. 55, 15 May 1983.

82. "Madereros con la soga al cuello," *Noticiero de la Realidad Agraria,* no. 3, January 1983, 3.

83. Union leader, Confederación de Trabajadores Forestales, interview.

84. Instituto Forestal, "Volumen exportaciones principales productos forestales, 1962–1997," statistical bulletin no. 61, *Estadísticas Forestales 1997,* 77.

85. "Madereros con la soga al cuello" *Noticiero de la Realidad Agraria,* no. 3, January 1983, 3.

86. García Elizalde and Leyton, *El desarrollo frutícola y forestal en Chile y sus derivaciones sociales,* 213.

87. Peter Winn, "The Pinochet Era," in *Victims of the Chilean Miracle: Workers and Neoliberalism in the Pinochet Era, 1973–2002,* ed. Peter Winn (Durham, NC: Duke University Press, 2004), 32–36.

88. Otero Durán, *El problema social detrás de los bosques,* 16.

89. Otero Durán, *El problema social detrás de los bosques,* 16.

90. Luis Otero Durán, "Los trabajadores y el sistema de contratistas en el sector forestal," *Boletín de Estudios Agrarios,* no. 7, March 1981, 52–88.

91. *El Pino Insigne,* no. 11, 1983. See also *Páginas Sindicales,* no. 43, 9 December 1981.

92. *Hoy,* no. 165, 17–23 September 1980, 35–36; Otero Durán, *El problema social detrás los bosques;* "Primer Taller Forestal," *Boletín de Estudios Agrarios,* no. 22, May 1988, 10–12.

93. Otero, *El problema social detrás los bosques,* 24.

94. "Madereros con la soga al cuello," *Noticiero de la Realidad Agraria,* no. 3, January 1983, 3.

95. "¿Qué hay tras esas nuevas aldeas rurales?" *Noticiero de la Realidad Agraria,* no. 4, February 1982, 17.

96. Union leader, Federación Forestal Liberación, interview.

97. *El Pino Insigne,* October 1984.

98. *Páginas Sindicales,* no. 55, 15 May 1983.

99. Caupolicán Pávez, interview by the author, Concepción, April 2000.

100. *El Pino Insigne,* n.d.

101. Confederación de Trabajadores Forestales, Grupo de Estudios Agro-Regionales, Centro para el Desarrollo Forestal, "Propuestas forestales y laborales para la democracia," October 1989, Concepción; "Informe central al quinto Congreso Nacional Eleccionario de la Confederación Nacional de Trabajadores Forestales de Chile," Constitución, 5–6 June 1998, Confederación de Trajabadores Forestales, Concepción.

102. Caupolicán Pávez interview.

103. Confederación de Trabajadores Forestales, Grupo de Estudios Agro-Regionales, Centro para el Desarrollo Forestal, "Propuestas forestales y laborales para la democracia."

104. Caupolicán Pávez interview.

105. Confederación de Trabajadores Forestales, Grupo de Estudios Agro-Regionales, Centro para el Desarrollo Forestal, "Propuestas forestales y laborales para la democracia."

9 The Mapuche Challenge

1. *The New York Times,* 12 December 2004; *La Tercera,* 3 April 1999 Mónica Olivia Monckeberg, *El Saqueo de los grupos economícos al estádo Chileno* (Santiago: Ediciones B Grupo Zeta, 2001).

2. *Chile Forestal,* no. 215, 1994, quoted in Roger A. Clapp, "Waiting for the Forest Law: Resource-Led Development and Environmental Politics in Chile," *Latin American Research Review* 33, no. 2 (1998), 19.

3. I interviewed a number of forestry workers and residents of Neltume with the journalist Mauricio Durán in March 2002. I do not include their names here to protect their anonymity. See also see an investigative report by Arnaldo Pérez Guerra, "Neltume; los esclavos del bosque," *Punto Final,* 1 April 1999.

4. Interviews with forestry workers by the author, Neltume, March 2002; Luis Rosales, interview by the author, Panguipulli, March, 2002.

5. Instituto Forestal, "Exportaciones chilenas por sectores de la economía," and "Volumen exportaciones principales productos forestales, 1962–1997," statistical bulletin no. 61, *Estadísticas Forestales, 1997* (Santiago: INFOR, 1978), 73, INFOR Library, Santiago.

6. Patricio Escobar S. and Diego López F., *El sector forestal en chile: Crecimiento y precarización del empleo* (Santiago: Programa de Economía del Trabajo, 1996), 56.

7. *Punto Final,* January 1997; Stephanie Rosenfeld and Juan Luis Marre, "How Chile's Rich got Richer," NACLA *Report on the Americas,* May–June 1997, 20–26; Diane Haughney, *Neoliberal Economics, Democratic Transition, and Mapuche Demands for Rights in Chile* (Gainesville: University Press of Florida, 2006), 163; Pablo Camus Gayán, *Ambiente, bosques y gestión forestal en Chile, 1541–2005* (Santiago: LOM Ediciones and Dirección de Bibliotecas, Archivos y Museos, 2006), 270.

8. Ricardo Carrere and Larry Lohmann, *El papel del Sur: Plantaciones forestales en la estrategia papelera internacional* (Montevideo: World Rainforest Movement, 1997), 156–57. See also Haughney, *Neoliberal Economics, Democratic Transition, and Mapuche Demands for Rights in Chile,* 163–64; Camus Gayán, *Ambiente, bosques, y gestión forestal en Chile,* 272–74.

9. *El Diario Austral,* 23 July 1997; Carrere and Lohmann, *El papel del Sur,* 158; Camus Gayán, *Ambiente, bosques, y gestión forestal en Chile,* 274.

10. *El Diario Austral,* 12 February, 14 February 1992; Carrere and Lohmann, *El papel del Sur.*

11. Carrere and Lohmann, *El papel del Sur,* 159.

12. Carrere and Lohmann, *El papel del Sur,* 156.

13. Instituto Forestal, "Volumen exportaciones principales productos forestales."

14. *Análisis,* no. 20, July 1992.

15. Marcel Claude, *Una vez más la miseria: Es Chile un país sustentable?* (Santiago: LOM Ediciones, 1997), 109–12; Marlene Gimpel Madariaga, "El sector forestal ante la apertura económica: Exportaciones y medio ambiente," in *El tigre sin selva: Consecuencias ambientales de la transformación económica de Chile,* ed. Rayén Quiroga Martínez, Saar van Hauwermeiren, Jorge Berghammer Vega, Patricio Del Real Jaramillo, and Marlene Gimpel Madariaga (Santiago: Instituto de Ecología Política, 1994), 313.

16. Guillermo Castillejo, "Changing Trends in Forests Policy in Latin America: Chile, Nicaragua and Mexico," *Unasylva*, vol. 44, no. 175, 1993–94.

17. For Marcel Claude's orginal study and calculations of forest conversión see Claude, *Una vez más la miseria;* "Las miserias del desarrollo chileno (una mirada desde la sustentabilidad" in *El modelo chileno: Democracia y desarrollo en los noventa,* ed. Paul Drake and Iván Jaksic (Santiago: LOM Ediciones, 1999), 155–66; and Banco Central de Chile, "Proyecto de cuentas ambientales y bosque nativo," Banco Central de Chile, Santiago, 1995. For the revised and scaled down figures by the Banco Central see "Estudio cuentas ambientales: Metodología de medición de recursos forestales en unidades físicas, 1985–1996," report, Banco Central de Chile, Santiago, 2001, BCN. See also *The Economist,* 3 February 1996; Carola Fernández, *La conveniente oscuridad del bosque Chileno* (Santiago: Fundacion Terram, 2003), 3–4, 6. For a study that reaffirms Claude's orginal conclusions see Consuelo Espinosa Proaño, "El bosque nativo de Chile: Situación actual y proyecciones" (Santiago: Fundación Terram, 2002).

18. Fernández, *La conveniente oscuridad del Bosque Chileno,* 14–15. Antonio Lara and Cristian Echeverria, *Certificacion forestal: Una necesidad para la conservacion de los bosques en Chile* (Valdivia: Universidad Austral de Chile, n.d.). For a discussion of the wood chip industry and central bank report, see Camus Gayán, *Ambiente, bosques y gestión forestal en Chile,* 290–93, 320–21.

19. "Alarm in Llanquihue: Depredation of the Forests," *El Diario Austral,* 12 March 1990, 6–9.

20. *El Diario Austral,* 11 March 1991, 5.

21. *El Diario Austral,* 30 November 1992, B13. For a discussion of the debate around the wood chip industry and the native forests see Clapp, "Waiting for the Forest Law."

22. Eduardo Silva, "Democracy, Market Economics, and Environmental Policy in Chile," *Journal of Interamerican Studies and World Affairs* 38, no. 4 (Winter 1996): 13–14; Eduardo Silva, "People, Forests, and Politics in Costa Rica and Chile: The Struggle for Grassroots-Development Friendly Initiatives," paper presented at the 20th International Congress, Latin American Studies Association, Guadalajara, 17–19 April 1997; Guillermo Castillejo, "Changing Trends in Forests Policy in Latin America: Chile, Nicaragua and Mexico," *Unasylva,* vol. 44, no. 175, 1993–94; Clapp, "Waiting for the Forest Law."

23. For analyses of the "authoritarian enclaves" during the transition, see Paul Drake and Iván Jaksic, "El 'modelo' chileno: Democracia y desarrollo en los noventa" 11–38; Andrés Allamand, "Las paradojas de un legado" 169–190; and Peter M. Siavelis, "Continuidad y transformación del sistema de partidos en una transición 'modelo,'" 223–256, in *El modelo chileno: Democracia y desarrollo en los noventa,* ed. Paul Drake and Iván Jaksic.

24. Fernández, *La conveniente oscuridad del Bosque Chileno,* 14–15. For a discussion of conflicts between environmentalist organizations and the forestry industry during the 1990s, see Camus Gayán, *Ambiente, bosques y gestión forestal en Chile,* 311–31. See also Silva, "People, Forests, and Politics in Costa Rica and Chile."

25. Silva, "People, Forests, and Politics in Costa Rica and Chile," 14–15.

26. *El Diario Austral,* 24 March 1997, A8, 6.

27. *El Diario Austral,* 30 June 1997, 10–11. See also *El Diario Austral,* 2 July 1992, A5 for a description of CONAF's Program for Forestation on Peasant and Indigenous Land.

28. Peter Winn, "The Pinochet Era," in *Victims of the Chilean Miracle: Workers and Neoliberalism in the Pinochet Era, 1973–2002,* ed. Peter Winn (Durham, NC: Duke University Press, 2004), 32–36.

29. Volker Frank, "Politics without Policy: The Failure of Social Concertation in Democratic Chile, 1990–2000" in Winn, *Victims of the Chilean Miracle,* 73–74.

30. Union leader, Confederación de Trabajadores Forestales, interview by the author, Concepción, April 2000.

31. Interview with Caupolicán Pávez, "Hay mucho que recorrer todavía . . ." in Fernando Echeverría Bascuñán and Jorge Rojas Hernández, *Añoranzas, sueños, realidades: Dirigentes sindicales hablan de la transición* (Santiago: Ediciones Sur, 1992), 137–143.

32. Escobar S. and López F., *El sector forestal en Chile,* 126.

33. Confederación Nacional de Trabajadores Forestales, "Informe central al quinto congreso nacional eleccionario de la Confederación Nacional de Trabajadores Forestales de Chile," Constitución, 5–6 June 1998, Confederación Nacional de Trabajadores Forestales, Concepción; former textile union leader and regional labor organizer, Departamento Pastoral Obrera, interview by the author, Concepción, April 2000.

34. Union leader, Federación Forestal Liberación, interview by the author, Concepción, April 2000.

35. Union leader, Confederación de Trabajadores Forestales, interview.

36. Director, Federación Forestal Liberación, interview by the author, Concepción, April 2000.

37. Confederación Nacional de Trabajadores Forestales, "Informe central al quinto congreso nacional eleccionario de la Confederación Nacional de Trabajadores Forestales de Chile."

38. Union leader, Confederación de Trabajadores Forestales, interview.

39. Also important were conflicts over hydroelectric dam projects in Alto Bío Bío involving Mapuche-Pehuenche communities, especially the Ralco community discussed in previous chapters: see Domingo Namuncura, *Ralco: Represa o Pobreza?* (Santiago: LOM Ediciones, 1999); Haughney, *Neoliberal Economics, Democratic Transition, and Mapuche Demands for Rights in Chile,* chap. 5.

40. For an important account of this history, see José Bengoa, *Quinquén: Cien años de historia pehuenche* (Santiago: Ediciones Chile América-CESOC, 1992).

41. Bengoa, *Quinquén.*

42. *El Diario Austral,* 23 September 1990, A7, A19. For accounts of the case, see also *El Mercurio,* 21 September, 23 September 1990; *La Tercera,* 24 September 1990.

43. Don Sergio, Comunidad Quinquén, interview by the author, Lonquimay, August 2010.

44. *El Mercurio,* 23 November 1990.

45. *El Mercurio,* 4 January 1991.

46. *El Diario Austral,* 24 September 1990, A6.

47. *El Diario Austral,* 27 September, A7, 29 September 1990, A5, 29 April, 1991, A5.

354 = *Notes for Chapter 9*

48. Florencia E. Mallon, *Courage Tastes of Blood: The Mapuche Community of Nicolás Ailío and the Chilean State, 1906–2001* (Durham, NC: Duke University Press, 2005), 174; José Bengoa, *Historia de un conflicto: El estado y los mapuches en el siglo XX* (Santiago: Planeta, 1999), 159–64, 171–75.

49. Mallon, *Courage Tastes of Blood*, 180–81; Haughney, *Neoliberal Economics, Democratic Transition, and Mapuche Demands for Rights in Chile*, 71–74. For an excellent history of the different Mapuche organizations during this period, see Christian Martínez Neira, "Transición a la democracia, militancia, y proyecto étnico: La fundación de la organización mapuche Consejo de Todas las Tierras (1978–1990)," *Estudios Sociológicos* 37, no. 80 (May–August 2009): 595–618.

50. *El Diario Austral*, 26–29 September 1990.

51. *La Nación*, 23 April 1991.

52. Bengoa, *Quinquén*, 114.

53. *El Diario Austral*, 16 April 1991, A6; 28 April 1991, A7; 29 April 1991, A5.

54. *La Época*, 9 May 1991; *La Nación*, 9 May 1991; *El Diario Austral*, 28 April 1991, A7; 4 May 1991, A8; 9 May 1991, B7; 10 May 1991, A19; 11 May 1991, A7; 12 May 1991, A7.

55. *El Diario Austral*, 9–11 May 1991.

56. *El Diario Austral*, 14 May 1991, A5; 16 May 1991, A10.

57. Bengoa, *Quinquén*, 62. See also *El Diario Austral*, 19 May 1991, A16; 31 January 1992, A5; 24 February 1992, A10; 25 February 1992, A5. It is notable that the CORA decree of 1974 that revoked Quinquén's expropriation also describes the property as 6,769 hectares: see Quinquén folder, CORA Archive, SAG.

58. *El Diario Austral*, 3 March 1992, 8, B7; El *Mercurio*, 3 March 1992; Bengoa, *Quinquén*; Raúl Molina and José Aylwin, "Entrega de las tierras a la Comunidad de Quinquén," *El Observador* 5, no. 2 (2007): 9–10.

59. Molina and Aylwin, "Entrega de las tierras a la Comunidad de Quinquén."

60. World Wildlife Fund, "WWF Chile y comunidad indígena se unen para crear un parque de Araucarias y promover el turismo en la Araucanía andina," 12 March 2009, available at http://wwf.panda.org/es/sala_redaccion/noticias/?163301/lanzamientoquinquen (accessed 2 June 2011).

61. David Tecklin and Rodrigo Catalán, "La gestión comunitaria de los bosques nativos en el sur de Chile: Situación actual y temas in discussion"; Rodrigo Catalán, "Aspectos fundamentales en la gestión comunitaria de bosques en Chile: La experiencia del Fondo Bosque Templado"; and Carmen Gloria Reyes, "Ecoturismo para la protección de la araucaria: Un desafío para la asociación pehuenche Quimpe Wentru de Lonquimay," in *Bosques y comunidades del sur de Chile*, ed. Rodrigo Catalán, Petra Wilken, Angelika Kandzior, David Tecklin, and Heinrich Burschel eds., (Santiago: Editorial Universitaria, 2005), 19–40, 107–18, 300–7.

62. Human Rights Watch and Observatorio de Derechos de los Pueblos Indígenas, "Undue Process: Terrorism Trials, Military Courts, and the Mapuche in Southern Chile," *Human Rights Watch* 16, no. 5 (October 2004): 13. Haughney also provides an excellent analysis of the law in *Neoliberal Economics, Democratic Transition, and Mapuche Demands for Rights in Chile*.

63. *El Diario Austral*, 14 October 1992, A5.

64. *El Diario Austral*, 21 October 1992, A5.

65. *El Diario Austral*, 18 April 1992, A5; 20 April 1992, 6–7.

66. Raúl Molina and Martín Correa, *Territorio y comunidades pehuenches del Alto Bío-Bío* (Santiago: Corporación Nacional de Desarrollo Indígena, 1996), 109.

67. Don Segundo, Comunidad Bernardo Ñanco, interview by the author, Lonquimay, August 2010.

68. Don Aladino, Comunidad Bernardo Ñanco, interview by the author, Lonquimay, August 2010; Frida Schweitzer, interview by the author, Lonquimay, August 2010.

69. Don Aladino interview; Don Segundo interview.

70. Angela Loncoñanco, interview by the author, Panguipulli, June 2012.

71. Interviews with Angela Loncoñanco, Humberto Manquel Millañguer and Pedro Cardyn, interviews by the author, Panguipulli, June 2012.

72. Rosales interview.

73. *El Mercurio Electrónico*, 4 May 1999, available at http://www.mapuche.info/lumaco/merc990505.htm, accessed 14 October, 2013.

74. *El Mercurio Electrónico*, 24 April 1999, available at http://www.mapuche.info/lumaco/merc990424.htm, accessed 14 October 2013.

75. *El Mercurio Electrónico*, 24 April 1999.

76. Angela Loncoñanco interview.

77. Observatorio de Derechos de los Pueblos Indígenas, "Parques Nacionales y exclusión indígena," *Revista Ser Indígena*, 18 April 2006; José Aylwin, "De quién es el Parque Nacional Villarrica?" *La Nación.cl*, 9 December 2008, available at http://www.mapuexpress.net/content/publications/print.php?id=1797, accessed 14 October 2013; José Luis Vargas, "Comunidades mapuches y locales: Rechazan inscripción de Parque Nacional Villarrica a favor del fisco," *Observatario Ciudadano*, 5 February 2010, available at http://www.observatorio.cl/2011/comunidades-mapuche-y-locales-rechazan-inscripcion-de-parque-nacional-villarrica-favor-del, accessed 14 October 2013.

78. For an excellent overview of the Mapuche movement during this period, see Haughney, *Neoliberal Economics, Democratic Transition, and Mapuche Demands for Rights in Chile*; Florencia E. Mallon, "Cuando la amnesia se pone con sangre el abuso se hace costumbre: El pueblo mapuche y el estado chileno, 1881–1998," in Drake and Jaksic, *El modelo chileno*, 435–84.

79. Tecklin and Catalán, "La gestión comunitaria de los bosques nativos en el sur de Chile," 20, 38.

80. For the ecological impact of forestry development on Mapuche communities, see Sara McFall, ed., *Territorio mapuche y expansión forestal* (Temcuo: Ediciones Escaparate, 2001); Haughney, *Neoliberal Economics, Democratic Transition, and Mapuche Demands for Rights in Chile*, 177–80. An excellent description of ecological degradation and social dislocation in Lumaco is provided in Senén Conejeros, Manuel Baquedano, José Bengoa, Nelson Caucoto, et al., "Informe de la Comisión Especial de Observadores de la Sociedad Civil para conocer los hechos ocurridos en las comunidades mapuche de Lumaco," 23 December 1997, Santiago. For a superb study of Lumaco, see René Montalba Navarro, Noelia Carrasco Hen-

ríquez, and José Araya Cornejo, *The Economic and Social Context of Monocultural Tree Plantations in Chile* (Montevideo: World Rainforest Movement, 2005).

81. See *El Diario Austral*, 6–21 December 1997.

82. For articles on El Rincón, see *El Diario Austral*, 18–25 February 1998.

83. *La Época*, 18 February, 21 February 1998.

84. *El Mercurio Electrónico*, 4 May 1999, available at http://www.mapuche.info/lumaco/merc990505.htm, 24 September 1999, available at http://www.mapuche.info/lumaco/merc990924.html, 14 December 1999, available at http://www.mapuche.info/lumaco/merc991214.html, accessed October 15, 2013.

85. See, e.g., the case of the Pidenco estate, *La Tercera en* Internet, 6 December 1997, available at http://www.mapuche.info/lumaco/terc971206d.htm, accessed 15 October 2013. For an investigation of these conflicts in Lumaco, see Conejeros et al., "Informe de la Comisión Especial de Observadores de la Sociedad Civil para conocer los hechos ocurridos en las comunidades mapuche de Lumaco"; Coordinadora de Comunidades en Conflicto Arauco-Malleco, "Informe de derechos humanos en las comunidades Mapuches en Conflicto de Arauco y Malleco," report presented to the United Nations Human Rights Commission, 55th sess., Geneva, Switzerland, April 1999.

86. Mallon, "Cuando la amnesia se impone con sangre el abuso se hace costumbre."

87. Mauricio Buendía, "Forestales en guerra contra los mapuches," *Punto Final*, 5 March 1999. For an extended discussion of the Temulemu case, see Conejeros et al. "Informe de la Comisión Especial de Observadores de la Sociedad Civil para conocer los hechos ocurridos en las comunidades mapuche de Lumaco." See also the detailed report by Chilean anthropologists, "Informe Comunidad Temulemu," Colegio de Antropólogos, Santaigo, August 1999. Haughney discusses Temulemu in *Neoliberal Economics, Democratic Transition, and Mapuche Demands for Rights in Chile*.

88. Daniel Ramos Fuentes, "Conflicto indígena en Traiguén: Una nueva pacificación," *El Sur en Internet*, 1998, available at http://www.mapuche.info/fakta/dsur990315.html, accessed 14 June 2011.

89. Fuentes, "Conflicto indígena en Traiguén,." See also "Informe Comunidad Temulemu."

90. "Informe Comunidad Temulemu." See also articles in *El Diario Austral*, 22–23 September 1998, 6 October 1998.

91. Ramos Fuentes, "Conflicto indígena en Traiguén." See also "Informe Comunidad Temulemu"; *El Diario Austral*, 21 April 1998, 23 April 1998, 22–23 September 1998, 6 October 1998.

92. *El Diario Austral*, 29 December 1997, A6.

93. *El Siglo*, 9–15 April 1999. For similar accounts, see also Haughney, *Neoliberal Economics, Democratic Transition, and Mapuche Demands for Rights in Chile*, 176; "Informe Temulemu."

94. Ramos Fuentes, "Conflicto indígena en Traiguén." As Haughney notes, during the 1990s, forestry companies also sprayed with a number of highly carcinogenic chemicals containing dioxin: Haughney, *Neoliberal Economics, Democratic Transition, and Mapuche Demands for Rights in Chile*, 263, n. 69.

95. *El Mercurio Electrónico,* 4 June 2002, avalable at http://www.mapuche.info/news/merco20604.html, accessed 15 October 2013; Human Rights Watch and Observatorio de Derechos de los Pueblos Indígenas, "Undue Process," 13–14.

96. Members of the Parlamento de Coz Coz, interviews by the author, Panguipulli, June 2012. For a description of this process in the Nicolás Ailío community, see Mallon, *Courage Tastes of Blood.*

97. *El Diario Austral,* 29 March, A13, 19 May, A16, 1991.

98. Human Rights Watch-Observatorio de Derechos de los Pueblos Indígenas, "Undue Process," 13.

99. Human Rights Watch, "Undue Process," 1–3.

100. Human Rights Watch, "Undue Process," 22–23.

101. Human Rights Watch, "Undue Process," 3.

102. Human Rights Watch, "Undue Process," 19.

103. Human Rights Watch, "Undue Process." See also Coordinadora de Comunidades en Conflicto Arauco-Malleco, *Informe de derechos humanos en la comunidades mapuches en conflicto de Arauco y Malleco.*

104. Francisco J. Fuentes, "Mapuches pactan con firma forestal explotación de predios adquiridos," *La Tercera en Internet,* 19 November 2009, available at http://www.latercera.com/contenido/654_202160_9.shtml, accessed 15 October 2013.

105. Quoted in Ramos Fuentes, "Conflicto Indígena en Traiguén." See also Haughney, *Neoliberal Economics, Democratic Transition, and Mapuche Demands for Rights in Chile,* 207.

106. Haughney provides an excellent overview of the Mapuche movement's demands for territorial autonomy and control of natural resources in *Neoliberal Economics, Democratic Transition, and Mapuche Demands for Rights in Chile.*

107. Roger A. Clapp, "The Unnatural History of the Monterey Pine," *Geographical Review* 85, no. 1 (January 1995): 16.

108. Clapp, "The Unnatural History of the Monterey Pine."

109. Clapp argues that, rather than re-create native forest systems on already cleared and degraded soil, the best solution in certain regions is to introduce different kinds of exotic species that work well in agroforestry or silvopastoralism: "The Unnatural History of the Monterey Pine," 16–17.

Conclusion

1. Diana K. Davis, *Resurrecting the Granary of Rome: Environmental History and French Colonial Expansion in North Africa* (Athens: Ohio University Press, 2007); Ramachandra Guha, *The Unquiet Woods: Ecological Change and Peasant Resistance in the Himalaya* (Berkeley: University of California Press, 2000).

2. For environmental history, see esp. Alfred W. Crosby, *The Columbian Exchange: Biological and Cultural Consequences of 1492* (Westport, CT: Praeger, 2003); Alfred W. Crosby, *Ecological Imperialism: The Biological Expansion of Europe, 900–1900* (Cambridge: Cambridge University Press, 2004).

3. Gunther Peck, "The Nature of Labor: Fault Lines and Common Ground in Environmental and Labor History," *Environmental History* 11, no. 2 (2006): 214.

4. Nancy Jacobs, *Environment, Power, and Injustice: A South African History* (Cambridge: Cambridge University Press, 2003), 16–21.

5. Donald Wooster, "Transformations of the Earth: Toward an Agroecological Perspective in History," *Journal of American History* 76, no. 4 (March 1990): 1087–1106; Donald Wooster, *Dust Bowl: The Southern Plains in the 1930s* (Oxford: Oxford University Press, 2004).

6. William Cronon, *Nature's Metropolis: Chicago and the Great West* (New York: W. W. Norton, 1992). See the critical essays on *Nature's Metropolis* in *Antipode* 26, no. 2 (1994).

7. See the essays in and introduction to William Cronon, ed., *Uncommon Ground: Rethinking the Human Place in Nature* (New York: W.W. Norton, 1996), esp. Richard White, "Are You an Environmentalist or Do You Work for a Living?" 171–85. See also Richard White, *The Organic Machine: The Remaking of the Columbia River* (New York: Hill and Wang, 1995). See also William Cronon, "Modes of Prophecy and Production: Placing Nature in History," *Journal of American History* 76, no. 4 (March 1990): 1122–131.

8. William Cronon, "The Trouble with Wilderness; or, Getting Back to the Wrong Nature" in Cronon, *Uncommon Ground*. See also White, "Are You an Environmentalist or Do You Work for a Living?"; Peck, "The Nature of Labor"; Douglas C. Sackman, " 'Nature's Workshop:' The Work Environment and Workers' Bodies in California's Citrus Industry, 1900–1940," *Environmental History*, vol. 5, no. 1 (January 2000): 27–53; Steve Marquardt, " 'Green Havoc': Panama Disease, Environmental Change, and Labor Process in the Central American Banana Industry," *American Historical Review* 106, no. 1 (February 2001): 49–80. A number of recent histories take up the challenge of bringing social, cultural, and environmental history together. See, e.g., Jacobs, *Environment, Power, and Injustice*. On Latin America, see John Soluri, *Banana Cultures: Agriculture, Consumption, and Environmental Change in Honduras and the United States* (Austin: University of Texas Press, 2005); Thomas D. Rogers, *The Deepest Wounds: A Labor and Environmental History of Sugar in Northeast Brazil* (Chapel Hill: University of North Carolina Press, 2010). On the United States, see Thomas G. Andrews, *Killing for Coal: America's Deadliest Labor War* (Cambridge, MA: Harvard University Press, 2008).

9. Two exceptions are Pablo Camus Gayán, *Ambiente, bosques, y gestión forestal en Chile, 1541–2005* (Santiago: LOM Ediciones and Dirección de Bibliotecas, Archivos y Museos, 2006), and Luis Otero Durán, *La huella del fuego: Historia de los bosques nativos, poblamiento, y cambios en el paisaje del sur de Chile* (Santiago: Corporación Nacional Forestal and Pehuén, 2006). For contemporary environmental crises and conflicts, see Rayén Quiroga Martínez, Saar van Hauwermeiren, Jorge Berghammer Vega, Patricio Del Real Jaramillo, and Marlene Gimpel Madariaga, eds., *El tigre sin selva: Consecuencias ambientales de la transformación económica de Chile, 1974–1993* (Santiago: Instituto de Ecología Política, 1994).

10. Richard Grove, *Green Imperialism: Colonial Expansion, Tropical Island Edens and the Origins of Environmentalism, 1600–1860* (Cambridge: Cambridge University Press, 1996).

11. See Peck, "The Nature of Labor," to which I am greatly indebted for this

understanding of the link between frontier commons and laborers' critique of commodification.

12. For a discussion of the community forestry approach in contrast to Chile's current model of forestry development, see Eduardo Silva, "Democracy, Market Economics, and Environmental Policy in Chile," *Journal of Interamerican Studies and World Affairs* 38, no. 4 (Winter 1996): 1–33; Eduardo Silva, "Forests, Livelihood, and Grassroots Politics: Chile and Costa Rica Compared," *European Review of Latin American and Caribbean Studies,* 66 (June 1999): 39–73.

13. For the case for mixed forestry, agriculture, and pastoralism (agroforestry) and silvopastoralism, see Roger A. Clapp, "The Unnatural History of Monterey Pine," *Geographical Review* 85, no. 1 (January 1995): 1–19. Also see David Tecklin and Rodrigo Catalán, "La gestión comunitaria de los bosques nativos en el sur de Chile: Situación actual y temas en discussion," 67–72, and other essays in *Bosques y comunidades del sur de Chile,* ed. Rodrigo Catalán, Petra Wilken, Angelika Kandzior, David Tecklin, and Heinrich Burschel (Santiago: Editorial Universitaria, 2006).

14. Tecklin and Catalán, "La gestión comunitaria de los bosques nativos en el sur de Chile." See also Juan J. Armesto, "Biodiversidad, ética y manejo sustentable del bosque nativo en Chile" in Defensores del Bosque Chileno, *La tragedia del bosque chileno* (Santiago: Ocho Libros Editores, 1998).

15. In 2009, roughly 70 percent of all farms (195,000) were 2 hectares or smaller and owned only 6 percent of Chile's basic irrigated hectares. At the other end of the spectrum, 5,331 farms of 60 hectares or more controlled about 66 percent of all basic irrigated hectares. Another study found that 87 percent of Chile's farms were small family farms that controlled in total only 13 percent of farmland: Julio A. Berdegué and Ricardo Fuentealba, "Latin America: The State of Smallholders in Agriculture," paper presented at the Conference on New Directions for Smallholder Agriculture, International Fund for Agricultural Development, Rome, 24–25 January 2011. The authors of that paper cite Jorge Echenique and Lorena Romero, *Evolución de la agricultura familiar en Chile en el período 1997–2007* (Santiago: United Nations Food and Agriculture Organization, 2009).

Bibliography

Archives and Government Records

Archivo Nacional, Santiago (AN)
   Intendencia de Cautín
Archivo Nacional de la Administración, Santiago (ARNAD)
   Ministerio de Agricultura
   Ministerio de Relaciones Exteriores (MRREE)
   Ministerio de Tierras y Colonización (MTC)
   Ministerio del Interior
   Archivo de la Corporación de Fomento de la Producción de Chile (CORFO Archive)
Archivo Regional de la Araucanía, Temuco (ARA)
   Archivo General de Asuntos Indígenas (CONADI Archive), Temuco
      Carpetas Administrativas
      Juzgados de Indios
   Intendencia de Cautín
Biblioteca del Congreso Nacional, Santiago (BCN)
   Ministerio del Fomento, Memorias anuales
   Inspección General de Tierras y Colonización, Memorias anuales
   Ministerio de Relaciones Exteriores, Culto i Colonización, Memorias anuales
   Oficina Menura de Tierras, Memorias anuales
Biblioteca Nacional, Santiago (BN)
   Archivo Alessandri
   Inspección General de Tierras y Colonización, Memorias anuales
Servicio Agrícola y Ganadero, Santiago (SAG)
   Archivo de la Ex-Corporación de Reforma Agraria (CORA Archive)

Newspapers and Periodicals

Agricultura Austral (Llanquihue)
El Agricultor (Temuco)
Análisis (Santiago)
Boletín de Bosques, Pesca i Caza (Santiago)
Boletín de Estudios Agrarios (Santiago)
Boletín de la Sociedad Agrícola del Sur (Osorno)

El Campesino (Santiago)
Chile Maderero (Santiago)
El Comercio (Lonquimay)
El Correo de Valdivia (Valdivia)
El Diario Austral (Temuco)
La Época (Santiago)
Hoy (Santiago)
Justicia (Santiago)
Lignum (Santiago)
El Malleco (Malleco)
El Mercurio (Santiago)
El Mercurio Electrónico (Santiago)
La Nación (Santiago)
Noticiero de la Realidad Agraria (Santiago)
La Opinión (Santiago)
Pacífico Magazine (Santiago)
Páginas Sindicales (Santiago)
Panorama Económico (Santiago)
Pensamiento iberoamericano: Revista de Economía Política
Periódico Forestal (Santiago)
El Pino Insigne (Concepción)
Poder Campesino (Santiago)
Punto Final (Santiago)
Qué Pasa? (Santiago)
Ránquil (Santiago)
El Siglo (Santiago)
El Sur (Concepción)
El Sur en Internet (Concepción)
La Tercera (Santiago)
La Tercera en Internet (Santiago)
Unasylva

Published Sources

Aagesen, David L. "Burning Monkey-Puzzle: Native Fire Ecology and Forest
    Management in Northern Patagonia," Agriculture and Human Values 21 (2004):
    232–44.
Aagesen, David L. "Indigenous Resource Rights and Conservation of the
    Monkey-Puzzle Tree (Araucaria araucana, Araucariaceae): A Case Study from
    Southern Chile." Economic Botany 52, no. 2 (1998): 146–160.
Aagesen, David L. "On the Northern Fringe of the South American Temperate
    Forest: The History and Conservation of the Monkey-Puzzle Tree," Environ-
    mental History 3, no. 1 (January 1998): 64–85.
Agrawal, Arun. Environmentality: Technologies of Government and the Making of
    Subjects. Durham, NC: Duke University Press, 2005.

Andrews, Thomas G. *Killing for Coal: America's Deadliest Labor War.* Cambridge, MA: Harvard University Press, 2008.

Armesto, Juan J. "Biodiversidad, ética y manejo sustentable del bosque nativo en Chile." In *La tragedia del bosque chileno,* ed. Defensores del Bosque Chileno. Santiago: Ocho Libros Editores, 1998.

Armesto, Juan J., Carolina Villagrán, and Mary Kalin Arroyo. *Ecología de los bosques nativos de Chile.* Santiago: Editorial Universitaria, 1995.

Astorga Schneider, Eduardo. "Factores que han afectado la tasa de reforestación en Chile." *Actas: Jornadas Forestales* 7 (1971): 35–39.

Astorquiza, Octavio, and Oscar Galleguillos V. *Cien años del carbón de Lota, 1852–septiembre–1952.* Santiago: Compañía Carbonífera e Industrial de Lota, 1952.

Baezas, Eduardo. "Monografía del fundo Poco a Poco de Temuco." Licenciatura thesis, Instituto Agronómico, Universidad de Chile, Santiago, 1910.

Balogh, Brian. "Scientific Forestry and the Roots of the Modern American State: Gifford Pinchot's Path to Progressive Reform." *Environmental History* 7, no. 2 (April 2002): 198–225.

Barrientos, Claudio. "Emblems and Narratives of the Past: The Cultural Construction of Memories and Violence in Peasant Communities of Southern Chile, 1970–2000." PhD diss., University of Wisconsin, Madison, 2003.

Bauer, Arnold. *Chilean Rural Society from the Spanish Conquest to 1930.* Cambridge: Cambridge University Press, 1975.

Bay-Smith, Troels. "Algunas observaciones sobre ensayos de especies forestales en la provincia de Arauco." *Seminario Forestal,* no. 10 (July 1965): 4–14.

Bengoa, José. *Haciendas y campesinos: Historia social de la agricultura chilena,* vol. 2. Santiago: Ediciones Sur, 1990.

Bengoa, José. *Historia del pueblo mapuche, siglo XIX y XX.* Santiago: LOM Ediciones, 2000.

Bengoa, José. *Historia de un conflicto: El estado y los mapuches en el siglo XX.* Santiago: Planeta, 1999.

Bengoa, José. *Quinquén: Cien años de historia pehuenche.* Santiago: Ediciones Chile América-CESOC, 1992.

Berniger, Otto. "Bosque y tierra despejada en el sur de Chile desde la conquista española." Thesis: Universidad de Chile, Santiago, 1966.

Bianchi, Hugo. "Política forestal frente al problema de tenencia y uso de la tierra." *Actas: Jornadas Forestales* 6 (1970): 33–38.

Bussi Soto, Renato. "Monografía económico-cultural del fundo 'Puello,' provincia de Cautín, departamento de Temuco." Licenciatura thesis, Universidad de Chile, Santiago, 1941

Cabeza Monteira, Angel. *Aspectos históricos de la legislación forestal vinculada a la conservación, la evolución de areas silvestres protegidas de la zona de Villarrica y la creación del primer parque nacional de Chile.* Santiago: Corporación Nacional Forestal, 1988.

Camus Gayán, Pablo. *Ambiente, bosques, y gestión forestal en Chile, 1541–2005.* Santiago: LOM Ediciones and Dirección de Bibliotecas, Archivos y Museos, 2006.

Camus Gayán, Pablo. "Federico Albert: Artífice de la gestión de los bosques de Chile." *Revista de Geografía Norte Grande*, no. 30 (2003): 55–63.

Carrere, Ricardo, and Larry Lohmann. *El papel del Sur: Plantaciones forestales en la estrategia papelera internacional*. Montevideo, Uruguay: World Rainforest Movement, 1997.

Catalán Labarías, Rodrigo, and Ramos Ruperto Antiqueo. *Pueblo Mapuche, bosque nativo y plantaciones forestales: Las causas de las deforestación en en el sur de Chile*. Temuco: Ediciones Universidad Católica de Temuco, 1999.

Chaparro Ruminot, Leoncio. *Colonias de medieros*. Santiago: Multitud, 1941.

Chonchol, Jacques. *Sistemas agrarios en América Latina*. Santiago: Fondo de Cultura Económica, 1994.

Clapp, Roger Alex. "Creating Comparative Advantage: Forest Policy as Industrial Policy in Chile." *Economic Geography* 71, no. 3 (July 1995): 273–96.

Clapp, Roger Alex. "The Unnatural History of the Monterey Pine." *Geographical Review* 85, no. 1 (January 1995): 1–19.

Clapp, Roger Alex. "Waiting for the Forest Law: Resource-Led Development and Environmental Politics in Chile." *Latin American Research Review* 33, no. 2 (1998): 3–36.

Claude, Marcel. *Una vez más la miseria: ¿Es Chile un país sustentable?* Santiago: LOM Ediciones, 1997.

Comité Interamericano de Desarrollo Agrícola. *Chile: Tenencia de la tierra y desarrollo socio-económico del sector agrícola*. Santiago: Comité Interamericano de Desarrollo Agrícola, 1966.

Congreso Nacional. *Comisión Parlamentaria de Colonización: Informe, proyectos de ley, actas de las sesiones, y otros antecedents*. Santiago: Sociedad "Imprenta y Litografía Universo," 1912.

Coroníl, Fernando. *The Magical State: Nature, Money, and Modernity in Venezuela*. Chicago: University of Chicago Press, 1997.

Correa, Martín, Raúl Molina, and Nancy Yáñez. *La reforma agraria y las tierras mapuches: Chile, 1962–1975*. Santiago: LOM Ediciones, 2005.

Corte de bosques: Informe de la comisión nombrada para dictaminar esta materia i reglamento dictado por el Presidente de la República. Santiago: Imprenta Nacional, 1873.

Cronon, William. *Changes in the Land: Indians, Colonists, and the Ecology of New England*. New York: Hill and Wang, 1983.

Cronon, William. "Modes of Prophecy and Production: Placing Nature in History." *Journal of American History* 76, no. 4 (March 1990): 1122–131.

Cronon, William. *Nature's Metropolis: Chicago and the Great West*. New York: W. W. Norton, 1992.

Cronon, William, ed. *Uncommon Ground: Rethinking the Human Place in Nature*. New York: W. W. Norton, 1996.

Crosby, Alfred W. *The Columbian Exchange: Biological and Cultural Consequences of 1492*. Westport, CT: Praeger, 2003.

Crosby, Alfred W. *Ecological Imperialism: The Biological Expansion of Europe, 900–1900*. Cambridge: Cambridge University Press, 2004.

Davis, Diana K. *Resurrecting the Granary of Rome: Environmental History and French Colonial Expansion in North Africa*. Athens: Ohio University Press, 2007.

Defensores del Bosque Chileno, ed. *La tragedia del bosque chileno*. Santiago: Ocho Libros Editores, 1998.

Díaz Meza, Aureliao. *En la Araucanía: Breve relación del último parlamento araucano de Coz-Coz en 18 de enero de 1907*. Santiago: Imprenta "El Diario Ilustrado," 1907.

Domeyko, Ignacio. *Araucanía y sus habitantes*. Santiago: Imprenta Chilena, 1846.

Donoso Zegers, Claudio. "Bosques nativos de Chile: Patrimonio de la tierra." In *La tragedia del bosque chileno*, ed. Defensores del Bosque Chileno, 83–99. Santiago: Ocho Libros Editores, 1998.

Donoso Zegers, Claudio. *Bosques templados de Chile y Argentina: Variación, estructura y dinámica*. Santiago: Editorial Universitaria, 1993.

Drake, Paul, and Iván Jaksic. *El modelo chileno: Democracia y desarrollo en los noventa*. Santiago: LOM Ediciones, 1999.

Echenique, Jorge, and Lorena Romero. *Evolución de la agricultura familiar en Chile en el período 1997–2007*. Santiago: United Nations Food and Agriculture Organization, 2009.

Echeverría Bascuñán, Fernando, Jorge Rojas Hernández, and Manuel Bustos. *Añoranzas, sueños, realidades: Dirigentes sindicales hablan de la transición*. Santiago: Ediciones Sur, 1992.

Elizalde MacClure, Rafael. *La sobrevivencia de Chile: La conservación de sus recursos naturales renovables*. Santiago: Ministerio de Agricultura, 1970.

Elliott, Lilian E. *Chile: Today and Tomorrow*. New York: Macmillan, 1922.

Escobar S., Patricio, and Diego López F. *El sector forestal en Chile: Crecimiento y precarización del empleo*. Santiago: Programa de Economía del Trabajo, 1996.

Espinoza Cuevas, Víctor, Paz Rojas Baeza, and María Luisa Ortíz Rojas. *Derechos humanos: Sus huellas en el tiempo: Una experiencia de trabajo en derechos humanos y salud mental en una zona rural del sur de Chile*. Santiago: Comité de Defensa de los Derechos del Pueblo, 1999.

Fahrenkrog Reinhold, Harry. *La verdad sobre la revuelta de Ránquil: Memorias de Harry Fahrenkrog Reinhold*. Santiago: Editorial Universitaria, 1985.

Fazio R., Hugo. *Mapa actual de la extrema riqueza en Chile*. Santiago: LOM Ediciones and Universidad de Arte y Ciencias Sociales, 1997.

Fernández, Carola. *La conveniente oscuridad del bosque Chileno*. Santiago: Fundación Terram, 2003.

Figueroa Garavagno, Consuelo. "De rastros y extravíos: Guerras en exhibición, Chile, 1880s–1930s." In *Entre el humo y la niebla: Guerra y cultura en América Latina*, ed. Felipe Martínez P. and Javier Uriarte. Pittsburgh: University of Pittsburgh Press, forthcoming.

Figueroa Garavagno, Consuelo. "Geografías en disputa: La construcción del Chile territorial." *Revista 180*, no. 27 (2011): 10–13.

Figueroa Garavagno, Consuelo. *Revelación del subsole: Las mujeres en la sociedad minera del carbón, 1900–1930*. Santiago: Dirección de Bibliotecas, Archivos y Museos, 2009.

Flores Chávez, Jaime. "Europeos en la Araucanía: Los colonos del Budi a prinicipios del siglo xx." In *Emigración centroeuropea a América Latina*, ed. Josef Opatrny, 313–29. Prague: Editorial Karolinum, 2000.

Flores Chávez, Jaime. "Tierra y territorio Mapuche-Pehuenche, 1860–1997." In *Estudios introductorios en relaciones interétnicas*, ed. Belén Lorente Molina and Carlos Vladimir Zambrano, 79–104. Bogotá: Corporación Colombiana de Investigaciónes Humanísticas, 1992.

Flores Chávez, Jaime. "Un episodio en la historia social de Chile: Ránquil, una revuelta campesina." Master's thesis, Universidad de Santiago de Chile, 1993.

García Elizalde, Pedro, and José Ignacio Leyton. *El desarrollo frutícola y forestal en Chile y sus derivaciones sociales*. Santiago: Comisión Económica para América Latina, 1986.

González, Mauro E., Thomas T. Veblen, and Jason S. Sibold. "Fire History of *Araucaria-Nothofagus* Forests in Villarrica National Park, Chile." *Journal of Biogeography* 32 (2005): 1187–202.

Grove, Richard. *Green Imperialism: Colonial Expansion, Tropical Island Edens and the Origins of Environmentalism, 1600–1860*. Cambridge: Cambridge University Press, 1996.

Grupo de Investigaciones Agrarias. *Región forestal: Empresas y trabajadores*. Santiago: Academia de Humanismo Cristiano, n.d.

Guha, Ramachandra. *The Unquiet Woods: Ecological Change and Peasant Resistance in the Himalaya*. Berkeley: University of California Press, 2000.

Guha, Ranajit. "The Prose of Counter-Insurgency." In *Selected Subaltern Studies*, ed. Ranajit Guha and Gayatri Chakravorty Spivak, 45–88. Oxford: Oxford University Press, 1988.

Haig, Irvine T. *Forest Resources of Chile as a Basis for Industrial Expansion*. Washington, D.C.: U.S. Department of Agriculture, Forest Service, 1946.

Harambour-Ross, Albert. "Borderland Sovereignties: Postcolonial Colonialism and State Making in Patagonia, Argentina, and Chile, 1840s–1992." PhD diss., Stony Brook University, Stony Brook, NY, 2012.

Hartwig Carte, Fernando. *Chile, desarrollo forestal sustentable: Ensayo de política forestal*. Santiago: Editorial Los Andes, 1991.

Hartwig Carte, Fernando. *Federico Albert, pionero del desarrollo forestal en Chile*. Talca: Editorial Universidad de Talca, 1999.

Harvey, David. *The New Imperialism*. Oxford: Oxford University Press, 2003.

Haughney, Diane. *Neoliberal Economics, Democratic Transition, and Mapuche Demands for Rights in Chile*. Gainesville: University Press of Florida, 2006.

Hepp Dubiau, Ricardo, and Juan Jirkal H. *El problema forestal en Chile: Generalidades sobre el problema de la erosión del suelo en Chile*. Santiago: Litografía Concepción, 1943.

Huenchullán, Arturo. *Los sucesos del Alto Bío-Bío y el diputado Huenchullán*. Santiago: Selectam, 1934.

Human Rights Watch and Observatorio de Derechos de los Pueblos Indígenas. "Undue Process: Terrorism Trials, Military Courts, and the Mapuche in Southern Chile." *Human Rights Watch* 16, no. 5 (October 2004): 1–63.

Inostroza Córdova, Iván. *Etnografía mapuche del siglo xx*. Santiago: Dirección de Bibliotecas, Archivos y Museos, 1998.

Jacobs, Nancy J. *Environment, Power, and Injustice: A South African History*. Cambridge: Cambridge University Press, 2003.

Jacoby, Karl. *Crimes against Nature: Squatters, Poachers, Thieves, and the Hidden History of American Conservation*. Berkeley: University of California Press, 2003.

Jefferson, Mark. *Recent Colonization in Chile*. New York: Oxford University Press, 1921.

Keller, Carlos. *Revolución en la agricultura*. Santiago: Zig-Zag, 1956.

Keun, Ricardo Fernando. *Y así nació la frontera*. Santiago: Editorial Antartica, 1986.

Kunstmann, Germán. "Explotación de los bosques autóctonos del fundo Trafún desde el punto de vista de sus propietarios." *Seminario Forestal*, no. 10 (July 1965): 19–22.

Lara, Antonio, and Cristian Echeverria. *Certificacion forestal: Una necesidad para la conservación de los bosques en Chile*. Valdivia: Universidad Austral de Chile, n.d.

Lara, Antonio, and Thomas T. Veblen. "Forest Plantations in Chile: A Successful Model?" In *Afforestation: Policies, Planning, and Progress*, ed. Alexander S. Mather, 118–139. London: Belhaven, 1993.

Lavín, Joaquin. *Chile: Revolución silenciosa*. Santiago: Zig-Zag, 1987.

Leiva, Sebastián. "El partido comunista de Chile y levantamiento de Ránquil." *Cyber Humanitatis*, no. 28 (2003). Available at http:// www.cyberhumanitatis .uchile.cl/index.php/RCH/article/viewArticle/5711/5579.

León, Leonardo. *Maloqueros y conchavadores en la Araucanía y las pampas, 1700–1800*. Temuco: Ediciones Universidad de la Frontera, 1990.

León, Leonardo, Patricio Herrera, Luis Carlos Parentini, and Sergio Villalobos. *Araucanía: La frontera mestiza, siglo xix*. Santiago: LOM Ediciones, 2003.

Limerick, Patricia Nelson. *The Legacy of Conquest: The Unbroken Past of the American West*. New York: W. W. Norton, 1987.

Limerick, Patricia Nelson. "Turnerians All: The Dreams of a Helpful History in an Unintelligible World," *American Historical Review* 100 (1995): 697–716.

Linebaugh, Peter. "Karl Marx, the Theft of Wood, and Working Class Composition: A Contribution to the Current Debate." *Crime and Social Justice* 6 (Fall–Winter 1976): 5–16.

Lira, Elizabeth, and Brian Loveman. *Las ardientes cenizas del olvido: Via chilena de reconciliación política, 1932–1994*. Santiago: LOM Ediciones, 2000.

Loveman, Brian. *Chile: The Legacy of Hispanic Capitalism*. Oxford: Oxford University Press, 2001.

Loveman, Brian. *Struggle in the Countryside: Politics and Rural Labor in Chile, 1919–1973*. Bloomington: Indiana University Press, 1976.

Maige A., José. *Plantaciones forestales en las colonias agrícolas*. Santiago: n.p., 1937.

Mallon, Florencia E. *Courage Tastes of Blood: The Mapuche Community of Nicolás*

Ailío and the Chilean State, 1906–2001. Durham, NC: Duke University Press, 2005.

Mallon, Florencia E., ed. *Decolonizing Native Histories: Collaboration, Knowledge, and Language in the Americas.* Durham, NC: Duke University Press, 2011.

Mamalakis, Markos J. *Historical Statistics of Chile: Forestry and Related Activities,* vol. 3. Westport, CT: Greenwood, 1982.

Marquardt, Steve. " 'Green Havoc': Panama Disease, Environmental Change, and Labor Process in the Central American Banana Industry." *American Historical Review* 106, no. 1 (February 2001): 49–80.

Martínez Neira, Christian. "Transición a la democracia, militancia, y proyecto étnico: La fundación de la organización mapuche Consejo de Todas las Tierras (1978–1990)." *Estudios Sociológicos* 37, no. 80 (May–August 2009): 595–618.

Mathei, Adolfo. *La agricultura en Chile y la política agraria chilena.* Santiago: Imprenta Nacimento, 1939.

McBride, George McCutchen. *Chile: Land and Society.* New York: American Geographical Society, 1936.

McFall, Sara, ed. *Territorio mapuche y expansión forestal.* Temcuo: Ediciones Escaparate, 2001.

Merrriman, John M. "The Demoiselles of the Ariege, 1829–1831." In *1830 in France,* ed. John M. Merriman, 87–118. New York: New Viewpoints, 1975.

Miller, Char. *Gifford Pinchot and the Making of Modern Environmentalism.* Washington, DC: Island Press, 2001.

Molina, Raúl, and José Aylwin. "Entrega de las tierras a la Comunidad de Quinquén." *El Observador* 5, no. 2 (2007): 9–10.

Molina, Raúl, and Martín Correa. *Territorio y comunidades pehuenches del Alto Bío-Bío.* Santiago: Corporación Nacional de Desarrollo Indígena, 1996.

Montalba Navarro, René, Noelia Carrasco Henríquez, and José Araya Cornejo. *The Economic and Social Context of Monocultural Tree Plantations in Chile: The Case of the Commune of Lumaco.* Montevideo, Uruguay: World Rainforest Movement, 2006.

Mora Venegas, Alfonso. "Los bosques en la economía chilena." Licenciatura thesis, Universidad de Chile, Santiago, 1950.

Mosbach, Ernesto Wilhelm de. *Botanica indígena de Chile.* Santiago: Editorial Andrés Bello, 1992.

Namuncura, Domingo. *Ralco: Represa o pobreza?* Santiago: LOM Ediciones, 1999.

Neira, Eduardo, Hernán Verscheure, Carmen Revenga. *Chile's Frontier Forests: Conserving a Global Treasure.* Valdivia: World Resources Institute, Comité Nacional por la Defensa de la Fauna y Flora, and Universidad Austral de Chile, 2002.

Neíra, Paz, Samuel Línker, and Irene Romero. *Memorias de Llaima: Historias de las comunidades mapuches de Melipeuco.* Santiago: Talleres Designofpasten, 2011.

Opazo, Roberto. *Desarrollo agricola de los territorios situados al sur del río Bio-Bio.* Temuco: Sociedad Cooperativa y de Fomento Agricola, 1920.

Ortega Martínez, Luis, Cármen Norambuena Carrasco, Julio Pinto Vallejos, and

Guillermo Bravo Acevedo. *Corporación de Fomento de la Producción: 50 Años de Realizaciones, 1939–1989.* Santiago: Corporación de Fomento de la Producción, 1989.

Otero Durán, Luis. *La huella del fuego: Historia de los bosques nativos, poblamiento, y cambios en el paisaje del sur de Chile.* Santiago: Corporación Nacional Forestal and Pehuén, 2006.

Otero Durán, Luis. *El problema social detrás de los bosques.* Concepción: Vicaría de la Pastoral Obrera, 1984.

Palacios, Germán. *Ránquil: La violencia en la expansión de la propiedad agrícola.* Santiago: Instituto de Ciencias Alejandro Lipschutz, 1992.

Palacios, Nicolás. *La raza chilena: Libro escrito por un chileno y para los chilenos.* Valparaíso: Imprenta y Litografía Alemana, 1904.

Peck, Gunther. "The Nature of Labor: Fault Lines and Common Ground in Environmental and Labor History." *Environmental History* 11, no. 2 (2006): 212–38.

Peña, Alberto. "Reflexiones en torno a la extensión forestal en Chile." In *Bosques y comunidades del sur de Chile,* ed. Rodrigo Catalán, Petra Wilken, Angelika Kandzior, David Tecklin, and Heinrich Burschel, 67–72. Santiago: Editorial Universitaria, 2005.

Peñaloza Wagenknecht, Rubén, Federico Schlegel Sachs, and Claudio Donoso Zegers. "Importancia del recurso forestal del Complejo Forestal y Maderero Panguipulli." Report, Facultad de Ciencias Forestales, Universidad Austral, Valdivia, 1985.

Pérez Rosales, Vicente. *Memoria sobre colonización de la provincia de Valdivia.* Valparaíso: Imprenta del Diario, 1852.

Pérez Rosales, Vicente. *Times Gone By: Memoirs of a Man of Action,* trans. John H. R. Polt. Oxford: Oxford University Press, 2003.

Peters, Konrad. *Estudio experimental sobre selvicultura en Chile.* Santiago: Imprenta Nascimento, 1938.

Philippi, Rudolph. "Valdivia en 1852." *La Revista de Chile,* no. 74 (May 1901): 297–300, 329–35, 355–61.

Pinto Rodríguez, Jorge. "Redes indígenas y redes capitalistas: La Araucanía y las pampas en el siglo XIX" in *Los pueblos campesinos de las Américas. Etnicidad, cultura e historia en el siglo XIX.* Heraclio Bonilla and Amado A. Guerrero eds. Bucaramanga, Colombia: Universidad Industrial de Santander, Escuela de Historia, 1996: 137–144.

Pinto Rodríguez, Jorge. *La formación del estado y la nación, y el pueblo mapuche: De la inclusión a la exclusión.* Santiago: Dirección de Bibliotecas, Archivos y Museos and Centro de Investigaciones Diego Barros Arana, 2003.

Pinto Vallejos, Julio ed. *Cuando hicimos historia: La experiencia de la Unidad Popular.* Santiago: LOM Ediciones, 2005.

Poeppig, Eduard. *Un testigo en la alborada de Chile (1826–1829).* Santiago: Zig-Zag, 1960.

Quiroga Martínez, Rayén, and Saar van Hauwermeiren. *Globalización e insustentabilidad: Una mirada desde la economía ecológica.* Santiago: Instituto de Ecología Política, 1996.

Quiroga Martínez, Rayén, Saar van Hauwermeiren, Jorge Berghammer Vega, Patricio del Real Jaramillo, and Marlene Gimpel Madariaga, eds. *El tigre sin selva: Consecuencias ambientales de la transformación económica de Chile, 1974–1993.* Santiago: Instituto de Ecología Política, 1994.

Rivera A., Rigoberta, and María Elena Cruz. *Pobladores rurales.* Santiago: Grupo de Investigaciones Agrarias and Academia de Humanismo Cristiano, 1984.

Rogers, Thomas D. *The Deepest Wounds: A Labor and Environmental History of Sugar in Northeast Brazil.* Chapel Hill: University of North Carolina Press, 2010.

Rojas Baeza, Paz, Víctor Espinoza Cuevas, and María Luisa Ortíz Rojas. *Chile, recuerdos de la guerra: Valdivia, Neltume, Chihuio, Liquiñe.* Santiago: Comité de Defensa de los Derechos del Pueblo, 1991.

Rojas Magallanes, Victorino. *Informe sobre bosques presentado al Ministerio de Industria I Obras Públicas.* Santiago: Imprenta Litografía y Encuadernación Barcelona, 1904.

Ros Vera, Rafael. "Plantaciones: Sus efectos sobre el ambiente." *Actas: Jornadas Forestales* 14 (1992): 21–37.

Ruiz-Esquide, Andrea. *Los indios amigos en la frontera araucana.* Santiago: Dirección de Bibliotecas, Archivos y Museos and Centro de Investigaciones Diego Barros Arana, 1993.

Ruiz-Esquide, Andrea. "Migration, Colonization, and Land Policy in the Former Mapuche Frontier, Malleco, 1850–1900." PhD thesis, Columbia University, New York, 2000.

Saavedra, Cornelio. *Documentos relacionados a la ocupación de Arauco.* Santiago: Imprenta de la Libertad, 1870.

Saavedra Peláez, Alejandro. *Los mapuches en la sociedad chilena actual.* Santiago: LOM Ediciones, 2002.

Sackman, Douglas C. "'Nature's Workshop': The Work Environment and Workers' Bodies in California's Citrus Industry, 1900–1940." *Environmental History* 5, no. 1 (January 2000): 27–53.

Sahlins, Peter. *Forest Rites: The War of the Demoiselles in Nineteenth-Century France.* Cambridge, MA: Harvard University Press, 1994.

Sánchez Guerrero, Juan. *Hijo de las piedras.* Santiago: Zig-Zag, 1962.

Schilling, Juan. *Industria de papel: Ante-proyecto de instalación de una fábrica de papel i celulosa en Chile.* Santiago: Sociedad de Fomento Fabríl and Imprenta Universo, 1911.

Schneider, Teodoro. *La agricultura en Chile en los últimos cincuenta años.* Santiago: Imprenta Barcelona, 1904.

Scott, James C. *Seeing like a State: How Certain Schemes to Improve the Human Condition Have Failed.* New Haven, CT: Yale University Press, 1998.

Silva, Eduardo. "Democracy, Market Economics, and Environmental Policy in Chile." *Journal of Interamerican Studies and World Affairs* 38, no. 4 (Winter 1996): 1–33.

Silva, Eduardo. "Forests, Livelihood, and Grassroots Politics: Chile and Costa Rica Compared." *European Review of Latin American and Caribbean Studies,* 66 (June 1999): 39–73.

Silva, Patrico. *Estado, neoliberalismo y política agraria en Chile, 1973–1981*. Dordrecht: Centro de Estudios y Documentación Latinoamericanos, 1987.

Smith, Edmond Reuel. *The Araucanians; or, Notes of a Tour among the Indian Tribes of Southern Chili*. New York: Harper and Brothers, 1855.

Solberg, Carl E. "A Discriminatory Land Policy: Chile, 1870–1914." *The Americas* 26, no. 2 (October 1969): 115–33.

Soluri, John. *Banana Cultures: Agriculture, Consumption, and Environmental Change in Honduras and the United States*. Austin: University of Texas Press, 2005.

Steenland, Kyle. *Agrarian Reform under Allende: Peasant Revolt in the South*. Albuquerque: University of New Mexico Press, 1977.

Tecklin, David, and Rodrigo Catalán. "La gestión comunitaraia de los bosques nativos en el sur de Chile: Situación actual y temas en discusión." In *Bosques y comunidades del sur de Chile*, ed. Rodrigo Catalán, Petra Wilken, Angelika Kandzior, David Tecklin, and Heinrich Burschel, 19–40. Santiago: Editorial Universitaria, 2005.

Téllez Lúgaro, Eduardo, Cristian Arancibia, Juan Canales, Larisa de Ruit, Rodrigo Quinteros, and Yuri Quintupirray. "El levantamiento del Alto Bíobio y el Soviet y la República Araucana de 1934." *Annales de la Universidad de Chile*, sixth series, no. 13, August 2001. Available at http://www.revistas.uchile.cl/index.php/ANUC/article/viewArticle/2530/2448.

Thompson, E. P. *Whigs and Hunters: The Origins of the Black Act*. New York: Pantheon, 1975.

Tinsman, Heidi. *Partners in Conflict: The Politics of Gender, Sexuality, and Labor in the Chilean Agrarian Reform, 1950–1973*. Durham, NC: Duke University Press.

Tohá, Jaime. "Plan nacional de reforestación." *Actas: Jornadas Forestales* 2 (1966): 30–39.

Toledo Martel, Gonzalo. "La muerte y sus interpretaciones: Represión política en el Complejo Forestal y Maderero Panguipulli Ltda., 1973." Licenciatura thesis, Universidad Austral de Chile, Valdivia, 1994.

Treutler, Paul. *Andanzas de un alemán en Chile, 1851–1863*. Santiago: Editorial del Pacífico, 1958.

Uliánova, Olga. "Levantamiento campesino de Lonquimay y la internacional comunista," *Estudios Públicos*, no. 89 (Summer 2003). http://www.cepchile.cl/1_3185/doc/levantamiento_campesino_de_lonquimay_y_la_internacional_comunista.html#.UfZjPY3VCjo.

Verniory, Gustave. *Diez años en Araucanía, 1889–1899*. Santiago: Pehuén, 2001.

Villalobos, Sergio. *Vida fronteriza en la Araucanía: El mito de la Guerra de Arauco*. Santiago: Editorial Andrés Bello, 1995.

Villalobos, Sergio, Carlos Aldunate, Horacio Zapater, Luz María Méndez, and Carlos Bascuñan. *Relaciones fronterizas en la Araucanía*. Santiago: Ediciones Universidad Católica de Chile, 1982.

White, Richard. *The Organic Machine: The Remaking of the Columbia River*. New York: Hill and Wang, 1995.

Wilcox, Ken. *Chile's Native Forests: A Conservation Legacy*. Redway, CA: Ancient Forest International, 1996.

Winn, Peter, ed. *Victims of the Chilean Miracle: Workers and Neoliberalism in the Pinochet Era, 1973–2002.* Durham, NC: Duke University Press, 2004.

Winn, Peter. *Weavers of Revolution: The Yarur Workers and Chile's Road to Socialism.* Oxford: Oxford University Press, 1986.

Wooster, Donald. *Dust Bowl: The Southern Plains in the 1930s.* Oxford: Oxford University Press, 2004.

Wooster, Donald. "Transformations of the Earth: Toward an Agroecological Perspective in History." *Journal of American History* 76, no. 4 (March 1990): 1087–106.

Zenteno Barros, Julio. *Recopilación de leyes y decretos supremos sobre colonización.* Santiago: Imprenta Nacional, 1892.

Zon, Raphael, and William N. Sparhawk. *Forest Resources of the World.* New York: McGraw-Hill, 1923.

Index

Memorial at, 268–69; ownership of, 285, 325n12; plywood factory of, 187, 213; temporary workers of, 268–69, 276–78
neoliberalism, 2–7, 23–24, 27, 238–67, 270–71, 275, 298
nguillatún (religious ceremony), 284
nitrate mines, 14, 15, 17, 57, 120; decline of, 150, 152
Nueva Escocia estate, 100
Nueva Italia concession, 47, 48, 51, 76
Ñiochas estate, 100

Observatorio de los Derechos de los Pueblos Indígenas (Observatory of the Rights of Indigenous Peoples), 281, 292
ocupantes (squatters), 3, 16, 49–50; colonization laws and, 173; *colonos* and, 49–57, 96–99, 110–11; conservationism of, 98–100; control of, 31; *inquilinos* and, 107, 162; in national parks, 21, 95–96; Popular Front and, 156–57; uprisings by, 113–17
Oelckers timber company, 159–60, 217, 305–6
Oficina Mensura de Tierras (Office of Land Measurement), 24, 32, 38, 40, 43–46
Opazo, Roberto, 66, 68, 70, 76, 87
Ortíz Rojas, María Luisa, 347n24
Osorno, 43–44
Otero, Luis, 341n96
Otero Durán, Luis, 358n9
Otto, Guillermo, 127

"Pacification" of the Araucanía, 14, 15, 195–96
Pan-American Highway, 261
Pantano community, El, 290
Papelera, La. *See* Compañía Manufacturera de Papeles y Cartones
paper manufacturing, 6, 20, 125–26, 288; development of, 138–39, 149, 197–200; government support of, 23, 184–86, 198; tree plantations of, 130; unions of, 262. *See also* cellulose industry
paramilitary groups, 90–91
parcelas-bosques companies, 128–31, 185, 201
Park of the Araucarias, 280–81
Parlamento de Coz Coz, 27, 285, 286, 291–92

Parque Pehuenche de Quinquén, 281, 285
Parsons and Whitmore company, 187, 198
Pávez, Caupolicán, 263, 264, 276
Peck, Gunther, 300
Pedro Calfuqueo community, 165, 228–29, 279–81
Pehuenche, 12, 27, 106, 281, 307; agrarian reform and, 225–31; diet of, 13; lands of, 114–15, 194–96. *See also* Mapuche
pehuenes. See araucaria forests
Pellahuén estate, 44–45
Pemehue estate, 245
pentachlorophenol, 291
Peñuelas estate, 206
peones (seasonal laborers), 2, 3, 7, 16, 19–21, 152
Pérez, Francisco, 104
Pérez Guerra, Arnaldo, 351n3
Pérez Rosales, Vicente, 59–60, 62
Peru: Mapuche trade with, 14; War of the Pacific and, 14–15; wood imports of, 124
Peruvian lily, 93
pesticides, 1, 2, 7, 257–58, 287, 291, 302
Petermann, Víctor, 285, 286
Peters, Konrad, 85, 334n3
Philippi, Bernard, 60
Philippi, Rudolph, 11, 60
Pichún, Pascual, 290, 294
Pidenco estate, 291, 356n85
Pillén Pillén estate, 173
Pillimpillim estate, 171
Pilmaiquén estate, 247
Pilpilco Coal Company, 202
Pilpilco estate, 169
Pinares estate, 126, 205
Pinchot, Gifford, 18, 81–82, 101, 323n87
pine plantations, 118–24, 179–84, 299–308; agrarian reform and, 190, 200–201; Allende and, 209–10, 216, 218; cereal production and, 125, 132–33; of coal companies, 59, 123, 131, 172–73, 184, 271; ecological impact of, 284–87, 290–91, 296–301; harvesting of, 181–82; Pinochet and, 2, 7, 23–24, 240–44, 256–57; of private landowners, 125–37, 141, 144; Sánchez Guerrero on, 145–46; unemployment on, 183–84. *See also* Monterey pine; reforestation
Piñera, José, 275